AS Chemistry for AQA

John Atkinson and Carol Hibbert

Heinemann Educational Publishers
Halley Court, Jordan Hill, Oxford, OX2 8EJ
a division of Reed Educational & Professional Publishing Ltd
Heinemann is a registered trademark of Reed Educational & Professional Publishing Ltd

OXFORD MELBOURNE AUCKLAND
JOHANNESBURG BLANTYRE GABORONE
IBADAN PORTSMOUTH NH (USA) CHICAGO

First published 2000

ISBN 0 435 58134 1

04 03 02 01
10 9 8 7 6 5 4 3

Development editor Paddy Gannon

Edited by Andrew Nash

Designed and typeset by Ian Foulis & Associates

Illustrated by Ian Foulis & Associates, Plymouth, Devon
Original Illustrations © Heinemann Educational Publishers, 2000
Printed and bound in Spain by Edelvives

Acknowledgements
The authors and publishers would like to thank the following for permission to use photographs.

Cover photo: Science Photo Library

p43 all Carol Hibbert; **p70** Peter Gould; **p82 all** Carol Hibbert; **p117** Environmental Images; **p120 both** Gareth Boden; **p121 both** Peter Gould; **p129** Peter Gould; **p130** Peter Gould; **p137 all** Peter Gould; **p140** Peter Gould; **p141** Peter Gould; **p142 all** Peter Gould; **p143** Peter Gould; **p145** Environmental Images; **p152** Science Photo Library; **p162 all** Environmental Images; **p166** Peter Gould; **p178** Peter Gould; **p183** Peter Gould; **p185** Environmental Images; **p191 both** Environmental Images; **p192** Environmental Images; **p207** Chris Honeywell **p220** Peter Gould; **p223** Peter Gould.

Picture research by Charlotte Lippmann

The authors and publishers would like to thank the following for permission to reproduce copyright material:

p78 Label reproduced by kind permission of The **"Buxton Mineral Water Company Ltd"**; **p146** Label reproduced by kind permission of Tesco Stores Ltd.

The publishers have made every effort to trace the copyright holders, but if they have inadvertently overlooked any, they will be pleased to make the necessary arrangements at the first opportunity.

Tel: 01865 888058 www.heinemann.co.uk

Contents

Module 3 Foundation organic chemistry

Introduction

To the student

This book has been written specifically for Advanced Subsidiary (AS) Chemistry and matches AQA Specification 5421. It covers all the material you will need written in a clear and concise way with no added extras.

Each section starts with GCSE knowledge and gradually builds up to the depth of coverage you will need for AS level. This should enable you to make a smooth transition from GCSE Double Award Science or GCSE Chemistry. Questions throughout the book will help you test your own knowledge and prepare for the exams.

How to use this book

- The book is divided into three **Modules**:

  Module 1 – Atomic structure, Bonding and Periodicity

  Module 2 – Foundation Physical and Inorganic Chemistry

  Module 3 – Introduction to Organic Chemistry

  At the beginning of each Module you will find an **introduction** which gives you an overview of the Module and a **concept map** which shows you the various links between the topics in that Module.
- Each Module is divided into units, each one of which covers a main topic area. Each double page covers one key aspect of the unit. You will find these titles at the top of each page for easy reference:

  Module 1 | 1 Atomic structure | Fundamental particles
- At the start of each double page there is a **summary** of the content covered and how it links back to GCSE. Within the text there are **questions** for you to answer about the information on the pages. You should do these questions as you work through each double page, as they will check your understanding. The answers to these questions are in the back of the book. At the end of each double page are the **key ideas** which you can use as a check list for revision.
- At different points in the text you will find **key words** in bold type. It is important that you understand these terms. Their meaning is given in the text and in the **glossary** at the end of the book.
- There are short answer examination questions at the end of each unit to test your knowledge of that topic and longer questions at the end of each module which link the topics within the module and across modules to test your application of knowledge. The answers to the end of unit questions are at the end of the book with mark allocations shown in brackets.
- Where appropriate we have included everyday applications of the chemistry covered in the book.

A2 book and further resources

An A2 book is also available for the next year of study if you are taking the full AQA Chemistry A level.

Both books have an accompanying resource pack with further questions and answers to all the end of module questions in the book.

Introduction to Module 1

Module 1 – Atomic structure, bonding and periodicity – is a very important module: all of the chemistry in this module provides the basic knowledge and understanding for the other modules.

Any study of chemistry requires an understanding of atomic structure and chemical bonding, together with calculations of reacting quantities in chemical reactions. This module develops some of the concepts that you will have studied at GCSE.

In Unit 1 ideas about the structure of atoms, and the arrangement of the electrons in the atoms, are developed further. This includes a study of the use of mass spectrometers to identify isotopes and to determine relative atomic masses of the elements. A study of the ionisation energies of the elements gives a fuller understanding of the electronic structure of the elements.

In Unit 2 the ideas of covalent and ionic bonding learned at GCSE are developed to give a deeper understanding. The study of covalent bonds is extended to include co-ordinate bond formation and the existence of polar covalent bonds. The principles of the shapes of simple molecules and molecular ions are introduced. The concepts of intermolecular forces studied at GCSE are extended to show that there are three different types of intermolecular forces, and to explain how these different forces influence the melting points and boiling points of simple covalent molecules. Several types of solid structure are studied, including examples of metallic, giant ionic, giant atomic and molecular solids.

In Unit 3 some periodic trends are studied. The study of the trends in chemical and physical properties requires a good understanding of atomic structure and bonding. In this unit some of the physical properties of the elements in the third Period are studied, as are the physical and chemical properties of the elements in Group II. This builds on the chemistry of Group I, which you will have studied at GCSE.

In Unit 4 calculations involving empirical formulae and reacting masses, studied at GCSE, are extended to include calculations involving reactions in solution and reactions with gases. The concept of the mole as the working unit in calculations is introduced. All of the calculations covered in this unit will be met again in later modules.

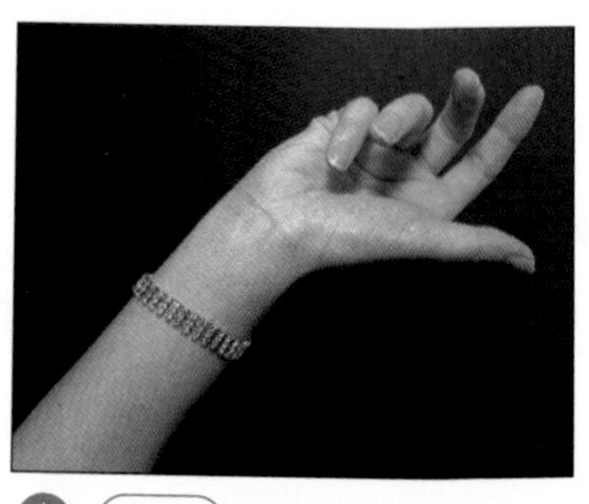

Fig. 1 *Diamond is a giant structure of carbon atoms*

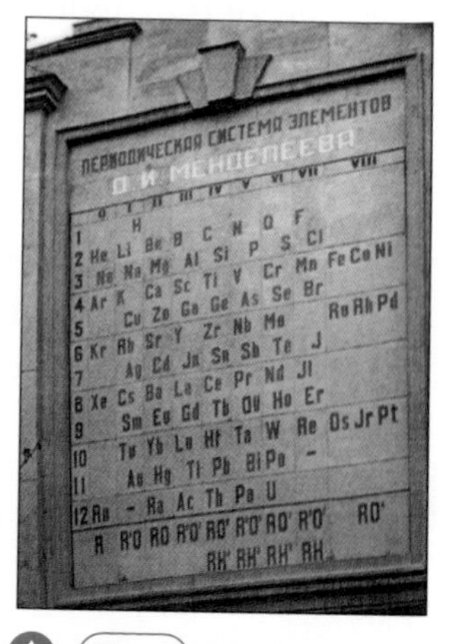

Fig. 2 *Mendeleev's Periodic Table*

Fig. 3 *Standard solutions are important in volumetric analysis*

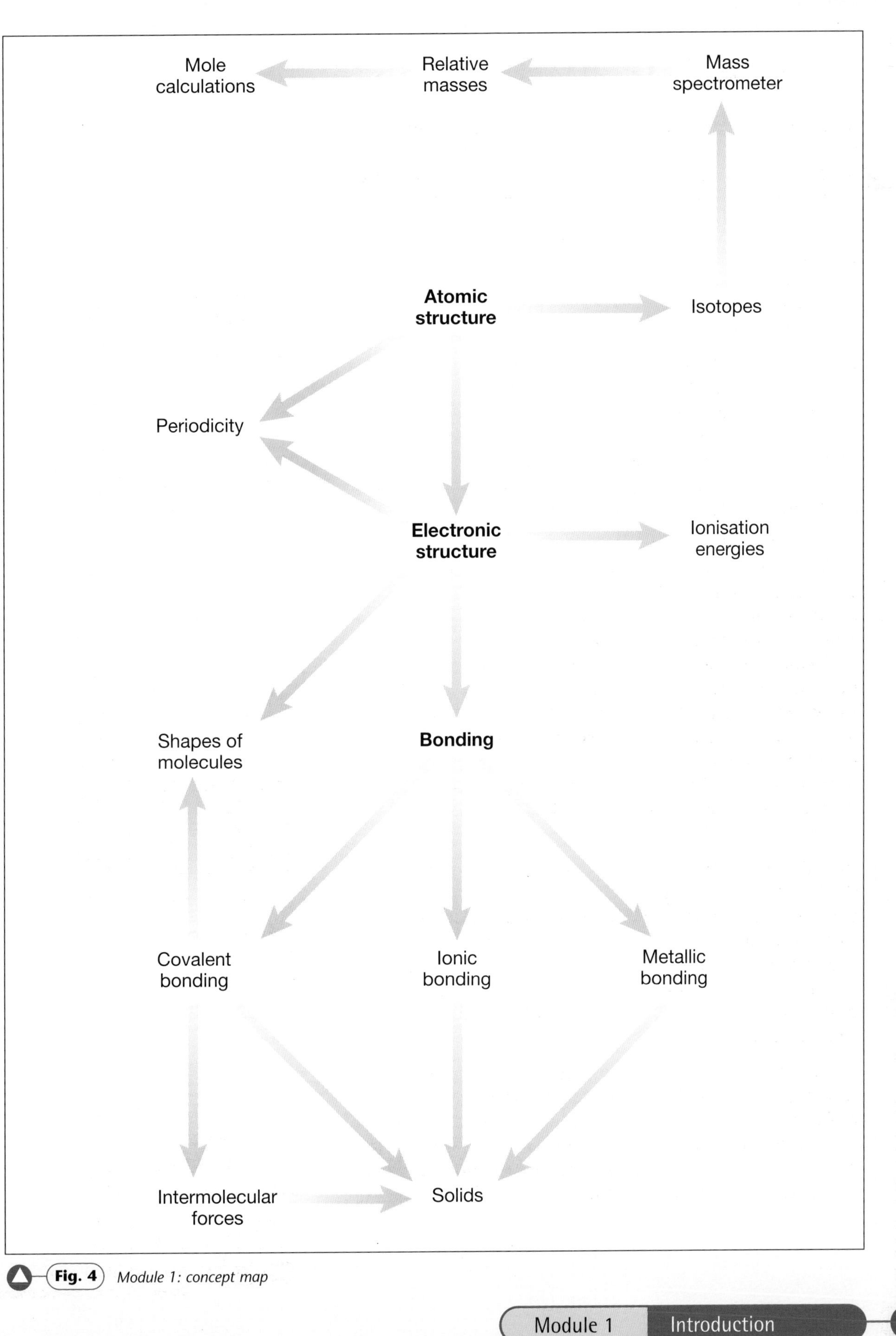

Fig. 4 *Module 1: concept map*

1 Where it all begins ...

At GCSE you learned that all matter in the Universe is made up from 92 naturally occurring elements. An **element** is made up of one sort of atom only.

All atoms are made up of three simpler **sub-atomic particles**, namely protons, neutrons and electrons. The protons and neutrons form a small, dense, positively-charged nucleus. The electrons are in **energy levels** outside the nucleus.

Atoms

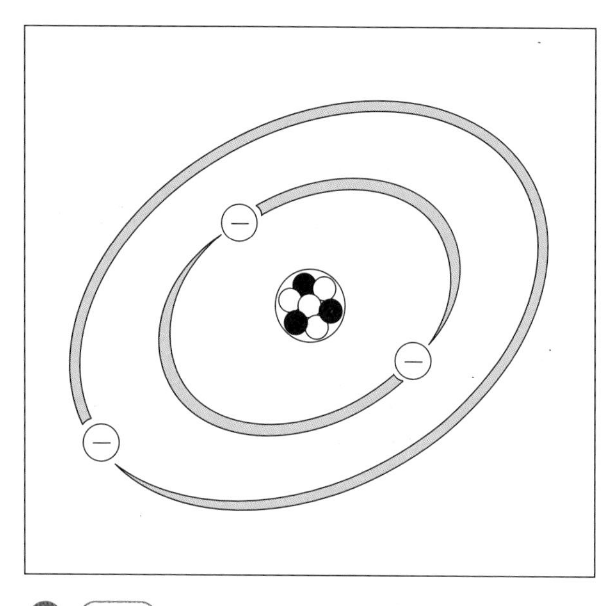

Fig. 1 *Representation of an atom*

Atoms are very small – the *radius* of a hydrogen atom is approximately 10^{-10} metres. As a consequence, the *mass* of an atom is very small. The radius of a hydrogen nucleus is approximately 10^{-15} metres, which demonstrates how small is the nucleus of an atom compared to the size of the atom. Most of an atom is empty space. If an atom were the size of a football pitch, then the nucleus would be about the size of a football on the centre spot of the pitch.

The actual values of the masses and charges of the sub-atomic particles are shown in Table 1. The values are so small that it is difficult to have a realistic understanding of the masses and charges of the sub-atomic particles.

	Proton (p)	Neutron (n)	Electron (e^-)
Mass/kg	1.672×10^{-27}	1.674×10^{-27}	9.109×10^{-31}
Charge/C	$+1.602 \times 10^{-19}$	0	-1.602×10^{-19}

Table 1

One meaningful way to consider the masses of the sub-atomic particles is to use **relative masses**. The proton is assigned a mass of 1, and the masses of other sub-atomic particles are given *relative* to the mass of one proton.

Electrons and protons have an equal but opposite **charge**. Neutrons have no charge. Table 2 gives the relative masses and charges of the three sub-atomic particles.

	Proton (p)	Neutron (n)	Electron (e^-)
Relative mass	1	1	1/1836
Relative charge	+1	0	–1

Table 2

Nearly all the mass of an atom is in the nucleus. The mass of an electron is negligible compared with that of the protons and neutrons. The atoms of different elements contain different numbers of sub-atomic particles.

The nucleus of an atom of an element can be described using the *atomic number* and the *mass number*. (See Appendix A.)

Atomic number

The **atomic number** (***Z***) is the number of protons in the nucleus of an atom. It is also known as the **proton number**.

The number of protons in the nucleus of an atom determines the element to which the atom belongs. If an atom has an atomic number of 6, the atom *must* be a carbon atom. All carbon atoms have 6 protons in the nucleus.

Atoms carry no overall charge. The number of protons must therefore be the same as the number of electrons. For example, each carbon atom contains 6 protons and 6 electrons.

Mass number

The **mass number** (***A***) is the sum of the number of protons and the number of neutrons in the nucleus of an atom. The number of neutrons can be found by subtracting the atomic number from the mass number.

Chemists use the following shorthand to represent an atom. The mass number is shown as a superscript (top number) and the atomic number is shown as a subscript (bottom number).

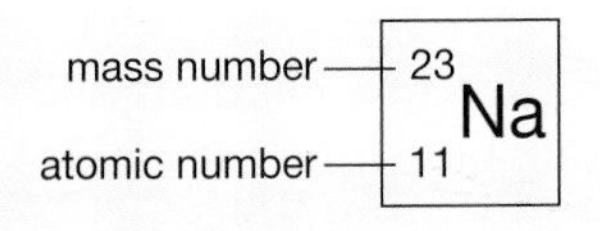

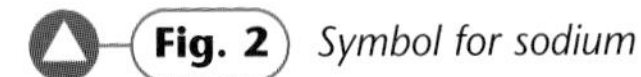
Fig. 2 *Symbol for sodium*

Each atom of sodium contains:

11 protons
11 electrons
12 neutrons

Q

1 State the numbers of protons, neutrons and electrons in an atom with an atomic number of 9 and a mass number of 19.

2 Copy and complete the following table.

Element	Atomic number	Mass number	Number of protons	Number of neutrons	Number of electrons
Potassium	19	40			
Oxygen			8	8	
Chlorine		37		20	
Phosphorus		31			15
Aluminium	13			14	

Key Ideas 4 – 5

- **Atoms are made up of protons, neutrons and electrons.**
- **Because atoms are so small, the protons, neutrons and electrons are given relative atomic masses and charges.**
- **The atomic number and the mass number of an atom gives information about the number of protons, neutrons and electrons in the atom.**

2 Different types of atoms ...

All of the atoms of any particular element contain the same number of protons in the nucleus, and the same number of electrons. However, the number of *neutrons* may vary, and this results in elements having different types of atoms. These different types of atoms are called *isotopes*. Isotopes have many uses in industry, medicine and scientific research.

Isotopes

Isotopes are atoms of the same element with the same *atomic* number, but different *mass* numbers. They have different numbers of neutrons.

1 State the number of protons, neutrons and electrons in:

a) one atom of carbon-12
b) one atom of carbon-14

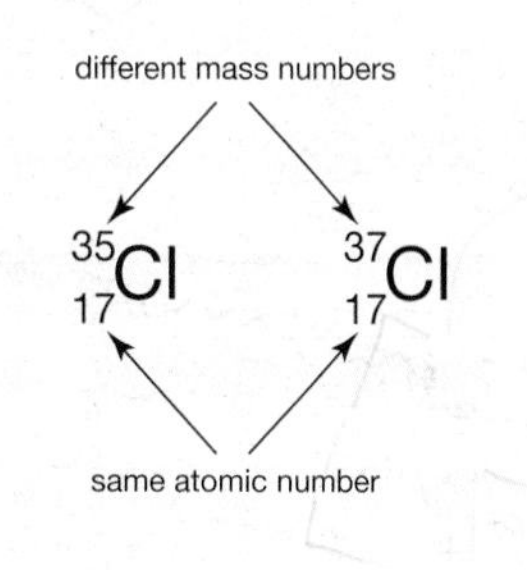

Fig. 1 *Isotopes of chlorine*

Each atom of chlorine contains the following:

$^{35}_{17}Cl$	$^{37}_{17}Cl$
17 protons	17 protons
17 electrons	17 electrons
18 neutrons	20 neutrons

The isotopes of chlorine are often referred to as *chlorine-35* and *chlorine-37*.

Isotopes of an element have the same *chemical* properties because they have the same number of electrons. When elements react, it is the *electrons* that are involved in the reactions. This means that the isotopes of an element cannot be differentiated by chemical reactions. For example, both isotopes of chlorine react with sodium to form sodium chloride, NaCl.

Because isotopes of an element have different numbers of neutrons, they have different *masses*, and isotopes of an element therefore have slightly different *physical* properties.

The isotopes of an element with *fewer* neutrons will have:

- *lower masses*
- *lower densities* – the atoms are the same size but have a smaller mass: the mass per unit volume of the isotope will therefore be less
- *faster rates of diffusion* – the rate of diffusion of a gas depends upon temperature and mass: if two gaseous isotopes of an element are at the same temperature, then the lighter isotope will diffuse more quickly
- *lower melting and boiling points* – the forces of attraction between particles increase as the mass of the particles increases.

Q

2 Explain why gaseous isotopes of an element can be separated by diffusion.

Here are some further examples of isotopes:

Each atom of uranium contains:

$^{235}_{92}U$	$^{238}_{92}U$
92 protons	92 protons
92 electrons	92 electrons
143 neutrons	146 neutrons

The isotopes of hydrogen are unusual in that the three isotopes have different names: *hydrogen* (H), *deuterium* (D) and *tritium* (T).

Each atom of hydrogen contains:

$^{1}_{1}H$	$^{2}_{1}D$	$^{3}_{1}T$
1 proton	1 proton	1 proton
1 electron	1 electron	1 electron
0 neutrons	1 neutron	2 neutrons

Hydrogen is the only element with atoms that have *no* neutrons.

Uses of isotopes

Some isotopes are unstable, and **decay** by giving off **radiation** which can be detected. This property of **radioactive** isotopes makes them very useful as tracers. Iodine-131 may be injected into the bloodstream to detect tumours. Iron-59 is used to monitor red blood cell production in bone marrow. Sodium-24 is used to monitor blood flow and to detect blood clots and obstructions in blood vessels. Leaks in underground water pipes can be found by adding small quantities of radioactive isotopes to the water.

The naturally occurring isotopes of carbon, carbon-12 and carbon-14, are used to date material made from wood or other plant matter. When a tree is cut down, the ratio of the two isotopes of carbon present in the wood steadily changes with time, because carbon-14 is radioactive and decays. If the *ratio* of the two isotopes is found, then the age of the wood can be determined.

Q

3 Define the term *isotope*.

4 State *two* ways in which the physical properties differ for the isotopes of an element.

Fig. 2 *Isotopes of carbon are used to date timbers*

Key Ideas 6 – 7

- **Isotopes are atoms of the same element that have different numbers of neutrons.**
- **Isotopes of an element have identical *chemical* properties. However, because isotopes have different masses, their *physical* properties are slightly different.**

3 Sorting out isotopes

If they have the same chemical properties, how do we know that isotopes exist? Isotopes have different masses, and this allows them to be separated and identified in a mass spectrometer.

Mass spectrometer

The **mass spectrometer** is an instrument used:

- to measure the relative masses of isotopes
- to find the relative abundance of the isotopes in a sample of an element.

The mass spectrometer works on the principle that when charged particles pass through a magnetic field, the particles are deflected by the magnetic field, and the *amount* of deflection depends upon the **mass/charge ratio** of the charged particle.

Once the sample of an element has been placed in the mass spectrometer, it undergoes five separate stages: vaporisation, ionisation, acceleration, deflection, and detection.

A mass spectrometer uses a vacuum pump to remove air from the apparatus, since the ions would otherwise undergo random collisions with air particles.

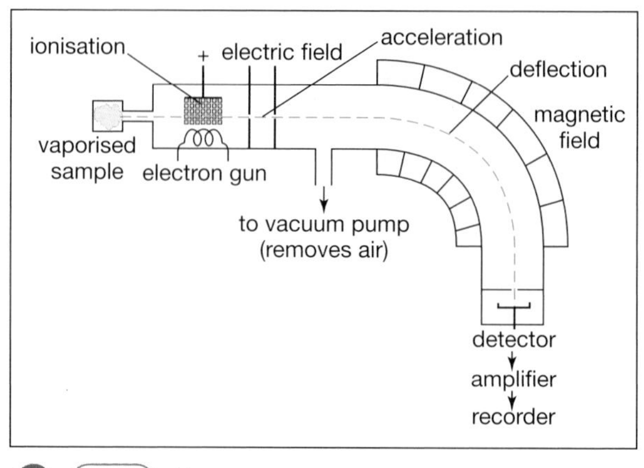

Fig. 1 *Mass spectrometer*

Provided that the energy of the electrons from the electron gun is not too high, only one electron per atom will be removed and the ions will each carry a single positive charge – they will be **unipositive** ions.

Vaporisation

The sample has to be in its **gaseous** form. If the sample is a liquid or solid, a heater is used to vaporise some of the sample.

$X(s) \rightarrow X(g)$

or $X(l) \rightarrow X(g)$

Ionisation

The sample is bombarded by a stream of high-energy electrons from an electron 'gun'. These high-energy electrons can 'knock' an electron from an atom. This produces a **positive ion**:

$X(g) \rightarrow X^+(g) + e^-$

Acceleration

An electric field is used to accelerate the positive ions towards the magnetic field. The accelerated ions are **focused** and passed through a slit: this produces a narrow beam of ions.

Deflection

The accelerated ions are deflected in the magnetic field. The amount of deflection is *greater* when:

- the mass of the positive ion is *less*
- the charge on the positive ion is *greater*
- the velocity of the positive ion is *less*
- the strength of the magnetic field is *greater*.

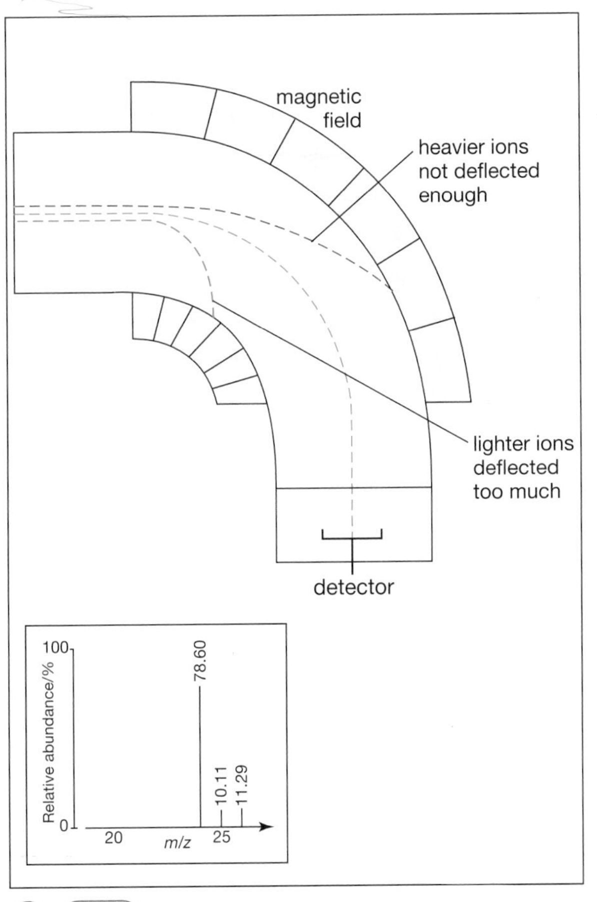

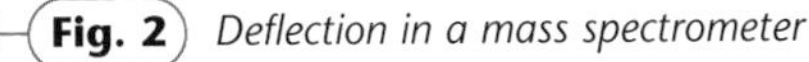
Fig. 2 *Deflection in a mass spectrometer*

If all the ions are travelling at the same velocity and carry the same charge, the amount of deflection in a given magnetic field depends upon the mass of the ion.

For a given magnetic field, only ions with a particular **relative mass (*m*) to charge (*z*) ratio** – the ***m/z* value** – are deflected sufficiently to reach the detector. Ions with too small an *m/z* value are deflected too much, and ions with too large an *m/z* value are not deflected enough.

Detection

Ions that reach the detector cause electrons to be released in an **ion-current detector**. The number of electrons released, and hence the current produced, is proportional to the number of ions striking the detector. The detector is linked to an amplifier and then to a recorder: this converts the current into a **peak** which is shown in a **mass spectrum**.

When the mass spectrometer is operating, the magnetic field is slowly increased until a stream of ions with a particular *m/z* value reaches the detector. The strength of the magnetic field is then further increased until a second stream of ions with a greater *m/z* value reaches the detector. The magnetic field is steadily increased until all of the different ions have been detected.

A mass spectrum gives information on the ions with different *m/z* values present in a sample, and on the relative abundance of each of these ions. In a sample of an element, each isotope has a different mass and hence each isotope produces a peak with a different *m/z* value.

Assuming that all the peaks in the spectrum are for unipositive ions (z=1), then the *m/z* value is the same as the ion's mass number. The height of each peak will be proportional to the relative abundance of that isotope.

Q

1 Why is the sample in a mass spectrometer ionised?

2 Explain how the ions in the mass spectrometer are accelerated and deflected.

Key Ideas 8 – 9

- **The mass spectrometer identifies the different isotopes present in an element.**
- **The isotopes are ionised, and the resulting positive ions undergo deflection in a magnetic field.**
- **In a mass spectrum, each isotope produces a separate peak with its own *m/z* value.**
- **The height of a peak is proportional to the relative abundance of that isotope.**

4 Working out masses

The mass spectrum of an element can be used to calculate the relative atomic mass of an element. Each isotope of the element will produce a separate peak. Using the *m/z* value and the peak height for each of the different isotopes, the relative atomic mass of the element can be calculated.

Relative atomic mass

Assuming that all the peaks in the spectrum are for unipositive ions ($z = 1$), then the *m/z* value is the same as the mass number of the ion and hence the relative mass of the atom. The height of each peak will be proportional to the relative abundance of that isotope.

The **relative atomic mass (A_r)** of an element is the average mass of the atoms, taking into account the natural relative abundance of that element's isotopes compared to one-twelfth of the mass of an atom of carbon-12.

$$\text{relative atomic mass } (A_r) = \frac{\text{average (mean) mass of an atom} \times 12}{\text{mass of one atom of carbon-12}}$$

Isotopes of boron

For boron there are two isotopes: boron-10, with an abundance of 18.7%, and boron-11, with an abundance of 81.3%. To calculate the relative atomic mass of boron, the natural abundance of each of the isotopes has to be taken into account.

$$A_r \text{ of boron} = \frac{(10 \times 18.7) + (11 \times 81.3)}{(18.7 + 81.3)}$$

$$= \frac{187 + 894.3}{100}$$

$$= \frac{1081.3}{100} = 10.8$$

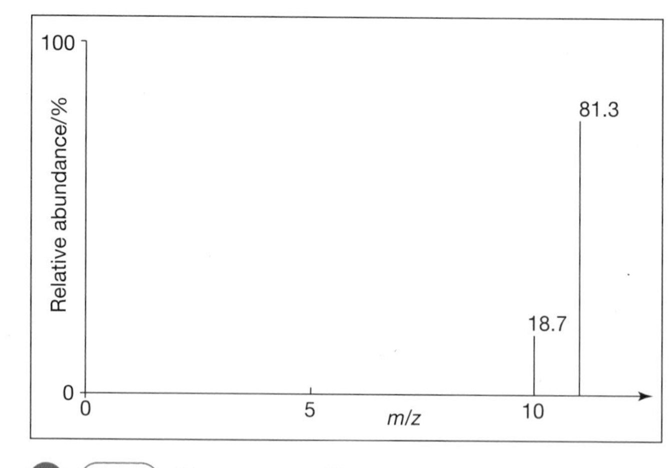

Fig. 1 Mass spectrum of boron

Isotopes of lead

m/z value	204	206	207	208
Relative abundance/%	1.55	23.6	22.6	52.3

$$A_r \text{ of lead} = \frac{(204 \times 1.55) + (206 \times 23.6) + (207 \times 22.6) + (208 \times 52.3)}{(1.55 + 23.6 + 22.6 + 52.3)}$$

$$= \frac{316.2 + 4861.6 + 4678.2 + 10878.4}{100.05}$$

$$= \frac{20734.4}{100.05} = 207.2$$

If an element exists as diatomic molecules, for example Cl_2, O_2, H_2, then the spectrum will contain peaks both for the separate atoms and for the molecules. For example, the mass spectrum of chlorine will have peaks for Cl^+, with m/z values of 35 and 37, and peaks for Cl_2^+, with m/z values of 70, 72 and 74.

$^{35}Cl—^{35}Cl^+$	$^{35}Cl—^{37}Cl^+$	$^{37}Cl—^{37}Cl^+$
$m/z = 70$	$m/z = 72$	$m/z = 74$

Q

1 Calculate the relative atomic mass of zinc using the following mass spectra information about its isotopes.

m/z value	64	66	67	68	70
Relative intensity	48.9	27.8	4.1	18.6	0.62

Q

2 Oxygen has two isotopes, oxygen-16 and oxygen-18. Give the m/z values of the three peaks in the mass spectrum caused by the O_2^+ ion.

Mass spectra for compounds

The mass spectrometer is used for analysis of compounds as well as elements. The ionisation of a gaseous molecule in the mass spectrometer produces a mass spectrum with a large number of peaks.

The peak with the greatest m/z value (this is the peak furthest to the right in the spectrum) is caused by the molecular ion. The molecular ion is formed when the high-energy electrons from the electron gun 'knock' an electron from the molecule. For example, ethanol, CH_3CH_2OH, can form the molecular ion $[CH_3CH_2OH]^+$, which has an m/z value of 46. Hence the relative molecular mass of ethanol is 46.

relative molecular mass (M_r)

$$= \frac{\text{average (mean) mass of the compound} \times 12}{\text{mass of one atom of carbon-12}}$$

The other peaks in the mass spectrum of a compound are caused by fragments from molecular ions. (This is dealt with in greater depth in A2 Chemistry.)

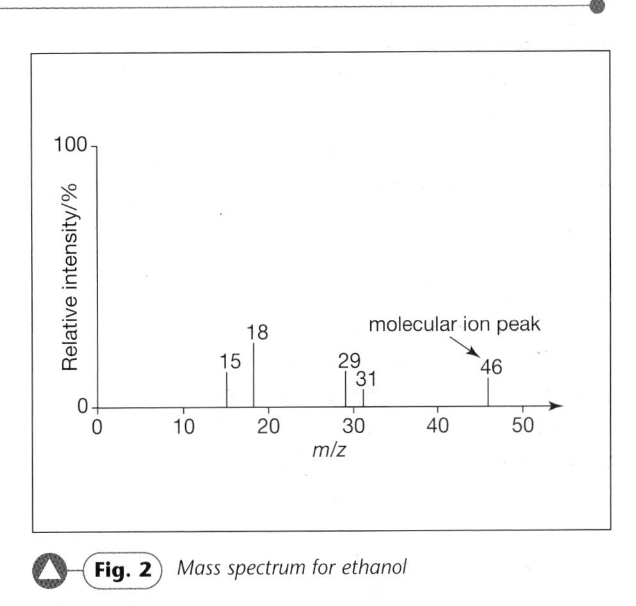

Fig. 2 *Mass spectrum for ethanol*

Q

3 Explain why the presence of $^{32}S^{2+}$ and $^{16}O^+$ in a sample produces only one peak in the mass spectrum.

Key Ideas 10 – 11

- **Relative atomic masses (A_r) of elements are calculated using the relative intensities of the peaks in the mass spectra of the elements.**
- **Relative molecular masses (M_r) of compounds can be determined from the mass spectra of compounds.**

5 Energy levels

It is the electrons and the *arrangement* of the electrons (electronic structure) that are responsible for chemical reactions. To be more precise, it is the number of **outer electrons** that control the chemical properties of the elements.

At GCSE we think of the electrons as being in **energy levels** or **shells**. The lowest energy level holds a maximum of two electrons, and the higher energy levels hold a maximum of eight electrons. The electrons are in the lowest available energy levels. For example, the sodium atom has 11 electrons and an electronic structure of 2,8,1. At AS level electronic structure is taken a step further.

Principal energy levels and sub-levels

The energy levels are called **principal energy levels** and are given the numbers 1, 2, 3, 4, 5, 6 … (**principal quantum numbers**).

There are at least seven different principal energy levels, but for most purposes at AS and A2 level only the first four principal energy levels are considered.

The principal energy levels contain **sub-levels**. Each principal energy level contains a different number of sub-levels.

Principal quantum number	Number of sub-levels
1	1
2	2
3	3
4	4

These sub-levels are assigned the letters **s**, **p**, **d** and **f**. (These letters were originally used for the different spectral series that provided evidence of electron energy levels in atoms.)

- The first principal energy level has a **1s** sub-level.
- The second principal energy level has a **2s** and a **2p** sub-level.
- The third principal energy level has **3s**, **3p** and **3d** sub-levels.

Within a principal energy level, each of the sub-levels has a different energy. The energy of the sub-levels *increases* from s to p to d to f (see Figure 1).

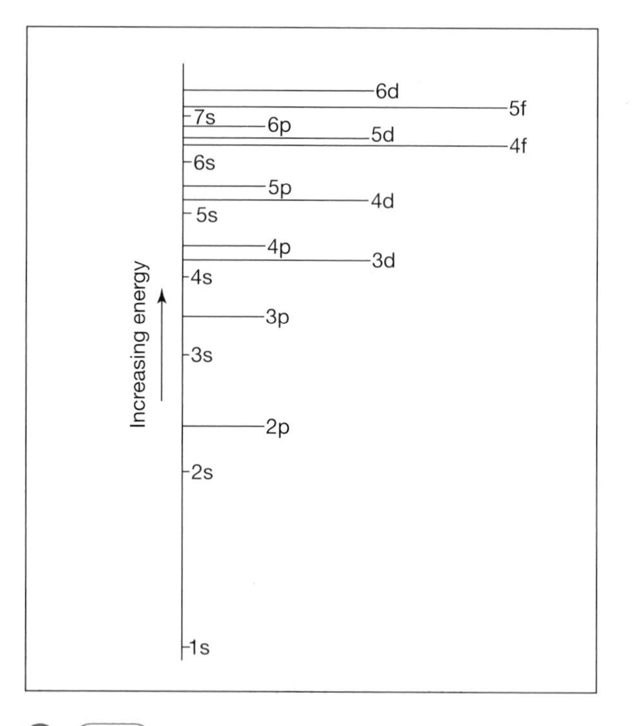

Fig. 1 *Energy levels*

Each type of sub-level can hold a different maximum number of electrons.

Sub-level	Maximum number of electrons
s	2
p	6
d	10
f	14

Putting all this information together, the first four principal energy levels contain the sub-levels shown in the table.

Principal energy level	Sub-levels	Maximum number of electrons
1	1s	2
2	2s2p	2 + 6 = 8
3	3s3p3d	2 + 6 + 10 = 18
4	4s4p4d4f	2 + 6 + 10 + 14 = 32

The Periodic Table consists of all the known elements arranged in order of increasing atomic number. But the arrangement of the elements also reflects the electronic structure of the elements. The Periodic Table is split into blocks called the **s block**, **p block**, **d block** and **f block**. The columns in the Periodic Table are called Groups; the rows are Periods.

The blocks correspond to the filling of sub-levels:

- the s block is two elements wide, corresponding to the filling of an s sub-level
- the p block is six elements wide, corresponding to the filling of a p sub-level
- the d block is ten elements wide, corresponding to the filling of a d sub-level.

Q

1 State the sub-levels in the third principal energy level.

2 In which block (s, p, d or f) in the Periodic Table are the following elements?

a) Na b) Fe c) S
d) Cl e) Ra f) Ne

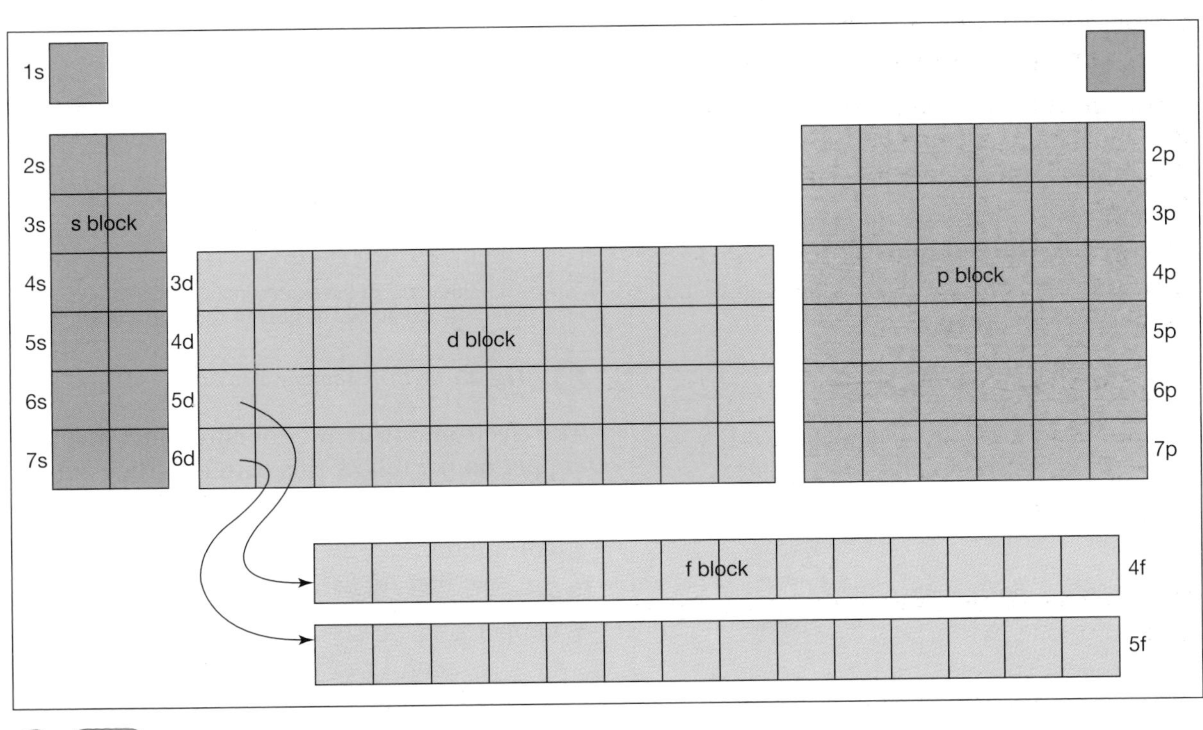

Fig. 2 *Periodic Table*

Key Ideas 12 – 13

- **Electrons are in principal energy levels. Each principal energy level is given a principal quantum number.**
- **Principal energy levels contain sub-levels that are of slightly different energies. Each different type of sub-level can hold a different maximum number of electrons.**
- **The shape of the Periodic Table depends upon the electronic structure of the elements. The blocks in the Periodic Table correspond to the filling of the different energy sub-levels.**

6 Filling the energy levels

The full electronic structure of an element gives information about the number of electrons in each occupied principal energy level and sub-level. The electrons are in the lowest available energy levels.

Electronic structure

The **electronic structure** follows a pattern which takes into account the filling of principal energy levels and sub-levels and the maximum number of electrons in each energy level.

The order of filling of the sub-levels is 1s 2s 2p 3s 3p. After this there is a break in the logical pattern, in that the 4s sub-level fills *before* the 3d sub-level. This is because the 4s sub-level is of lower energy than the 3d sub-level (see page 12). Once the 3d sub-level is full, the 4p sub-level is filled.

Elements with a partially filled d sub-level are called **transition elements**.

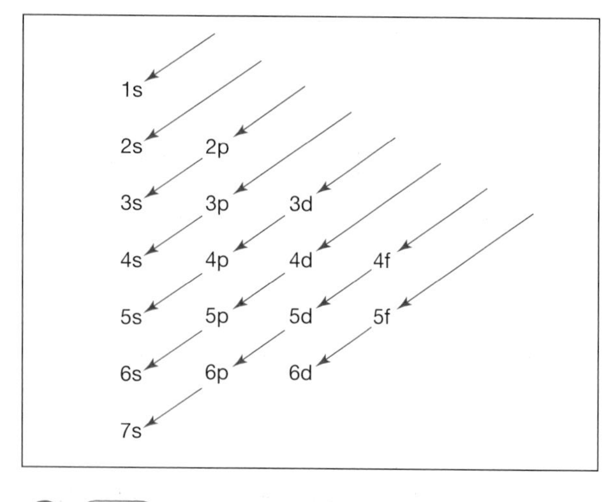

Fig. 1 *The order in which energy levels are filled*

This gives an overall order of filling of 1s 2s 2p 3s 3p 4s 3d 4p, for the elements hydrogen to krypton. (This building up of electronic structure is called the **Aufbau Principle**.) The full electronic structures of the first 36 elements in the Periodic Table are shown in Table 1.

From the electronic structure, we can see which sub-levels are occupied and the number of electrons in each occupied sub-level. The electronic structure of sodium – 2,8,1 at GCSE – becomes that shown in Figure 2.

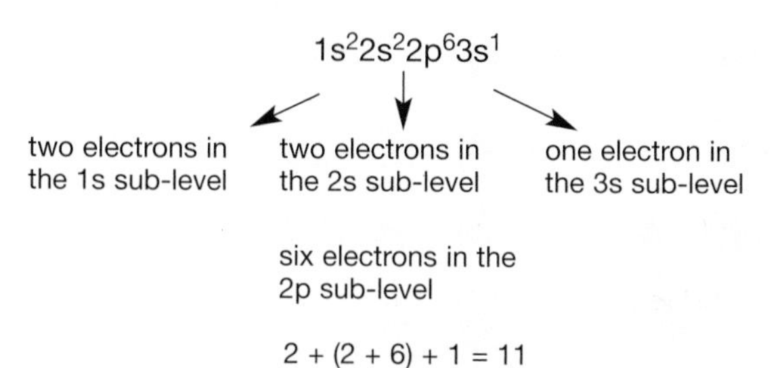

Fig. 2 *Electronic structure of sodium*

The electronic structures of chromium and copper do *not* follow this pattern – they are anomalous:

- chromium is $1s^2 2s^2 2p^6 3s^2 3p^6 \mathbf{3d^5 4s^1}$ (*not* $1s^2 2s^2 2p^6 3s^2 3p^6 3d^4 4s^2$)
- copper is $1s^2 2s^2 2p^6 3s^2 3p^6 \mathbf{3d^{10} 4s^1}$ (*not* $1s^2 2s^2 2p^6 3s^2 3p^6 3d^9 4s^2$)

The reason for these anomalies is that in both cases the outer sub-levels are either full or half-full, and the structures shown are of lower energy and represent more stable arrangements.

Since the *chemistry* of an element depends upon the *outer electrons*, a shortened version of the electronic structure is sometimes used. For example, the electronic structure of chromium can be shown as $[Ar]3d^5 4s^1$, where [Ar] represents the electronic structure of an argon atom, $1s^2 2s^2 2p^6 3s^2 3p^6$.

Element	Atomic number	Electronic structure
H	1	$1s^1$
He	2	$1s^2$
Li	3	$1s^22s^1$
Be	4	$1s^22s^2$
B	5	$1s^22s^22p^1$
C	6	$1s^22s^22p^2$
N	7	$1s^22s^22p^3$
O	8	$1s^22s^22p^4$
F	9	$1s^22s^22p^5$
Ne	10	$1s^22s^22p^6$
Na	11	$1s^22s^22p^63s^1$
Mg	12	$1s^22s^22p^63s^2$
Al	13	$1s^22s^22p^63s^23p^1$
Si	14	$1s^22s^22p^63s^23p^2$
P	15	$1s^22s^22p^63s^23p^3$
S	16	$1s^22s^22p^63s^23p^4$
Cl	17	$1s^22s^22p^63s^23p^5$
Ar	18	$1s^22s^22p^63s^23p^6$

Element	Atomic number	Electronic structure
K	19	$1s^22s^22p^63s^23p^64s^1$
Ca	20	$1s^22s^22p^63s^23p^64s^2$
Sc	21	$1s^22s^22p^63s^23p^63d^14s^2$
Ti	22	$1s^22s^22p^63s^23p^63d^24s^2$
V	23	$1s^22s^22p^63s^23p^63d^34s^2$
Cr	24	$1s^22s^22p^63s^23p^63d^54s^1$
Mn	25	$1s^22s^22p^63s^23p^63d^54s^2$
Fe	26	$1s^22s^22p^63s^23p^63d^64s^2$
Co	27	$1s^22s^22p^63s^23p^63d^74s^2$
Ni	28	$1s^22s^22p^63s^23p^63d^84s^2$
Cu	29	$1s^22s^22p^63s^23p^63d^{10}4s^1$
Zn	30	$1s^22s^22p^63s^23p^63d^{10}4s^2$
Ga	31	$1s^22s^22^6p3s^23p^63d^{10}4s^24p^1$
Ge	32	$1s^22s^22^6p3s^23p^63d^{10}4s^24p^2$
As	33	$1s^22s^22^6p3s^23p^63d^{10}4s^24p^3$
Se	34	$1s^22s^22^6p3s^23p^63d^{10}4s^24p^4$
Br	35	$1s^22s^22^6p3s^23p^63d^{10}4s^24p^5$
Kr	36	$1s^22s^22p^63s^23p^63d^{10}4s^24p^6$

Table 1 *Electronic structures of the elements H–Kr*

Q

1 Change the following GCSE electronic structures into full electronic structures:

a) chlorine 2,8,7 b) calcium 2,8,8,2 c) oxygen 2,6

2 Identify the elements with the following electronic structures:

a) $1s^22s^22p^2$ b) $1s^22s^22p^63s^23p^63d^24s^2$ c) $1s^22s^22p^63s^23p^63d^{10}4s^2$

3 Expand the following shortened electronic structures to show the full electronic structures:

a) $[He]2s^22p^5$ b) $[Ne]3s^23p^4$ c) $[Ar]3d^34s^2$

Key Ideas 14 – 15

- **The electrons are in the lowest available sub-levels.**
- **The order of filling the principal energy levels and sub-levels of the elements up to krypton is 1s 2s 2p 3s 3p 4s 3d 4p.**
- **The 4s sub-level fills *before* the 3d sub-level.**

7 Electrons in orbit ...

The electronic structure in terms of principal energy levels and sub-levels builds on the GCSE ideas of electronic structure. The sub-levels are themselves further divided into orbitals.

Orbitals

The energy sub-levels are made up of **orbitals**, each of which can hold a maximum of two electrons. The different sub-levels have different numbers of orbitals.

Sub-level	Number of orbitals	Maximum number of electrons
s	1	2
p	3	6
d	5	10
f	7	14

The orbitals in different sub-levels have different *shapes*. The shapes of the s and p sub-levels are shown in Figure 1.

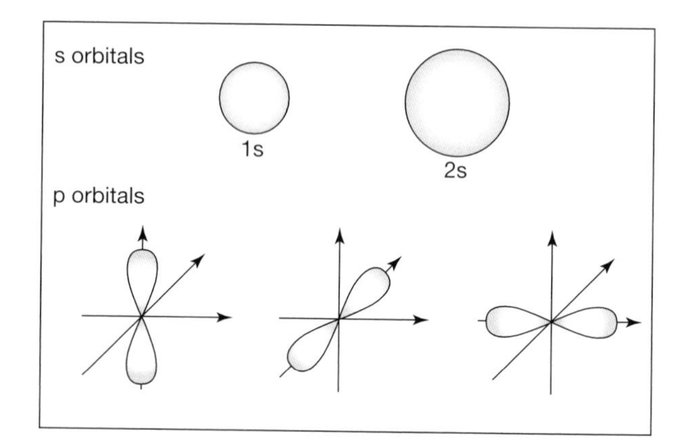

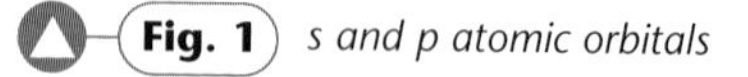
Fig. 1 *s and p atomic orbitals*

Within a sub-level, the electrons occupy orbitals as *unpaired* electrons rather than as *paired* electrons. (This is known as **Hund's Rule**.) The electrons are only paired when there are no more empty orbitals available within a sub-level.

Hund's rule is best illustrated by the use of examples. In a carbon atom the electrons are arranged as shown. The different orbitals are represented as boxes. The 1s and 2s electrons are paired electrons, but the 2p electrons are unpaired. The arrows represent the electrons in the orbitals. The *direction* of the arrows indicates the **spin** of the electron. Paired electrons will have opposite spin, as this reduces the **mutual repulsion** between the paired electrons.

Electronic structure of carbon, $1s^22s^22p^2$

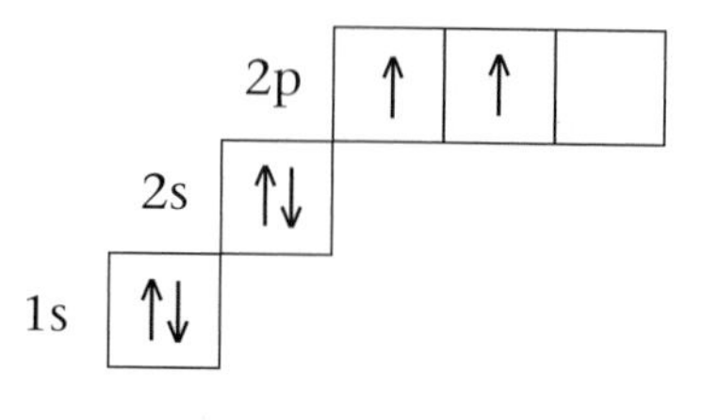

Electronic structure of nitrogen, $1s^22s^22p^3$

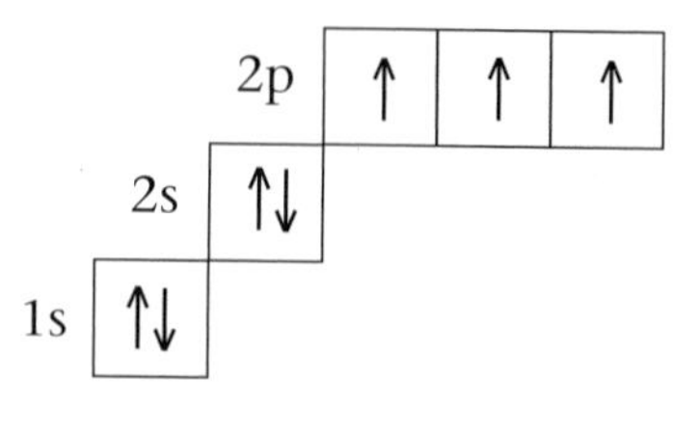

Electronic structure of oxygen, $1s^22s^22p^4$

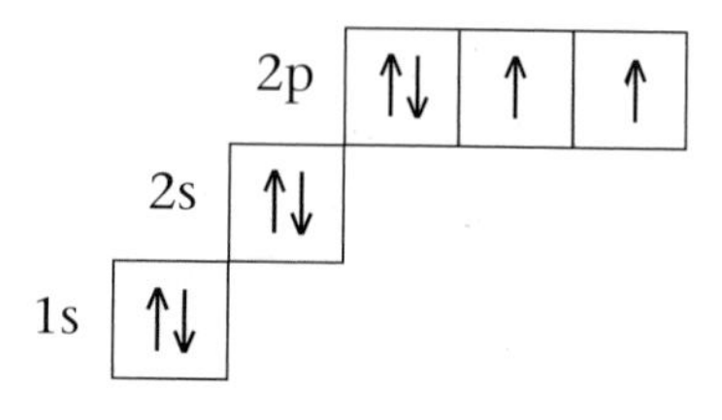

Q

1 Using boxes to represent orbitals, give the full electronic structure of the following atoms:

a) lithium b) fluorine c) potassium

Electronic structure of ions

When an atom gains or loses electrons to form an ion, the electronic structure changes.

Positive ions: formed by the loss of electrons

The sodium atom loses an electron from the highest occupied energy level to form the sodium ion, Na^+:

$1s^2 2s^2 2p^6 3s^1 \rightarrow 1s^2 2s^2 2p^6$

Na atom — Na^+ ion

Negative ions: formed by the gain of electrons

The oxygen atom gains two electrons to form the oxide ion, O^{2-}:

$1s^2 2s^2 2p^4 \rightarrow 1s^2 2s^2 2p^6$

O atom — O^{2-} ion

Notice that the Na^+ ion and the O^{2-} ion have the same electronic structure. Ions that have the same electronic structure are **isoelectronic**.

Transition metals

With the transition metals, it is the 4s electrons that are lost first when they form ions:

loss of two electrons:

Ti $1s^2 2s^2 2p^6 3s^2 3p^6 3d^2 \mathbf{4s^2}$

forming Ti^{2+} $1s^2 2s^2 2p^6 3s^2 3p^6 3d^2$

lost first

loss of three electrons:

Cr $1s^2 2s^2 2p^6 3s^2 3p^6 3d^5 \mathbf{4s^1}$

forming Cr^{3+} $1s^2 2s^2 2p^6 3s^2 3p^6 3d^3$

Q

2 Give the full electronic structure of the following positive ions:

a) Mg^{2+} b) Ca^{2+} c) Al^{3+}

3 Give the full electronic structure of the following negative ions:

a) Cl^- b) Br^- c) P^{3-}

4 Copy and complete the following table:

	Atomic number	Mass number	Number of protons	Number of neutrons	Number of electrons	Electronic structure
Mg				12		$1s^2 2s^2 2p^6 3s^2$
Al^{3+}		27			10	
S^{2-}			16	16		
Sc^{3+}	21	45				
Ni^{2+}				30	26	

5 Write out the full electronic structure of the Fe^{2+} and Fe^{3+} ions. Which of these ions has the greater stability? Explain why.

Key Ideas 16 – 17

- **Sub-levels are split into orbitals, each of which contains a maximum of two electrons.**
- **The s, p, d and f orbitals have different shapes.**
- **Within a sub-level, electron–electron repulsion is reduced by having unpaired electrons in preference to paired electrons.**
- **When the transition metals form ions, it is the 4s electrons that are lost first.**

8 Evidence for electronic structure

How do we *know* the electronic structure of atoms, as they are too small to be seen or studied separately? A study of the ionisation energies of the elements provides evidence for the existence of principal energy levels and sub-levels in their atoms.

Ionisation energy

Ionisation of an atom involves the loss of an electron to form a *positive* ion.

The **first ionisation energy** is defined as the energy required to remove *one* electron from a gaseous atom.

The first ionisation energy of an atom can be represented by the following general equation:

$$X(g) \rightarrow X^+(g) + e^- \quad \Delta H + ve$$

Since all ionisations require energy, they are **endothermic** processes, and have a *positive* **enthalpy change** (ΔH) value – see page 96.

In **mole** quantities, the first ionisation energy is defined as the energy required to remove one electron from each atom in a mole of gaseous atoms to produce a mole of gaseous unipositive ions.

The value of the first ionisation energy depends upon two main factors:

- the size of the nuclear charge
- the energy of the electron that is removed (this depends upon its distance from the nucleus).

As the size of the nuclear charge increases, the force of attraction between the negatively charged electrons and the positively charged nucleus increases. More energy is needed to overcome the stronger force of attraction, and hence the value of the ionisation energy increases too.

As the energy of the electron increases, the electron is farther from the nucleus. As a result, the force of attraction between the nucleus and the electron *decreases*. Less energy is needed to overcome this force of attraction, hence the value of the ionisation energy decreases.

Knowledge of the full electronic structure of the elements is therefore an essential aspect of understanding trends in ionisation energy.

1 Write an equation to represent the first ionisation of aluminium.

Trends across a Period

Look at Table 1. This shows that in the third Period in the Periodic Table, sodium to argon, the link between electronic structure and first ionisation energy is very important.

Going across a Period, the size of the first ionisation energy shows a general increase, as the electron removed comes from the same principal energy level but the size of the **nuclear charge** steadily increases.

There are two breaks in this pattern of general increase in ionisation energy going across a Period.

	Electronic structure	Ionisation energy /kJ mol^{-1}
Na	$1s^2 2s^2 2p^6 3s^1$	494
Mg	$1s^2 2s^2 2p^6 3s^2$	736
Al	$1s^2 2s^2 2p^6 3s^2 3p^1$	577
Si	$1s^2 2s^2 2p^6 3s^2 3p^2$	786
P	$1s^2 2s^2 2p^6 3s^2 3p^3$	1060
S	$1s^2 2s^2 2p^6 3s^2 3p^4$	1000
Cl	$1s^2 2s^2 2p^6 3s^2 3p^5$	1260
Ar	$1s^2 2s^2 2p^6 3s^2 3p^6$	1520

Table 1 *Electronic structures and first ionisation energies for the elements in the third Period*

Fig. 1 *Graph of first ionisation energies for the elements hydrogen to calcium*

In the third Period the first break in the pattern occurs between magnesium and aluminium. The first ionisation of aluminium is less than that of magnesium, despite the increase in the nuclear charge. The reason for this is that the outer electron removed from aluminium is in a higher sub-level: the electron removed from aluminium is a 3p electron, whereas that removed from magnesium is a 3s electron. A similar break in the pattern occurs in the second Period between beryllium and boron.

There is also a break in the pattern in the third Period between phosphorus and sulphur. The first ionisation energy of sulphur is less than that of phosphorus, despite the increase in the nuclear charge. In both cases the electron removed is from the 3p sub-level. However, the 3p electron removed from sulphur is a **paired electron**, whereas the 3p electron removed from phosphorus is an **unpaired electron**. When the electrons are paired the extra **mutual repulsion** results in less energy being required to remove an electron, and hence a reduction in the ionisation energy. The first ionisation energy of oxygen is less than that of nitrogen for the same reason.

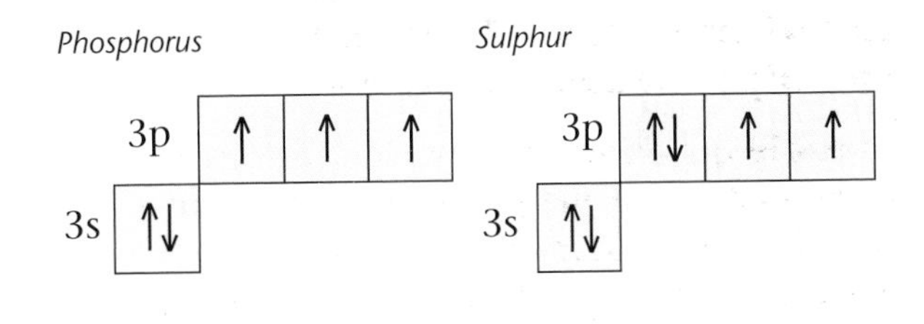

2 Explain why the first ionisation of boron is less than that of beryllium.

Key Ideas 18 – 19

- **Ionisation involves the loss of electrons from gaseous atoms, thereby forming gaseous positive ions.**
- **The value of the ionisation energy depends upon the nuclear charge and the energy of the electron that is removed.**
- **Trends in first ionisation energies across a Period provide evidence for principal energy levels and sub-levels.**

9 Trends down a Group

The trends in the first ionisation energies of the elements down a Group and the successive ionisation energies of an element provide evidence for principal energy levels.

Trends in ionisation energy

The ionisation energy *decreases* going down a Group.

For example, here are the first ionisation energies ($kJ\,mol^{-1}$) for the elements in Group II:

beryllium	900
magnesium	736
calcium	590
strontium	548
barium	502

Going down a Group in the Periodic Table, the electron removed during the first ionisation is from a successively higher principal energy level and hence is further from the nucleus. The nuclear charge also increases, but the effect of the increased nuclear charge is reduced by the inner electrons which **shield** or **screen** the outer electron from the nuclear charge. The extra mutual repulsion between the outer electrons and the greater number of inner electrons together reduce the effect of the nuclear charge.

The idea of the shielding or screening effect of inner electrons is an acceptable way of explaining the trend in first ionisation energies *down a Group* in the Periodic Table. It should not be used as an explanation of trends *across a Period* in the Periodic Table.

Q

1 Explain why sodium has a higher first ionisation energy than potassium.

Successive ionisation energies

Once an electron has been removed from an atom, the remaining electrons are more tightly held by the unchanged nuclear charge. There is less mutual repulsion between the remaining electrons, and they are drawn closer to the nucleus as the force of attraction between the nucleus and the electrons increases. Hence successive ionisation energies always show an increase in value.

These equations represent the first four ionisation energies of magnesium:

1st	$Mg(g) \rightarrow Mg^{+}(g) + e^{-}$	$\Delta H = +736\,kJ\,mol^{-1}$
2nd	$Mg^{+}(g) \rightarrow Mg^{2+}(g) + e^{-}$	$\Delta H = +1450\,kJ\,mol^{-1}$
3rd	$Mg^{2+}(g) \rightarrow Mg^{3+}(g) + e^{-}$	$\Delta H = +7740\,kJ\,mol^{-1}$
4th	$Mg^{3+}(g) \rightarrow Mg^{4+}(g) + e^{-}$	$\Delta H = +10\,500\,kJ\,mol^{-1}$

The *large* increase in the third ionisation energy compared to the first and second ionisation energies occurs because the third electron removed is from a lower principal energy level and is under greater influence from the nuclear charge.

A *logarithmic* scale is used in graphs of successive ionisation energies because of the large range in values.

The electronic structure of magnesium is $1s^2 2s^2 2p^6 3s^2$. The 3s electrons are removed first, followed by the 2p electrons, then by the 2s electrons, and finally the 1s electrons.

The two breaks in the graph are evidence of principal energy levels. The first break occurs after the second electron has been removed: the third electron is removed from a second principal energy level. The second break occurs after the tenth electron has been removed: the eleventh electron removed is from the first principal energy level.

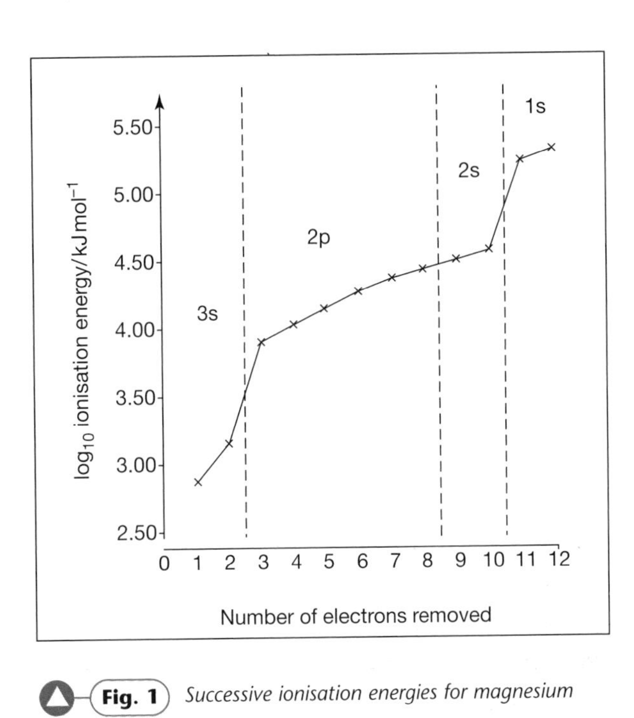

Fig. 1 *Successive ionisation energies for magnesium*

Examination questions

It is important that answers to questions comparing ionisation energy should refer to:

- the electronic structure
- the nuclear charge.

1 Why does helium have the highest first ionisation energy of all the elements?

The electron removed from helium ($1s^2$) is from the lowest principal energy level, and the nuclear charge is greater than that of hydrogen ($1s^1$), whose first ionisation energy also involves removing an electron from the first principal energy level.

2 Why is the first ionisation energy of sodium less than that of neon, but the second ionisation energy of sodium is greater than the second ionisation energy of neon?

For the first ionisation energies the electron removed from sodium ($1s^2 2s^2 2p^6 3s^1$) is from the third principal energy level, but the electron removed from neon ($1s^2 2s^2 2p^6$) is from the second principal energy level. For the second ionisation energies in both cases the electron removed is from the second principal energy level, but the sodium ion has the greater nuclear charge.

Q

2 Sketch a graph to show the successive ionisation energies for potassium.

3 Explain why the fourth ionisation energy of aluminium is much greater than the third ionisation energy.

Key Ideas 20 – 21

- **Ionisation energies decrease down a Group since the electron removed is from a higher principal energy level.**
- **Successive ionisation energies for an element increase, since the remaining electrons are pulled closer to the nucleus and are more tightly held.**

Unit 1 Questions

10

1. Copy and complete the following table, giving the relative masses and the relative charges of the three sub-atomic particles. (3)

	Relative mass	Relative charge
Proton		
Neutron		
Electron		

2. Define the terms *atomic number* and *mass number*. (2)

3. **a)** Define the term *isotopes*. (2)
   b) Explain why isotopes have identical chemical properties but different physical properties. (4)

4. Give the full electronic structure for each of the following atoms and ions:

   a) chlorine
   b) magnesium
   c) potassium
   d) copper
   e) argon
   f) B^{3+}
   g) O^{2-}
   h) Cr^{3+}
   (8)

5. **a)** Explain briefly the processes in a mass spectrometer by which a sample of an element is: (i) ionised; (ii) accelerated; (iii) deflected; (iv) detected. (8)
   b) A sample of iron was analysed in a mass spectrometer. Four peaks with the *m/z* values shown in the table were observed in the mass spectrum.

Relative abundance/%	5.8	91.6	2.2	0.33
m/z	54	56	57	58

   (i) Define the term *relative atomic mass* of an element.
   (ii) Calculate the relative atomic mass of iron from the mass spectrum data. (5)

6) The first ionisation energies of the elements neon to phosphorus are given in the table.

Element	Ne	Na	Mg	Al	Si	P	S
Ionisation energy/kJ mol^{-1}	2080	494	736	577	786	1060	

a) Explain the meaning of the term *first ionisation energy*. (2)

b) Write an equation for the process involved in the first ionisation energy of sodium. (2)

c) Explain why the first ionisation energy of sodium is much less than the first ionisation energy of neon. (2)

d) Explain why the first ionisation energy of magnesium is greater than that of sodium. (2)

e) Explain why the first ionisation energy of aluminium is smaller than that of magnesium. (2)

f) Predict, and explain your reasoning, whether the first ionisation energy of sulphur is smaller or larger than the first ionisation energy of phosphorus. (2)

7)

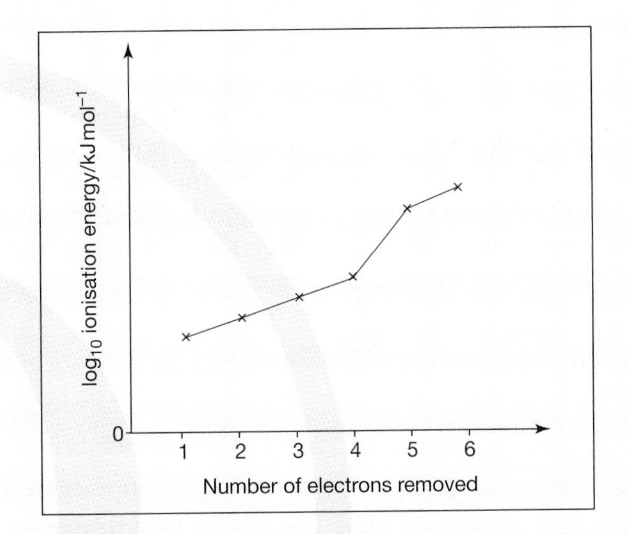

The graph shows the successive ionisation energies for the element carbon.

a) Explain why a logarithmic scale is used. (1)

b) Explain why successive ionisation energies increase in value. (3)

c) Explain the shape of the graph in terms of the electronic structure of carbon. (4)

1 How atoms join together ...

At GCSE you learned that compounds are formed when the atoms of elements combine chemically. The outer electrons of the atoms are involved in bonding. The atoms can lose, gain or share electrons when forming compounds. *Ionic* compounds are formed by the *transfer* of electrons between atoms, and *covalent* compounds are formed by the *sharing* of electrons between the atoms.

Ionic bonding

An **ionic bond** is the force of attraction between *oppositely charged ions*. The ions are formed by the *transfer* of electrons from metal atoms to non-metal atoms. It is only the *outer* electrons (those in the highest occupied principal energy level) that are involved in bonding.

The metal atoms lose electrons and form **positive ions** (**cations**); the non-metal atoms gain electrons and form **negative ions** (**anions**). The **charge** on the ion depends upon the number of electrons lost or gained.

When discussing the formation of the ions in an ionic compound, it is convenient to use the electronic structure from GCSE to show the outer electrons. An example of an ionic compound is sodium chloride. The sodium atom, Na, has an electronic structure of 2,8,1. Loss of an electron results in the formation of a sodium ion, Na^+, with electronic structure 2,8. The chlorine atom, Cl, has an electronic structure of 2,8,7. Gain of one electron forms the chloride ion, Cl^-, with electronic structure 2,8,8. Hence the formula of sodium chloride is NaCl.

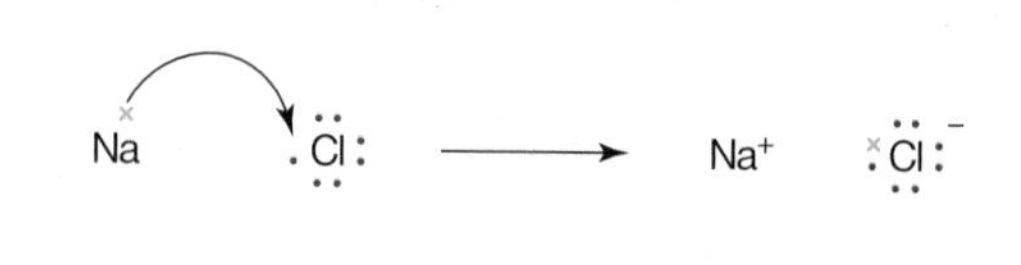

Fig. 1 *Formation of the ions in NaCl*

The number of electrons lost or gained when an atom forms an ion depends upon the electronic structure, and hence on the Group number in the Periodic Table. (The Group number is equal to the number of outer electrons in the atom.)

When the s block and p block elements gain or lose electrons to form ions, they achieve the electronic structure of a **noble gas**. (This does *not* apply to most of the positive ions formed by the transition metals.) The metal atoms *lose* their outer electrons to achieve a noble gas electronic structure; the non-metal atoms *gain* electrons to achieve a noble gas electronic structure.

- Group I metals form M^+ ions
- Group II metals form M^{2+} ions
- Group III metals form M^{3+} ions
- Group VII non-metals form X^- ions
- Group VI non-metals form X^{2-} ions
- Group V non-metals form X^{3-} ions

Since an ionic compound does not have a net charge, the charges on the positive and negative ions must be balanced. The number of positive and negative ions needed to balance the charges is reflected in the *formula* of the compound.

Figure 2 shows further examples of the formation of ions in ionic compounds.

Q

1 Use dot-cross diagrams to show the formation of the ions in:

a) K_2O b) $CaCl_2$ c) MgO d) AlF_3

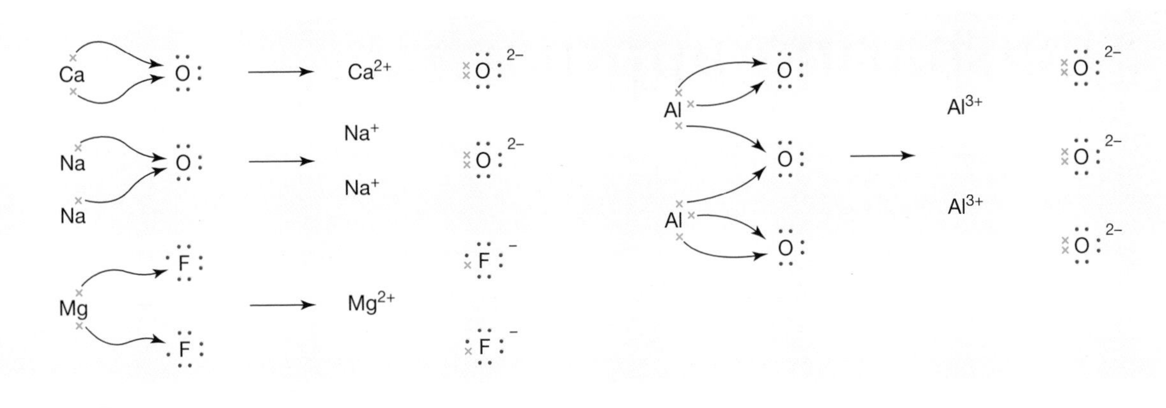

Fig. 2 *Formation of the ions in CaO, Na_2O, MgF_2 and Al_2O_3*

Properties of ionic compounds

There is a strong **electrostatic force** of attraction between the oppositely charged ions. It requires a lot of energy to overcome these strong forces of attraction, and as a result ionic compounds are solids with high melting and boiling points.

The strength of the electrostatic forces of attraction between oppositely charged ions depends upon the size and charge on the ions. The electrostatic forces increase in strength as

- the charge on the ions increases
- the size of the ions decreases.

The melting point of CaF_2 is greater than the melting point of $CaCl_2$ because the fluoride ion is smaller than the chloride ion: there is a greater force of attraction between the calcium ion and the the fluoride ion, hence more energy is needed to overcome the forces of attraction between the ions.

Ionic compounds exist as regular arrangements of ions in a **giant ionic lattice**. When an ionic compound is melted or in solution, the lattice holding the ions together breaks down and the ions are free to move. When the ions are free to move, the ionic compound will conduct electricity.

Q

2 Explain the differences in the melting points (given in K) of the following ionic compounds:

NaCl	KCl	MgO
1074	1043	3125

Key Ideas 24 – 25

- **The ionic bond is the force of attraction between oppositely charged ions.**
- **Ions are formed by atoms losing or gaining electrons, and as a result achieving the electronic structure of the nearest noble gas.**

2 The sharing approach ...

At GCSE covalent bonding is thought of as the sharing of a pair of electrons between two atoms. By sharing electrons each atom achieves the electronic structure of a noble gas. AS requires a deeper understanding of bonding.

Covalent bonding

Covalent bonding involves the *sharing* of a **pair of electrons** between two atoms. A covalent bond is formed when two atomic orbitals, each containing a single unpaired electron, overlap. Each electron is then attracted to the nuclei of both atoms. It is this attraction that holds the atoms together in a covalent bond.

The amount (degree) of overlap of the atomic orbitals determines the strength of the bond: the greater the amount of overlap, the stronger the bond.

The formation of a covalent bond is best shown by the simplest example, the hydrogen molecule. Each hydrogen atom has an unpaired 1s electron. When the two 1s orbitals overlap, the two electrons are now shared between the two atoms (Figure 1).

A dot-cross diagram is usually used to show covalent bonding (Figure 2). Dots and crosses are used to show the electrons from different atoms. Only the electrons in the outermost energy level are involved in bonding.

When atoms combine by covalent bonds, **molecules** are formed.

Non-bonded pairs of outer electrons in a compound are called **lone pairs** of electrons. There are lone pairs on the oxygen atom in water and on the nitrogen atom in ammonia.

In the structural formula, a line between the atoms is used to represent the covalent bond.

In *most* cases, when covalent bonds are formed in a compound the atoms achieve the **electronic structure of a noble gas**. However there are many examples where this does *not* occur.

- In PCl_5 each phosphorus atom has *ten* outer electrons (called the **expansion of the octet**: see Figure 3).
- In $AlCl_3$ each aluminium atom has only *six* outer electrons.

Q

1 Use dot-cross diagrams to show the bonding in the following molecules.

a) CCl_4 b) PF_5 c) H_2S d) BF_3 e) $SiCl_4$ f) $BeCl_2$

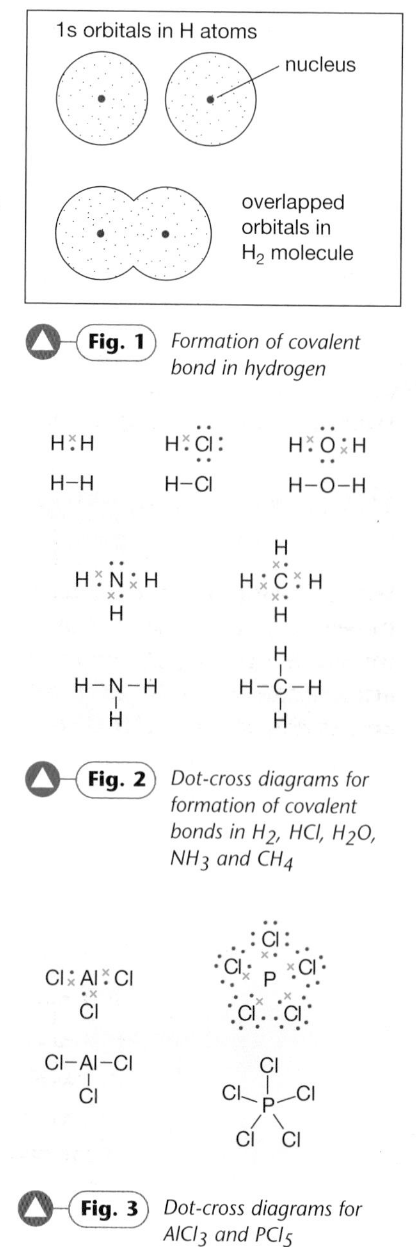

Fig. 1 Formation of covalent bond in hydrogen

Fig. 2 Dot-cross diagrams for formation of covalent bonds in H_2, HCl, H_2O, NH_3 and CH_4

Fig. 3 Dot-cross diagrams for $AlCl_3$ and PCl_5

Multiple bonds

Sometimes atoms can share *four* electrons to form a **double covalent bond**, or *six* electrons to form a **triple covalent bond**.

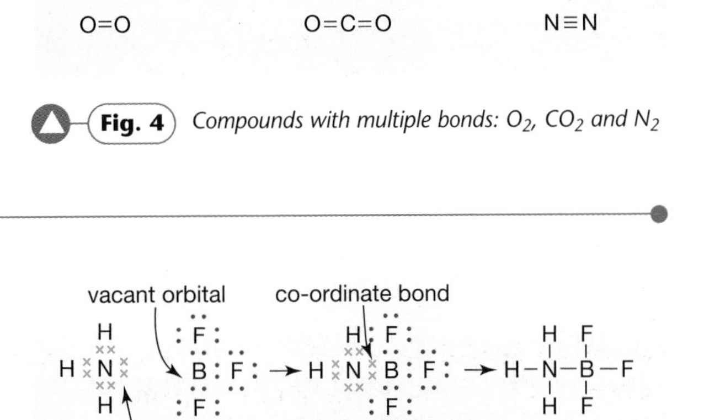

Fig. 4 *Compounds with multiple bonds: O_2, CO_2 and N_2*

Co-ordinate bonding

Co-ordinate or **dative covalent bonds** are formed when *one* atom contributes *both* of the electrons needed for the covalent bond. Once a co-ordinate bond has formed it is just the same as an ordinary covalent bond – the only difference is in the way that the bond has formed.

If two atoms are to form a co-ordinate bond between them, then one atom must have a lone pair of electrons and the other atom must have a vacant orbital.

NH_3 and H_2O have lone pairs of electrons; BF_3 and H^+ each have vacant orbitals. NH_4^+ and H_3O^+ are examples of **polyatomic ions**.

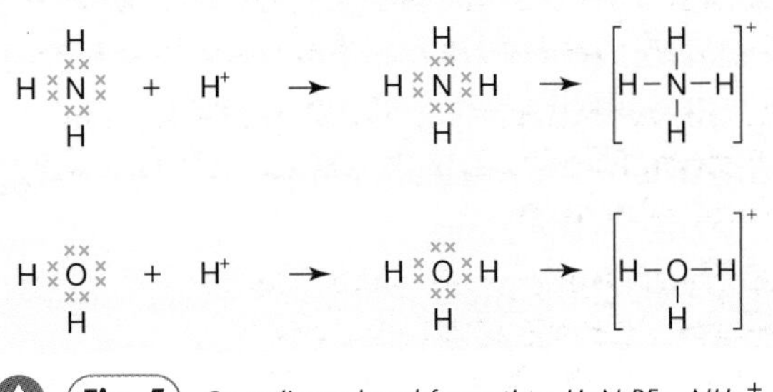

Fig. 5 *Co-ordinate bond formation: $H_3N.BF_3$, NH_4^+ and H_3O^+*

Properties of simple covalent compounds

Simple covalent compounds consist of small molecules. The covalent bonds *within* the molecules are strong, but the forces of attraction *between* the molecules are weak. The amount of energy needed to break these weak **intermolecular forces** is small, and simple covalent compounds therefore have low melting and boiling points, and are usually gases or volatile liquids.

There are no ions or free electrons present in simple covalent compounds, and they are **non-conductors** of electricity.

Q

2 Draw dot-cross diagrams to show the formation of a co-ordinate bond:

a) between PH_3 and H^+

b) between $AlCl_3$ and Cl^-

Key Ideas 26 – 27

- **A covalent bond is the sharing of a pair of electrons between two atoms.**
- **A covalent bond is formed by the overlap of atomic orbitals that contain unpaired electrons.**
- **Co-ordinate bonds are formed when one atom supplies the pair of electrons for a covalent bond. One atom has a lone pair of electrons; the other atom has a vacant orbital.**

3 Where have the electrons gone?

A covalent bond consists of a shared pair of electrons. In some cases the two electrons are shared equally between the two atoms, but in most cases the sharing is unequal. Unequal sharing of the pair of electrons in a covalent bond occurs when the two atoms have different electronegativities.

Electronegativity

Electronegativity is a measure of the relative ability of an atom to attract the pair of electrons in a covalent bond (to withdraw electron density).

Two factors help to determine the electronegativity value of an element:

1 the size of the nuclear charge
2 the size of the atom.

As the size of the nuclear charge increases, there will be an increased attraction between the nuclear charge and the pair of electrons in a covalent bond. Therefore the electronegativity *increases*.

As the size of the atom increases, the pair of electrons in the covalent bond will be further away from the nucleus, and there will be a decreased attraction from the nuclear charge. Therefore the electronegativity *decreases*. There will also be a shielding effect of the inner electrons.

Two trends are evident from the values given in Table 1.

1 Going across a Period, the electronegativity value *increases* as the nuclear charge increases, the size of the atoms decreases and hence there is a greater attraction between the nucleus and the pair of electrons in the covalent bond.
2 Going down a Group, the electronegativity value *decreases*. The effect of the increase in nuclear charge is less than the effect of increase in atomic radius and the shielding effect of the inner electrons.

If the electronegativity values of the two atoms forming a covalent bond are the same, then the pair of electrons will be equally shared and the covalent bond will be **non-polar**.

H							He
2.1							–
Li	Be	B	C	N	O	F	Ne
1.0	1.5	2.0	2.5	3.0	3.5	4.0	–
Na	Mg	Al	Si	P	S	Cl	Ar
0.9	1.2	1.5	1.8	2.1	2.5	3.0	–
K						Br	
0.8						2.8	
Rb						I	
0.8						2.5	
Cs						At	
0.7						2.2	

Table 1 *Electronegativity values for some selected elements. There are no values for the noble gases, since they do not form covalent bonds.*

Q

1 Explain why chlorine has a larger electronegativity value than bromine.
2 Explain why oxygen is more electronegative than nitrogen.

Polar bonds

When a covalent bond is formed between atoms with different electronegativity values, the *more* electronegative atom will have a *greater* attraction for the pair of electrons in the covalent bond. This causes the pair of electrons in the covalent bond to be pulled *towards* the more electronegative atom. This leads to a **polar bond** with an unequal sharing (**asymmetric distribution**) of the pair of electrons.

As a result the two atoms have partial charges, shown as δ^+ or δ^-. The *more* electronegative atom will have a δ^- charge and the *less* electronegative atom will have a δ^+ charge.

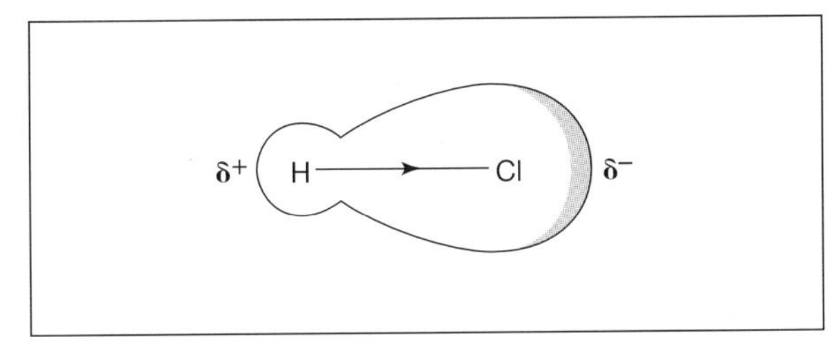

Fig. 1 *Polar bonds*

As the *difference* in electronegativity values of the atoms increases, the covalent bond formed between the two atoms becomes more polar.

> **Q**
> 3 Explain why the H–Cl bond is more polar than the H–Br bond.

A polar bond behaves as a small bar magnet and has a **dipole moment**. The size of the dipole moment depends upon the size of the partial charges on the atoms.

When there are polar bonds present it is important to consider the whole molecule. Some molecules that contain polar bonds have *no* overall polarity. This occurs when the dipole moments counteract each other, leaving no net dipole moment. This happens in molecules that have symmetry.

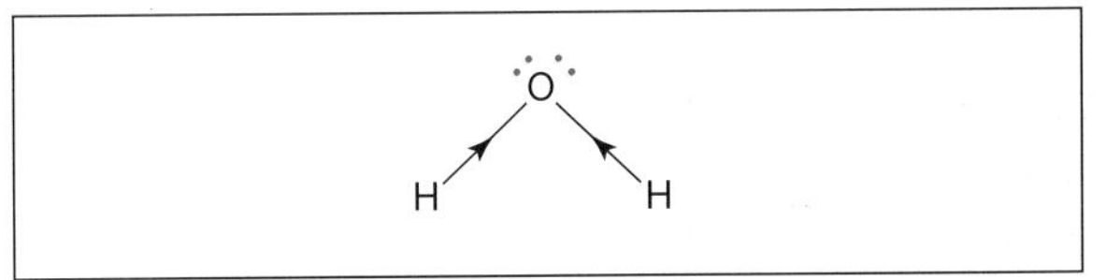

Fig. 2 *A molecule with overall polarity*

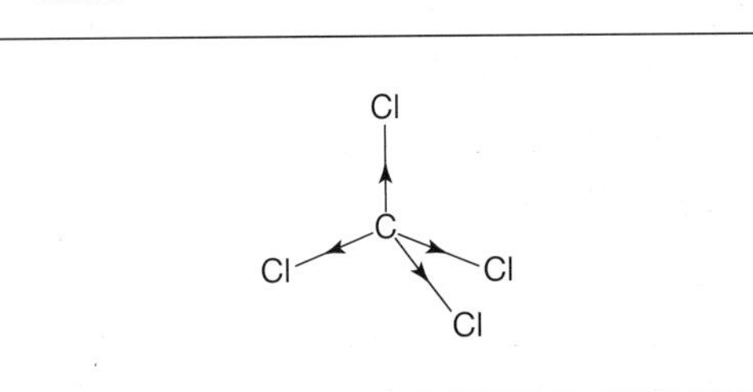

Fig. 3 *A molecule with no overall polarity*

> **Q**
> 4 Define the term *electronegativity.*
>
> 5 State and explain the trend in electronegativity across a Period.
>
> 6 Explain why some covalent bonds are polar.

Key Ideas 28 – 29

- **The pair of electrons in a covalent bond may not be shared equally, leading to the bond being polar.**
- **Electronegativity is a measure of the ability of an atom to attract the pair of electrons in a covalent bond.**
- **Electronegativity increases going across a Period and going up a Group.**
- **It is the difference in electronegativities that determines the polarity of a covalent bond.**

4 How some ions share electrons

When ions are formed by electron transfer we can picture the ions as acting as small charged spheres. The ions are held together by the strong force of attraction between the opposite charges on the ions. In some cases the electron clouds on the negative ions are *distorted* by the positive ions. The negative ions are polarised by the positive ions, and this can lead to the ionic compound having some covalent character.

Polarisation of ions

In simple ionic compounds the negative ion is often much larger than the positive ion (see Table 1).

Li^+	Be^{2+}		O^{2-}	F^-
0.060	0.020		0.140	0.136
Na^+	Mg^{2+}	Al^{3+}	S^{2-}	Cl^-
0.095	0.065	0.050	0.184	0.181
K^+	Ca^{2+}			Br^-
0.133	0.099			0.195
Rb^+				I^-
0.148				0.216

The ionic radii of some common ions (nanometres, 10^{-9} m)

Note that:

- positive ions are much *smaller* than the corresponding atoms
- negative ions are much *larger* than the corresponding atoms.

Q

1 Explain why a sodium ion is much smaller than a sodium atom.

2 Explain why Mg^{2+} has a smaller ionic radius than Na^+.

3 Explain why the O^{2-} ion is larger than an oxygen atom.

The electron cloud around the negative ion may be drawn towards the positive ion. The electron cloud on the negative ion is **polarised** by the positive ion. This polarisation effect *increases* as:

- the size of the positive ions *decreases*
- the charge on the positive ion *increases*
- the size of the negative ion *increases*.

Positive ions with a large charge-to-size ratio (high **charge density**) are highly polarising.

This effect can be illustrated by considering the bonding in the Group I metals. The atoms lose one electron to form an M^+ ion. The lithium ion, Li^+, is the smallest ion in the Group, and as a result is the most polarising. This results in lithium iodide, LiI, being predominantly covalent, whilst the other Group I halides are ionic.

The covalent character of lithium iodide arises from the fact that the Li^+ ion polarises the large iodide ion to such an extent that there is region of electron density between the lithium ion and the iodide ions (Figure 1). This results in some sharing of electrons between the ions.

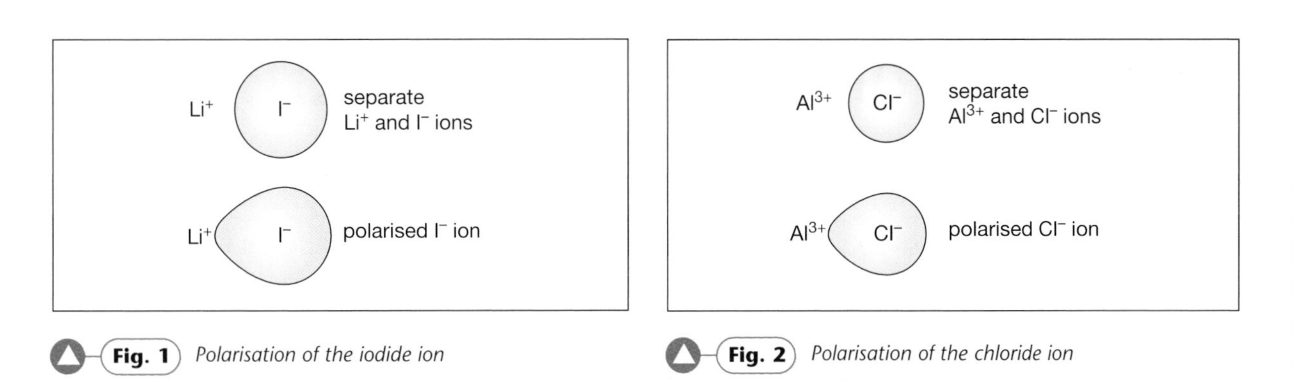

Fig. 1 *Polarisation of the iodide ion*

Fig. 2 *Polarisation of the chloride ion*

One example of the covalent character in lithium iodide is that it is *soluble* in non-polar organic solvents such as hexane or butane, and is *insoluble* in water. The other Group I halides are all ionic and dissolve in water.

It might be expected that aluminium chloride, $AlCl_3$, would be an ionic compound. However the Al^{3+} ion has a very high charge density and is very highly polarising. This results in the chloride ions being so polarised that in effect the electrons are shared with the aluminium ion, and aluminium chloride is covalent.

Electronegativity and bonding

In general we can use electronegativity values to predict the type of bonding in simple compounds. When the electronegativity differences are *small* then the bonding will be *covalent*. When the electronegativity differences are *large* then the bonding will be *ionic* (but if the positive ion is very highly polarising, the bonding may be predominantly covalent).

Q

4 The Group I halides are ionic compounds which dissolve in water. Explain why lithium iodide is insoluble in water but does dissolve in non-polar organic solvents.

Key Ideas 30 – 31

- **In some ionic compounds highly polarising positive ions attract the electrons in negative ions, causing the compound to have some covalent character.**
- **Polarising power depends upon the charge-to-size ratio of positive ions.**
- **Electronegativity differences can be used to predict the type of bonding in simple compounds.**

5 Let the force be with you ...

The covalent bonds within molecules are strong bonds, but the forces of attraction between molecules are weak. There are three different types of **intermolecular forces** that occur between the molecules of covalent compounds: van der Waals forces (induced-dipole forces), permanent dipole–dipole forces, and hydrogen bonding.

Van der Waals forces

The electrons in a molecule are constantly moving, and at any one instant there may be an unequal distribution of the electron cloud. This results in a **temporary dipole** on the molecule. **Van der Waals forces** occur when a temporary dipole on one molecule induces a dipole on a neighbouring molecule. There is an electrostatic force of attraction between the δ^+ charge on one molecule and the δ^- charge on a neighbouring molecule.

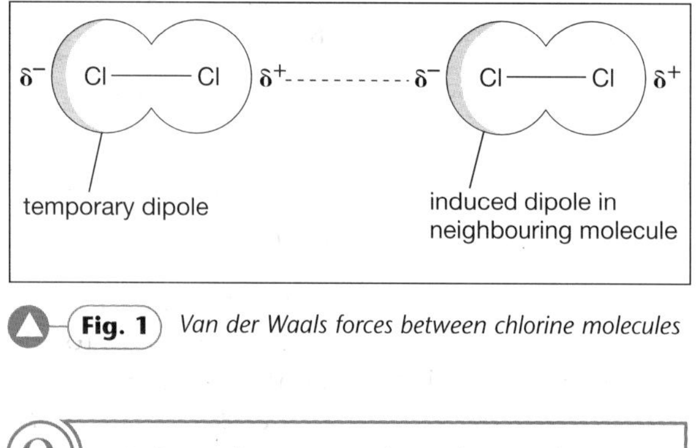

Fig. 1 *Van der Waals forces between chlorine molecules*

> **Q**
> 1 Draw diagrams to show the van der Waals forces between molecules of:
> a) hydrogen, H_2 b) oxygen, O_2

Van der Waals forces are the weakest intermolecular forces, and are easily broken. The strength of van der Waals forces depends upon the size of the molecules and the area of contact between molecules. The strength of van der Waals forces *increases* as:

- the size of the molecule *increases* (larger electron cloud)
- the points of contact between molecules *increase*.

This trend can illustrated by considering the boiling points of the halogens and the alkanes. The halogens all exist as diatomic molecules. Going down Group VII, the size of the halogen molecule increases. This means that the size of the electron cloud increases, and hence the amount of temporary distortion of the electron cloud increases, leading to stronger van der Waals forces.

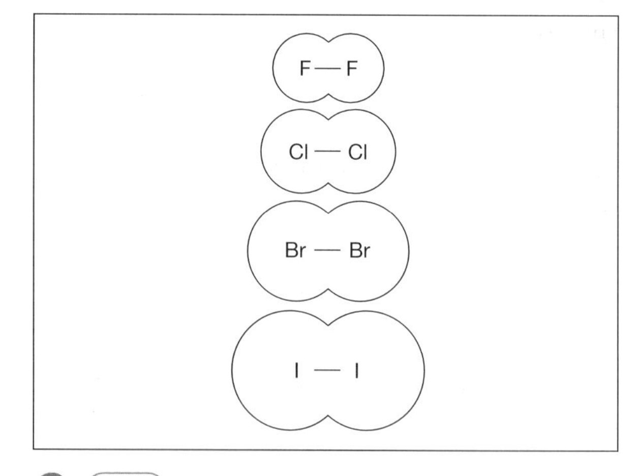

Fig. 2 *Halogen molecules*

The stronger the van der Waals forces between molecules, the higher the boiling point of the substance, as more energy is required to overcome these forces and release individual molecules as gases.

Halogen	F_2	Cl_2	Br_2	I_2
Boiling point/°C	–188	–35	59	184

A similar pattern occurs with the alkanes (see Module 3).

Name	Formula	Structural formula	Boiling point/°C
Methane	CH_4	CH_4	–164
Ethane	C_2H_6	CH_3CH_3	–89
Propane	C_3H_8	$CH_3CH_2CH_3$	–42
Butane	C_4H_{10}	$CH_3CH_2CH_2CH_3$	–1

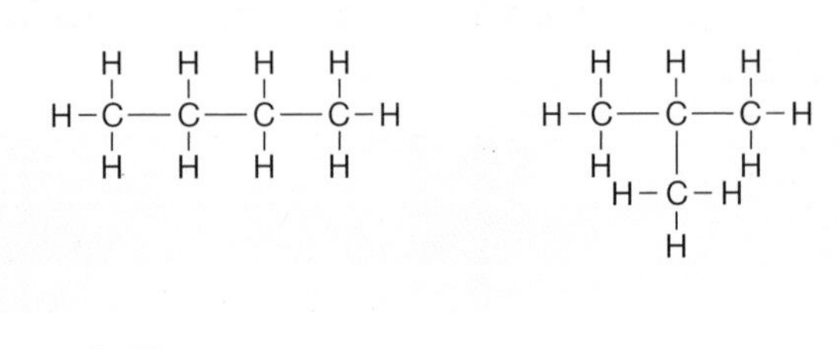

Fig. 3 *Isomers of butane*

As the size of the alkane molecules increases, the size of the van der Waals forces between molecules increases, and the boiling point increases.

Butane exists in two different forms. These two forms are called **isomers** (see page 171). Although the branched version is the same *size* as the straight-chain version, it has a more spherical shape and this results in there being fewer points of contact between the molecules. This results in weaker van der Waals forces and hence a lower boiling point.

Van der Waals forces are present between molecules of all covalent compounds.

Permanent dipole—dipole forces

A covalent bond is **polar** when the pair of electrons in the bond are unequally shared. Unequal sharing occurs when two atoms have different electronegativity values. This difference results in one atom having a δ^+ charge and the other atom having a δ^- charge. This in turn causes a **permanent dipole** (Figure 4).

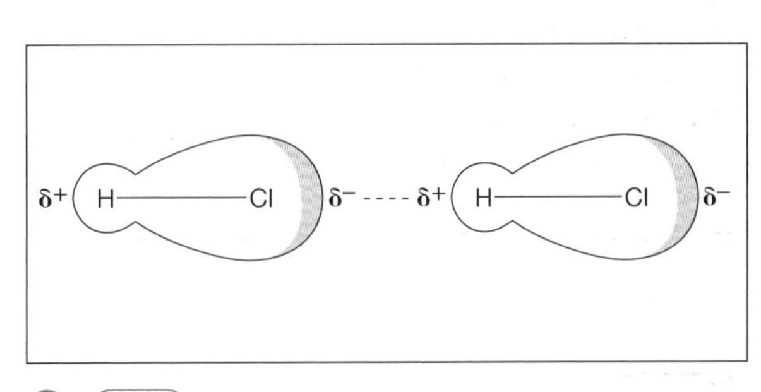

Fig. 4 *Dipole interaction between molecules of HCl*

Dipole–dipole forces occur between two molecules each of which has a permanent dipole. A dipole–dipole force is a force of attraction between the opposite charges on one molecule and a neighbouring molecule.

These permanent dipole–dipole forces are usually stronger than the *temporary* dipole forces in van der Waals forces. This leads to a higher boiling point for a compound with a permanent dipole compared to a similar-sized compound that only has van der Waals forces.

Q

2 Bromine, Br_2, and the interhalogen compound iodine monochloride, ICl, have different boiling points. Predict which species has the higher boiling point and explain why.

Key Ideas 32 – 33

- **The forces of attraction between simple covalent molecules are weak.**
- **There are three different types of intermolecular forces: van der Waals forces, permanent dipole forces, and hydrogen bonding.**

6 Let the force be with you ... (2)

The strongest of the intermolecular forces are hydrogen bonds. These are a special type of intermolecular force. They play a very important role in the physical chemistry of water, and in the chemistry of living systems including the structure of DNA and proteins.

Hydrogen bonding

Hydrogen bonding only occurs in compounds that have a hydrogen atom covalently bonded to an oxygen, a nitrogen or a fluorine atom. These are the three most electronegative atoms.

When a hydrogen atom is covalently bonded to a highly electronegative atom, the bond is polarised to such an extent that the δ^+ hydrogen atom can form a weak bond with a *different* fluorine, oxygen or nitrogen atom on a neighbouring molecule.

The oxygen, nitrogen and fluorine atoms are small enough to approach the δ^+ hydrogen atom closely enough for a hydrogen bond to be formed. Chlorine has the same electronegativity as oxygen, but the chlorine atom is too large for the chlorine atom to get close enough to a δ^+ hydrogen atom to form a hydrogen bond.

Q

1 Which of the following compounds have hydrogen bonding between molecules?

a) HBr b) CH_3OH c) NH_3 d) F_2O

Hydrogen bonding in water

Water can be used to illustrate hydrogen bonding. The pair of electrons in the oxygen—hydrogen covalent bond is pulled so close to the oxygen atom that the hydrogen nucleus is unshielded. Now the lone pair of electrons on a *different* oxygen atom can approach close to the nucleus of the hydrogen atom. The hydrogen bond is the force of attraction between the lone pair of electrons on the oxygen atom and the δ^+ charge on the hydrogen atom covalently bonded to a different oxygen atom.

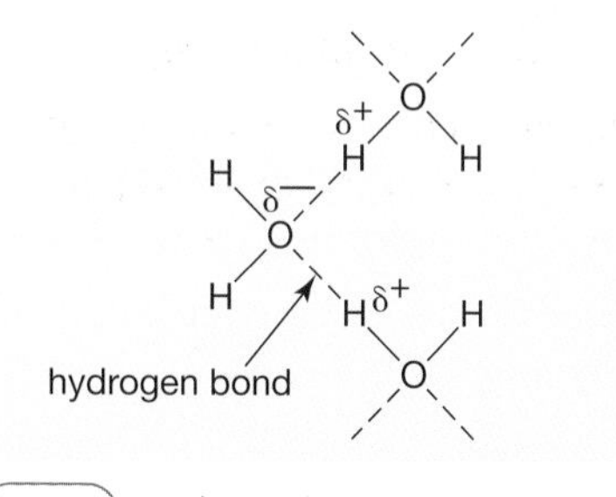

Fig. 1 *Hydrogen bonding in water*

Hydrogen bonds are directional in nature. In water, the lone pair of electrons on an oxygen atom is on the same axis as the adjacent hydrogen—oxygen covalent bond. In solid water (ice), the hydrogen bonds cause the water molecules to be slightly further apart than in liquid water. This results in ice having an 'open structure', and the density of solid water is less than that of liquid water.

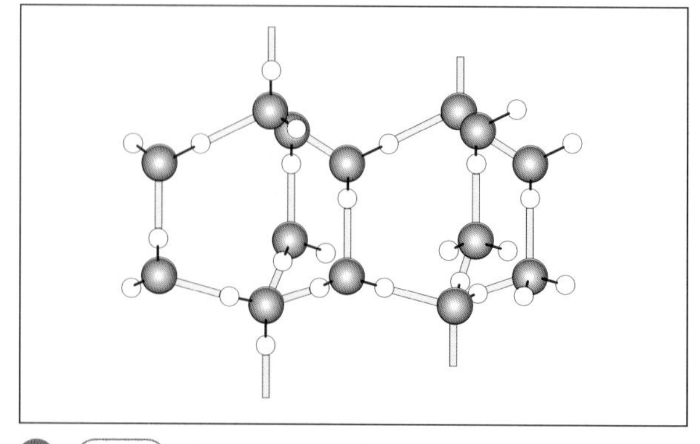

Fig. 2 *Open structure of ice*

Hydrogen bonding will occur with any compound that has a hydrogen atom covalently bonded to oxygen, nitrogen or fluorine. Examples include water, alcohols (ROH), carboxylic acids (RCOOH), hydrogen fluoride, ammonia, amines (RNH_2) and amino acids (see Module 3 and A2 Organic Chemistry).

Hydrogen bonding and boiling points

Type of bonding	Strength of bonds (kJ mol^{-1})
Covalent	150–900
Hydrogen bonding	20–40
Permanent dipole–dipole forces	5–20
van der Waals	1–20

Table 1 *Comparison of the strengths of intermolecular forces with the strength of covalent bonds*

Hydrogen bonds are the strongest intermolecular forces, and compounds that exhibit hydrogen bonding have higher than expected boiling points.

It might be expected that H_2S would have a higher boiling point than H_2O, since both covalent compounds have the same structure and bonding, but the H_2S molecule has a greater mass than the H_2O molecule. However the boiling point of H_2O is much higher than that of H_2S. This anomalous property is due to the fact that there is hydrogen bonding between water molecules and only dipole–dipole forces between H_2S molecules. Hydrogen bonding is also responsible for the anomalous boiling point of hydrogen fluoride, HF.

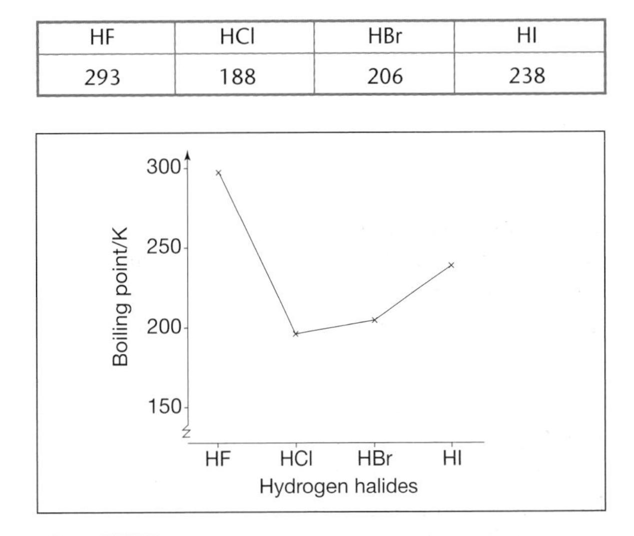

HF	HCl	HBr	HI
293	188	206	238

Fig. 3 *Boiling points of the hydrogen halides*

Hydrogen fluoride has a much higher boiling point than expected because there is hydrogen bonding between the HF molecules.

It is helpful to compare the boiling points of the hydrogen halides, as in Figure 3. Between molecules of HCl, HBr or HI are dipole—dipole forces, and these are weaker than hydrogen bonds (Table 1).

Q

2 Explain why the boiling point of ammonia, NH_3, is 240 K, but the boiling point of phosphine, PH_3, is 185 K.

Key Ideas 34 – 35

- **Hydrogen bonds are the strongest form of intermolecular forces.**
- **Hydrogen bonding occurs only with compounds that have a hydrogen atom covalently bonded to an atom of fluorine, oxygen or nitrogen.**

7 These are the states that matter ...

At GCSE you learned that there are three states of matter: solid, liquid and gas. At AS we concentrate on the behaviour of solids, liquids and gases in terms of the particles present, their motion and the forces acting between them.

States of matter

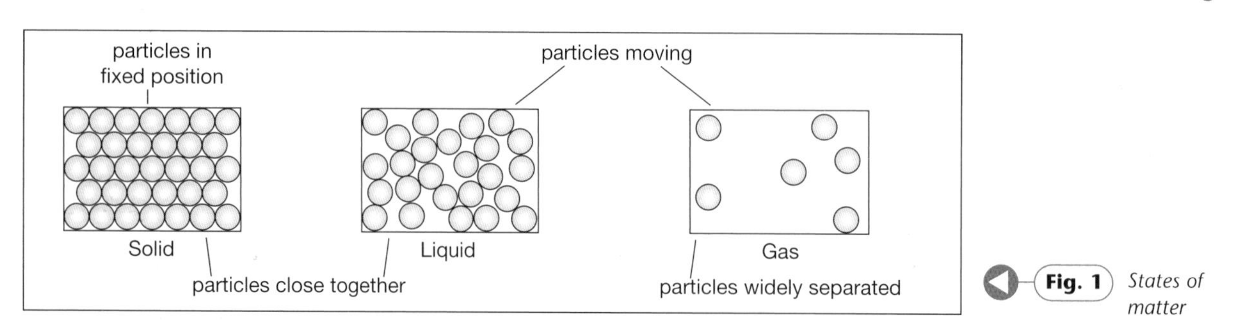

Fig. 1 *States of matter*

Solids

A **solid** has a fixed shape and a highly ordered structure.

In a solid the particles are close together and are densely packed. The particles are held together by the forces of attraction between the particles.

The particles in a solid **vibrate** about a fixed position, but are not free to move about (they have no translational kinetic energy).

Liquids

A **liquid** takes the shape of the containing vessel and has some degree of order.

In a liquid the particles are still close together, but the density is somewhat less than the density of a solid.

The particles in a liquid are free to move around.

Gases

A **gas** does not have a shape but occupies all of the volume available in its container.

The particles are widely spaced and the density of a gas is very low when compared to solids and liquids.

The particles in a gas are moving with rapid, random motion. It is this rapid random motion that allows a gas to spread out and occupy all of the available volume. A gas will **diffuse** by the particles moving randomly from an area of high concentration to an area of lower concentration.

Change of state

When a solid is heated, the temperature of the solid increases and the particles may gain sufficient energy to loosen the bonds between the particles and allow those particles to move. The solid melts and becomes a liquid.

When a liquid is heated, the particles may gain sufficient energy to overcome the forces of attraction between the particles and escape to form a gas.

When a solid is heated, the temperature will continue to rise until the substance reaches its melting point. The temperature remains constant until all of the solid has melted. The heat energy is used to break some of the forces between the particles.

When a liquid is heated, the temperature increases until it reaches its boiling point. The temperature remains constant until all of the liquid has been changed into a gas. The heat energy is used to break all of the forces between the particles, changing the liquid into gas.

A similar pattern occurs when a gas is cooled to form a liquid and then a solid. When the change of state occurs, the temperature remains constant.

Fig. 2 *Evaporation from a lake*

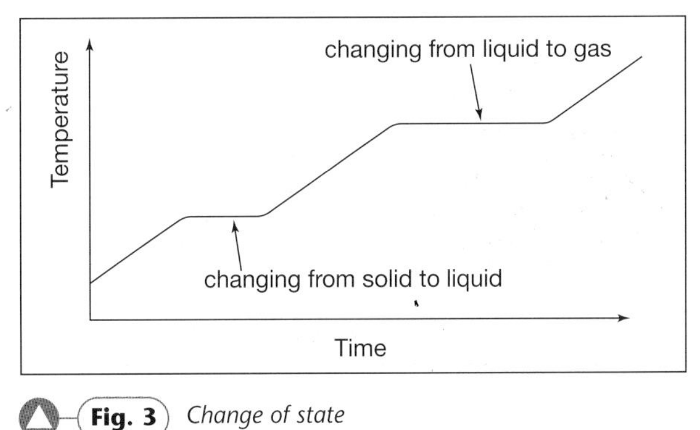

Fig. 3 *Change of state*

A *pure* substance has a sharp melting point and a sharp boiling point. Melting points and boiling points can be used as a test for the purity of a substance or as an aid to identification.

Evaporation

Evaporation is the change of a liquid into a gas without heating. The most energetic of the particles of a liquid may have sufficient energy to escape from the surface of a liquid and move into the gaseous state. Since the most energetic particles leave, this reduces the average energy of the remaining particles and hence evaporation causes the liquid to cool.

Q

1 Describe, in terms of the particles present, the changes that occur when a solid is heated until it becomes a liquid and then a gas.

Key Ideas 36 – 37

- **There are three states of matter: solid, liquid and gas.**
- **There are forces of attraction between the particles, and a change of state occurs when the particles have sufficient energy to loosen or break the forces of attraction.**
- **Pure substances have sharp melting and boiling points.**

8 Metallic and ionic solids

Many elements and compounds are solids at room temperature. There are different types of solids, each with different bonding and structure. The physical properties of solids depend upon their bonding and structure. Two examples of solids are metals and giant ionic lattices.

Metals

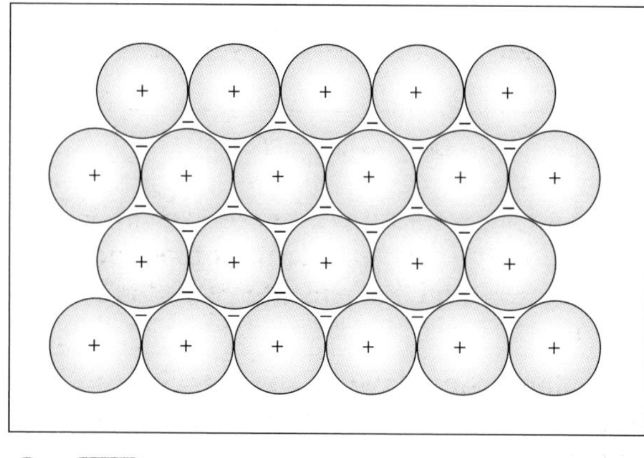

Fig. 1 *Metal structure*

In a metal solid the metal atoms are regularly arranged in a giant lattice (Figure 1). We can picture a metal structure as spherical atoms packed closely together. The outer electrons are **mobile** and are free to move throughout the metal structure: they are said to be **delocalised**. The delocalisation of the outer electrons leaves the metal atoms as ions or '**positive centres**'.

Metallic bonding is the force of attraction between the delocalised electrons and the 'positive centres'. We can picture the delocalised electrons as a 'sea of electrons' or 'glue' holding the positive centres together.

The strength of the metallic bonding depends upon the size and charge of the 'positive centres' and the number of mobile electrons per atom. The strength of metallic bonding *increases* as:

- the charge on the 'positive centre' *increases*
- the size of the 'positive centre' *decreases*
- the number of mobile electrons per atom *increases*.

Because metals have delocalised mobile electrons they are excellent conductors of electricity in both the solid and liquid states.

Q

1 Explain why magnesium is a better conductor of electricity than sodium.

2 Explain why aluminium has a higher melting point than sodium.

Giant ionic lattices

Ionic compounds form **giant ionic** crystals. In a giant ionic lattice the ions are arranged in a regular repeating pattern.

In sodium chloride each ion is surrounded by six of the oppositely charged ions to form a **face-centred cubic crystal** structure (Figure 2). In sodium chloride the **co-ordination number** (the number of nearest neighbours) for each ion is 6. The chloride ion is much larger than the sodium ion, and the sodium ions fit into the spaces between the chloride ions.

The type of ionic lattice formed by an ionic compound depends upon the relative sizes of the ions present. For example in caesium chloride each caesium ion is surrounded by eight chloride ions (Figure 3). The co-ordination number for each ion is 8.

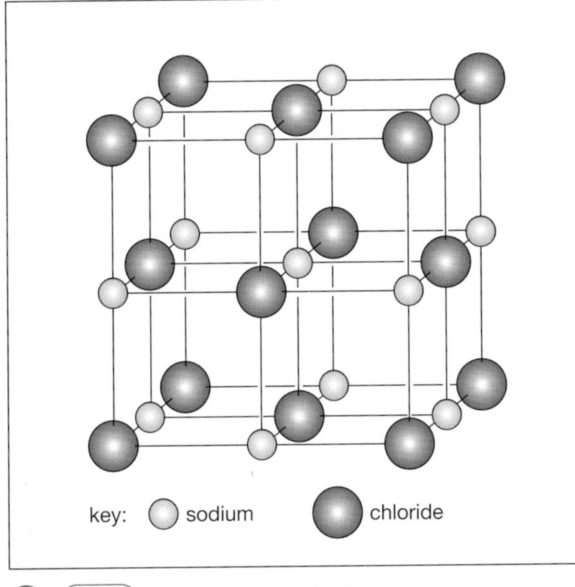

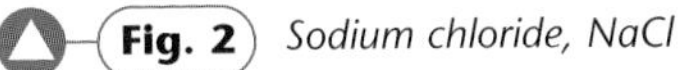
Fig. 2 *Sodium chloride, NaCl*

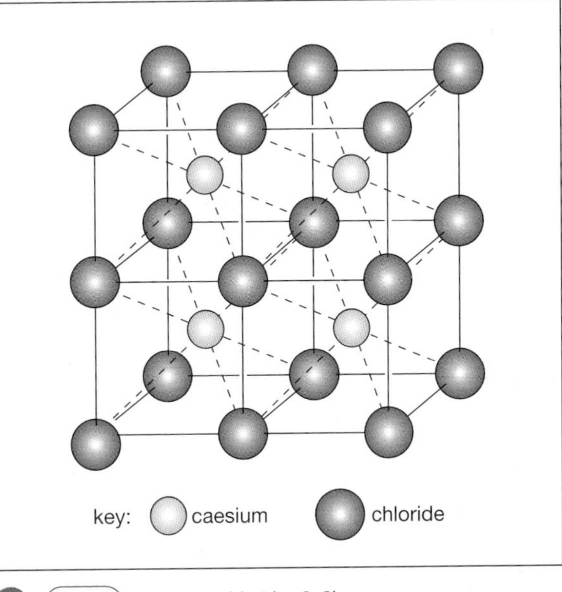

Fig. 3 *Caesium chloride, CsCl*

The reason why caesium chloride has a different ionic lattice structure from sodium chloride is that the caesium ion is larger than the sodium ion, which allows more chloride ions to 'pack' around the caesium ion. Caesium chloride has a **body-centred cubic crystal** structure.

A very large number of ionic bonds between the oppositely charged ions have to be weakened or broken in order to melt an ionic solid, and this results in ionic compounds having very high melting points and boiling points.

Solid ionic compounds do not conduct electricity because the ions are not free to move. When an ionic compound is melted or is in solution, the ions are free to move and the ionic compound can then conduct electricity.

Q

3 Explain the term *ionic bond*.

4 Draw dot-cross diagrams to show the formation of the ions in caesium chloride.

5 Explain why metals conduct electricity in the solid state, whereas ionic compounds will conduct electricity only when molten or dissolved in water.

Key Ideas 38 – 39

- **In metals the atoms are closely packed in a giant lattice.**
- **The outer electrons are free to move (delocalised) throughout the metal structure.**
- **Metallic bonding is the force of attraction between the positive centres and delocalised electrons.**
- **Ionic compounds form giant ionic lattices.**
- **When molten or dissolved in water, ionic compounds conduct electricity by the movement of ions.**

9 Covalently bonded solids

There are two types of solids that involve covalent bonding. Some simple molecules form molecular crystals in which the molecules are held together by van der Waals forces. Some elements and covalent compounds form giant covalent structures.

Molecular crystals

Some simple covalent compounds form **molecular crystals**. In solid iodine, the I_2 molecules are held together by van der Waals forces between the adjacent molecules. The distance between the iodine atoms *within* a molecule is less than the distance between iodine atoms in *adjacent* molecules.

The van der Waals forces are weak intermolecular forces: as a result, iodine has a low melting point and a low boiling point. When iodine is heated it changes directly from a black shiny solid into a purple vapour (**sublimation**), which consists of separate I_2 molecules.

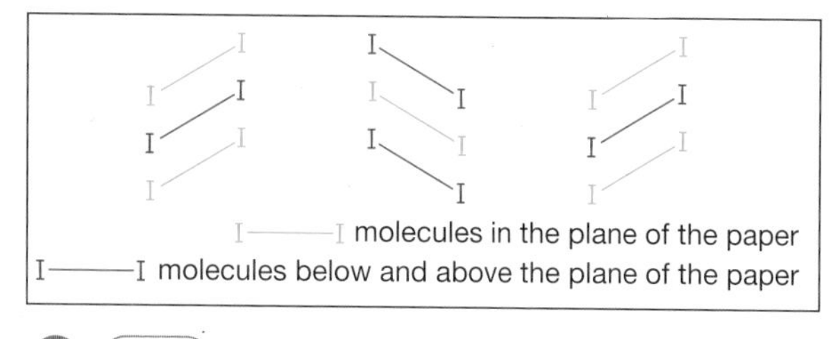

Fig. 1 *Structure of solid iodine*

Fig. 2 *Iodine crystals*

Macromolecular (giant covalent) crystals

Some elements form **giant covalent crystals**. Two examples are diamond and graphite (**allotropes** of carbon).

Diamond

In *diamond* each carbon atom is covalently bonded to four other carbon atoms in a giant covalent structure. The co-ordination number of the carbon atoms is 4.

The arrangement of the carbon atoms is tetrahedral and hence the structure is a **giant tetrahedral** structure. Diamond is a very rigid structure and is the hardest naturally occurring substance.

A great deal of energy is needed to break the large number of covalent bonds, so diamond has very high melting and boiling points.

All of the outer electrons in the carbon atoms are involved in the bonding and there are no free electrons. This results in diamond being a **non-conductor** of electricity.

Graphite

In *graphite* each carbon atom is covalently bonded to three other carbon atoms. The co-ordination number of the carbon is 3.

The arrangement of the carbon atoms is **trigonal planar**, leading to the formation of layers in which the carbon atoms form regular hexagons. The distance between the carbon atoms *within* the layers

Fig. 3 *Writing with a lump of graphite*

is less than the distance between the carbon atoms in *adjacent* layers.

Carbon has four outer electrons available for bonding. Each carbon atom in graphite uses three electrons in forming the covalent bonds. The remaining electron is **delocalised** between the plates of covalently bonded carbon atoms. The presence of the delocalised electrons results in graphite being an excellent electrical **conductor**.

There are weak van der Waals forces between the adjacent layers in graphite. The layers can easily slide over each other, and graphite is therefore soft.

As in diamond, the melting and boiling points of graphite are very high since a great deal of energy is needed to break the large number of covalent bonds.

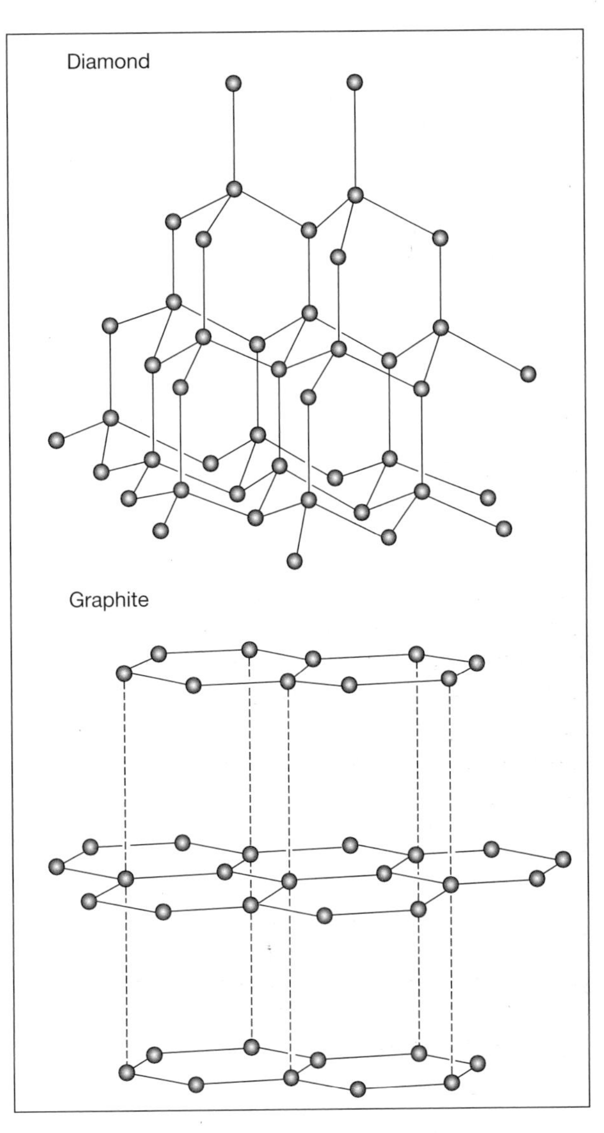

The structures of diamond and graphite **Fig. 4**

Q

1 Both carbon and iodine are non-metals. Explain why iodine has a low melting point but carbon has a very high melting point.

2 State *two* differences in the physical properties of diamond and graphite. Explain these differences in terms of the bonding and structure of diamond and graphite.

Key Ideas 40 – 41

- **Iodine forms molecular crystals in which the molecules are held together by van der Waals forces. These intermolecular forces are easily overcome and molecular crystals have low melting points and boiling points.**
- **Diamond and graphite form giant covalent structures. Because of the very large number of covalent bonds in their structures, diamond and graphite have very high melting points and boiling points.**

10 Chemistry gets you into shape ...

A covalent bond is formed by the overlap of atomic orbitals. An atomic orbital on one atom containing an unpaired electron overlaps with an atomic orbital on another atom containing an unpaired electron. Since atomic orbitals have a definite shape, this means that the covalent bond formed by atomic orbital overlap will be **directional**. This results in a covalent compound having a definite **shape**.

What determines the shape of a compound containing covalent bonds?

Because all electrons carry the same electrical charge, pairs of electrons will experience **mutual repulsion** with other pairs of electrons. This repulsion forces the pairs of electrons apart from other pairs of electrons. The mutual repulsion is minimised when the pairs of electrons are as far apart as possible.

The shape of a covalently bonded species will depend on the *number* of pairs of electrons around a central atom. These pairs of electrons may be **bond pairs** in covalent bonds or a **lone pair** of electrons. Lone pairs of electrons are non-bonded pairs of electrons in the highest occupied energy levels.

Table 1 gives details of the shapes and **bond angles** associated with different numbers of pairs of electrons around the central atom.

Number of pairs of electrons	Shape	Bond angles	Example
2	linear	180°	$BeCl_2$
3	trigonal planar	120°	BF_3
4	tetrahedral	109.5°	CH_4
5	triangular bipyramid	90° and 120°	PF_5
6	octahedral	90°	SF_6

Table 1

Q

1 Draw dot-cross diagrams for the following molecules and state the shape of each molecule.

a) $AlCl_3$ b) CCl_4 c) PCl_5

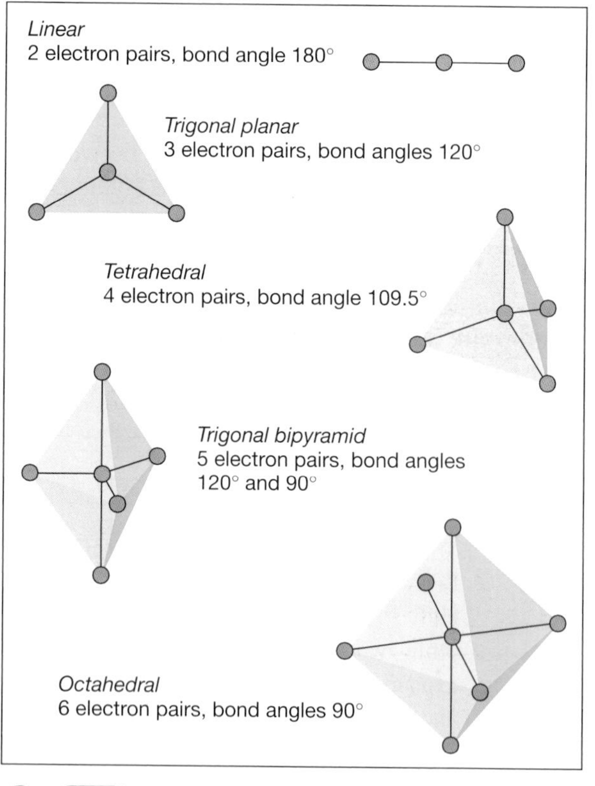

Fig. 1 *Shapes of molecules*

If all the pairs of electrons are bond pairs, the covalent compound will have a regular shape, with the bond angles as shown in Table 1. If, however, there are lone pairs present then the basic shape will be **distorted**. Lone pairs of electrons are more concentrated and are nearer the nucleus of an atom. This results in a different amount of mutual repulsion between the different pairs of electrons.

Order of increasing mutual repulsion: (↑ increasing)

- lone pair—lone pair
- lone pair—bond pair
- bond pair—bond pair

The *extra* mutual repulsion occurring when lone pairs are present results in the covalent bonds being pushed closer together. This results in a small reduction in the bond angle and a slight distortion to the shape.

- Methane, CH_4, has four bond pairs and hence is tetrahedral, with H—C—H bond angles of 109.5°.
- Ammonia, NH_3, has three bond pairs and a lone pair: the four pairs of electrons result in a tetrahedral shape, but the presence of the lone pair results in the covalent bonds being pushed closer together, giving a distorted tetrahedral shape with H—N—H bond angles of 107°. The shape of the ammonia molecule is referred to as a triangular pyramid.
- Water, H_2O, has two bond pairs and two lone pairs: the four pairs of electrons result in a tetrahedral shape, but the presence of two lone pairs results in the covalent bonds being pushed even closer, giving a distorted tetrahedral shape with an H—O—H bond angle of 105°. The water molecule is described as V-shaped.

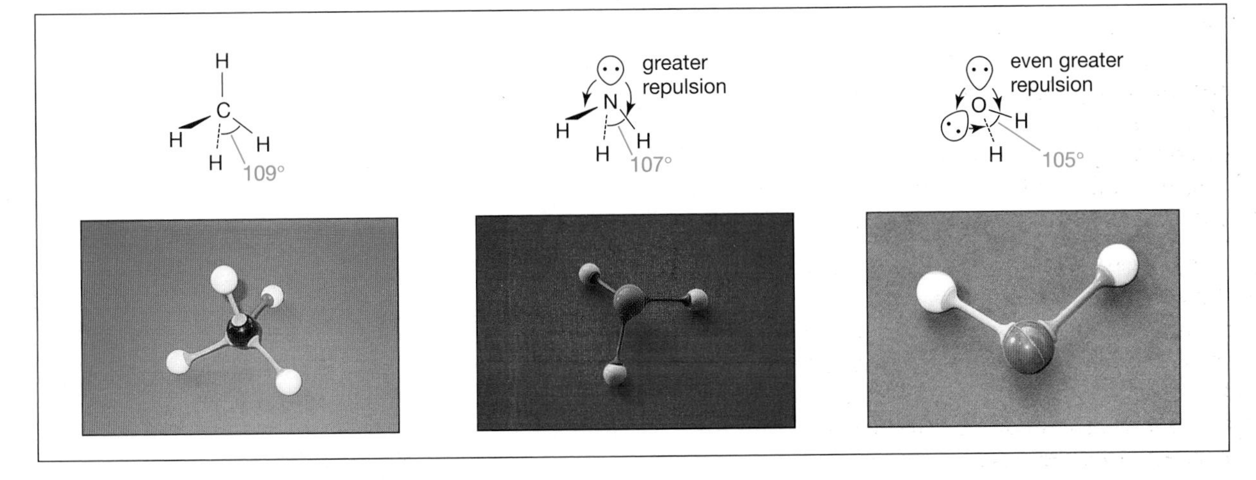

Fig. 2 *The shapes of CH_4, NH_3 and H_2O*

Q 2 Draw dot-cross diagrams of methane, ammonia and water to show the bond pairs and lone pairs present in the molecules.

Key Ideas 42 – 43

- **Since covalent bonds are directional in nature, covalently bonded compounds have a shape.**
- **The bonds are as far apart as possible to reduce the mutual repulsion between different pairs of electrons.**
- **The shape depends upon the number of pairs of electrons around the central atom.**
- **Lone pairs of electrons cause the shape to be distorted.**

11 Working out the shape ...

In order to determine the shape of a simple molecule or polyatomic ion, it is necessary to identify the central atom and work out how many pairs of electrons there are around it. This includes bond pairs and lone pairs. The total number of pairs of electrons (bond pairs and lone pairs) determines the shape. If any of the pairs of electrons are lone pairs, the shape will be distorted as lone pairs increase the mutual repulsion.

Determining the shape

One method of determining the shape of a simple molecule is to follow the sequence given below. This also applies to **polyatomic ions**, which are covalently bonded species carrying a charge.

1 Identify the Group number in the Periodic Table of the central atom: this gives the number of outer electrons on the central atom.
2 If the species is a polyatomic ion, add the negative charge or subtract the positive charge on the ion.
3 Count the number of single bonds formed by the central atom (one electron per bond). This gives the number of bond pairs present.
4 Any remaining outer electrons will be lone pairs of electrons.
5 The shape depends upon the number of bond pairs plus lone pairs.
6 The shape will be distorted if there are lone pairs present.

Some examples are shown in the table below.

	PCl_3	NH_4^+	PF_6^-	BF_4^-
Group number	5	5	5	3
Charge on ion	none	+	–	–
Outer electrons	5	4	6	4
Single bonds	3	4	6	4
Lone pairs	1	0	0	0
Total number of pairs of electrons	4	4	6	4
Shape	distorted tetrahedral	tetrahedral	octahedral	tetrahedral

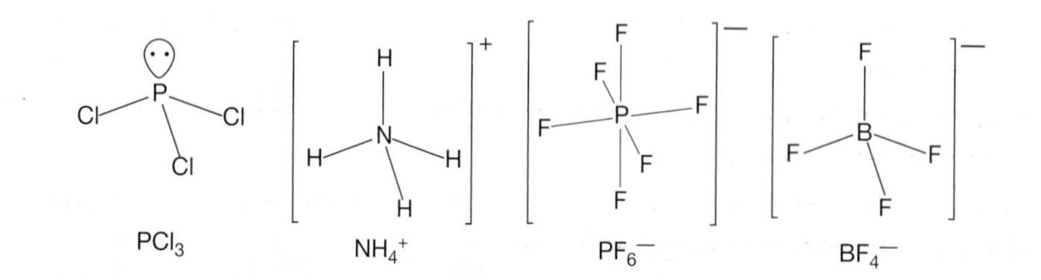

Fig. 1 *Shapes of molecules/polyatomic ions*

In compounds containing **double** or **triple covalent bonds** the same principles apply, with any double or triple bond behaving as a single covalent bond.

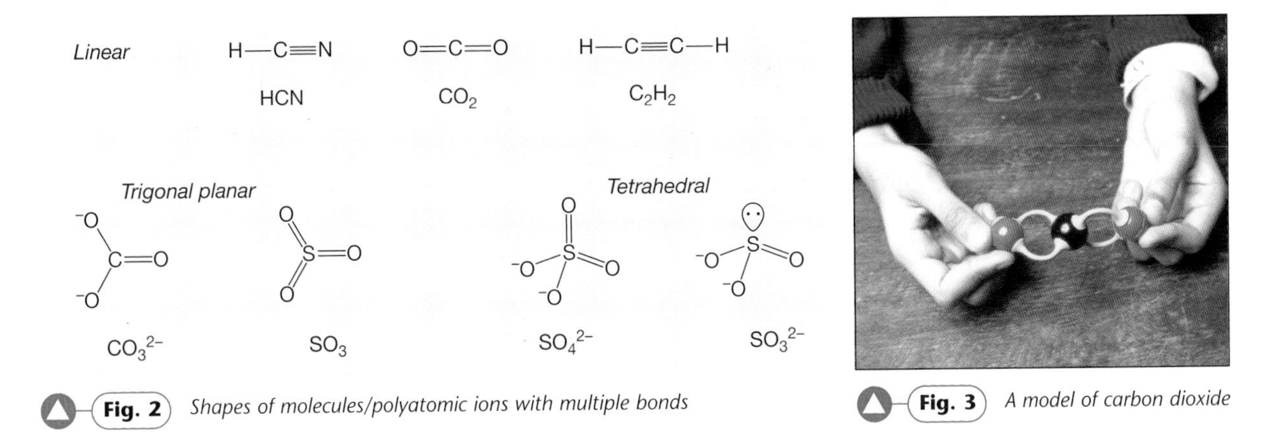

Fig. 2 *Shapes of molecules/polyatomic ions with multiple bonds*

Fig. 3 *A model of carbon dioxide*

All the shapes discussed have referred to the shape formed with respect to the number of pairs of electrons. The shape can also be referred to with respect to the 'positive centres' (atoms). If the shape is a regular shape then there is no difference between the two methods. If there are lone pairs present, however, then there are some differences. For example, ammonia, NH_3, is a pyramidal shape with respect to the 'positive centre', and water, H_2O, is a V-shape with respect to the positive centres.

Q

1 Work out the shapes of each of the following species.

a) $AlCl_3$
b) PCl_4^+
c) BeF_4^{2-}
d) H_3O^+
e) F_2O
f) PF_3
g) NH_2^-
h) ICl_4^+
i) ICl_5
j) ICl_3

Key Ideas 44 – 45

- **The shape of any simple molecule or polyatomic ion can be determined by following a simple method.**
  1. **Use the Group number of the central atom to identify the number of electrons involved in bonding, taking into account any charge on a polyatomic ion.**
  2. **Assuming that one electron from the central atom is used in the formation of each covalent bond, then any remaining electrons will be lone pairs.**
  3. **The total number of pairs of electrons (bond pairs and lone pairs) determines the shape.**
- **Double bonds and triple bonds behave as single bonds in determining the shape of a molecule or polyatomic ion.**

Unit 2 Questions

12

1. Draw dot-cross diagrams to show the formation of the ions in the following compounds.

   a) Li_2O
   b) $CaBr_2$
   c) AlF_3
   d) CaS
   e) Al_2O_3 (5)

2. **a)** Draw dot-cross diagrams to show the bonding in:
   (i) hydrogen chloride (HCl)
   (ii) oxygen (O_2)
   (iii) water (H_2O). (3)

   b) The relative molecular masses of oxygen and hydrogen chloride are similar, but the boiling point of oxygen is much less than that of hydrogen chloride. Explain why these two substances have such different boiling points. (4)

   c) Explain why the boiling point of water is much greater than that of hydrogen chloride, even though the relative molecular mass of hydrogen chloride is twice that of water. (4)

3. **a)** Define the term *electronegativity*. (1)
   b) State and explain the trend in electronegativity across a Period. (3)
   c) What is meant by the term *polar bond*? (2)
   d) Explain why the bond in Br_2 is non-polar but the bond in HBr is polar. (2)

4. **a)** Describe the structure and bonding in diamond and graphite. (6)
   b) Explain why graphite is a good conductor of electricity but diamond is a non-conductor. (2)

5. **a)** Describe the structure of sodium chloride. (2)
   b) Describe the motion of the particles in:
   (i) solid sodium chloride (2)
   (ii) liquid sodium chloride. (2)
   c) Describe the structure of solid iodine. (2)
   d) Explain why iodine forms a purple vapour when heated gently, but sodium chloride remains a solid when heated unless heated very strongly. (4)

6. **a)** Describe how a co-ordinate (dative covalent) bond is formed. (2)
   b) Draw a dot-cross diagram to show the formation of a co-ordinate bond between an ammonia molecule and a hydrogen ion. (2)

7 a) Name the three main types of intermolecular forces. (3)

b) Using H_2, NH_3 and CH_3Cl as examples, show how these three types of intermolecular force arise. (6)

8 Copy and complete the following table by stating the shape and giving the bond angles in each of the species included in the table. (14)

	Shape	Bond angle(s)
BF_3		
CCl_4		
PF_5		
SF_6		
NCl_3		
PCl_4^+		
ICl_4^+		

9 a) Draw the structure of an ammonia molecule, and state the shape of the molecule and the bond angle in the molecule. (2)

b) Draw the structure of a water molecule, and state the shape of the molecule and the bond angle in the molecule. (2)

c) Explain why the bond angle values are different in these two molecules. (2)

10 a) The bonding in aluminium chloride, $AlCl_3$, and phosphine, PH_3, is covalent.

(i) State the shape of, and give the bond angle in, a molecule of aluminium chloride.

(ii) State the shape of, and give the bond angle in, a molecule of phosphine. (4)

b) These two molecules can join together by the formation of a co-ordinate bond.

(i) Explain how the co-ordinate bond is formed.

(ii) Draw a 'dot-cross' diagram to show the bonding in the compound formed.

(iii) Give the Cl—Al—Cl and H—P—H bond angles in this compound, and explain why these angles are different from those in the separate molecules. (6)

11 For each of the following pairs of compounds, state which one of the pair has the higher boiling point. In each case give reasons for your answer in terms of the bonding involved.

a) MgO and CaO. (3)

b) NaCl and HCl. (3)

c) CO_2 and SO_2. (3)

1 Trends in physical properties

In this unit the trends in some of the physical properties of the elements of the third Period, sodium to argon, are studied.

Atomic radius, first ionisation energy and electronegativity depend upon the electronic structure of the elements; electrical conductivity and melting and boiling points depend upon the bonding and structure of the elements.

Physical properties

Atomic radius

Going across the Period, the atomic radius decreases. Each successive element has an extra proton in the nucleus and an extra electron. The extra electrons are added to the same principal energy level. The increasing nuclear charge pulls the electrons closer to the nucleus, resulting in a decrease in the atomic radius.

Element		Atomic number	Electronic structure
Sodium	Na	11	$1s^2 2s^2 2p^6 3s^1$
Magnesium	Mg	12	$1s^2 2s^2 2p^6 3s^2$
Aluminium	Al	13	$1s^2 2s^2 2p^6 3s^2 3p^1$
Silicon	Si	14	$1s^2 2s^2 2p^6 3s^2 3p^2$
Phosphorus	P	15	$1s^2 2s^2 2p^6 3s^2 3p^3$
Sulphur	S	16	$1s^2 2s^2 2p^6 3s^2 3p^4$
Chlorine	Cl	17	$1s^2 2s^2 2p^6 3s^2 3p^5$
Argon	Ar	18	$1s^2 2s^2 2p^6 3s^2 3p^6$

Table 1 *Electronic structures of the elements in Period 3*

First ionisation energy

Going across the Period, there is a general increase in the first ionisation energy, as although the electron removed is from the same principal *energy level*, the *nuclear charge* steadily increases.

There are two breaks in the pattern. The first break, between magnesium and aluminium, is evidence of sub-levels. The break between phosphorus and sulphur is evidence of the difference between unpaired and paired electrons (see page 19).

Electronegativity

The electronegativity increases across the Period. This is because the atomic radius *decreases* across the Period, and the nuclear charge *increases*. Therefore the pair of electrons in a covalent bond will be closer to the increasing nuclear charge, and the electrons will be more attracted to the nucleus of the atom (see page 28).

Electrical conductivity

Sodium, magnesium and aluminium are **metals**. The delocalised electrons in the metallic structure allow the metals to conduct electricity. The electrical conductivity increases from sodium to aluminium, as the number of delocalised electrons per atom increases.

The other elements in the third Period are all non-conductors of electricity since there are no free electrons in their structures.

Pure silicon is considered to be a **semiconductor** of electricity. However, when a small amount of a Group III or Group V element is added, then silicon becomes a good conductor: it is used in computer circuitry.

Melting and boiling points

In order to understand the trend in the melting points and boiling points across the Period, the structure and bonding in the elements must be known.

Na Mg Al	Si	P_4 S_8 Cl_2	Ar
metals	giant covalent	simple molecules	monatomic

Sodium, magnesium and aluminium are metals. The strength of the **metallic bonding** depends upon the number of delocalised electrons in the metal structure. Sodium has *one* mobile electron per atom, magnesium has *two* mobile electrons per atom, and aluminium has *three*. Hence the melting and boiling points increase from sodium to aluminium.

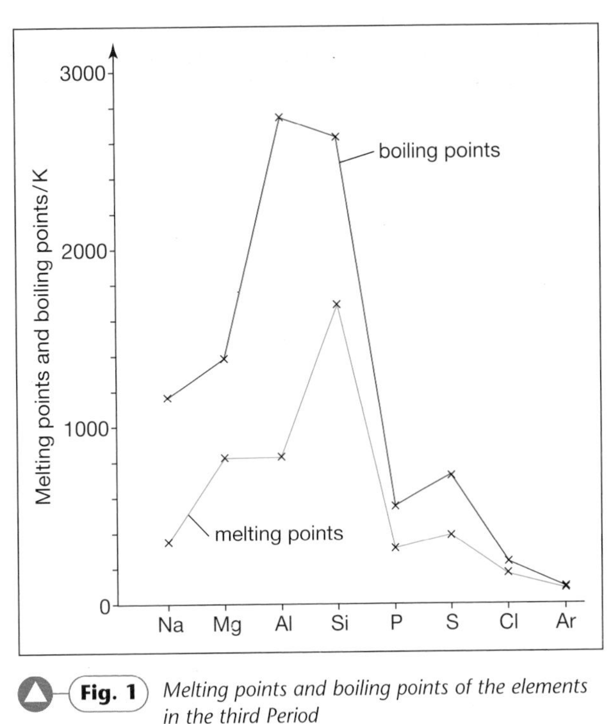

Fig. 1 *Melting points and boiling points of the elements in the third Period*

Silicon exists as a **giant covalent** structure. Each silicon atom is covalently bonded to four other silicon atoms in a giant tetrahedral structure (similar to diamond). A great number of covalent bonds have to be broken to break up the giant covalent structure of silicon, so it has very high melting and boiling points.

Phosphorus, sulphur and chlorine are all **simple molecular** species. Phosphorus consists of P_4 molecules; sulphur consists of S_8 molecules; chlorine consists of Cl_2 molecules. The forces between the molecules are weak van der Waals forces, so these elements have low melting and boiling points. The strength of the van der Waals forces increases as the size of the molecule increases, so the melting points and boiling points increase from Cl_2 to P_4 and to S_8.

Argon exists as **isolated atoms** (it is monatomic) with weak van der Waals forces between the atoms. Hence argon has the lowest melting and boiling points in the Period.

Q

1 Explain why a sulphur atom has a smaller atomic radius than phosphorus.

2 Explain why the first ionisation energy increases across the third Period.

3 Explain why electronegativity increases across the Period.

Key Ideas 48 – 49

- **The physical properties of the elements depend upon the electronic structure of the atoms, and upon the structure and bonding in the elements.**
- **Atomic radius decreases across the Period.**
- **Ionisation energy and electronegativity increase across the Period.**
- **The trends in melting and boiling points of the elements depend upon the structure and bonding of the elements.**

2 Group II elements

These pages consider the physical and chemical properties of the elements in Group II. The Group II elements are called the **alkaline earth metals.**
The physical properties and chemical reactivity depend upon the electronic structure of the elements.

Physical properties

Group II elements are s-block elements, since their outer electrons are in an s sub-level.

Beryllium	Be	$1s^2 2s^2$
Magnesium	Mg	$1s^2 2s^2 2p^6 3s^2$
Calcium	Ca	$1s^2 2s^2 2p^6 3s^2 3p^6 4s^2$
Strontium	Sr	$[Kr] 5s^2$
Barium	Ba	$[Xe] 6s^2$
Radium	Ra	$[Rn] 7s^2$

Atomic radius

Atomic radius increases down the Group, since more principal electron energy levels are occupied. These energy levels are further from the nucleus, hence the atomic radius increases.

First ionisation energy

First ionisation energy decreases down the Group, since the outer electron is removed more and more easily. This is because the electron removed is from a successively higher principal energy level, which is further from the nucleus. The nuclear charge also increases, but the extra inner electrons reduce the effect of the increased nuclear charge by shielding the outer electrons from the nucleus (see page 20).

Element	First ionisation energy/kJ mol^{-1}
Be	900
Mg	736
Ca	590
Sr	548
Ba	502
Ra	510

Electronegativity

Electronegativity *decreases* going down the Group. As the atomic radius *increases*, the pair of electrons in the covalent bond are further from the nucleus. With increasing distance, the pair of electrons are less attracted by the nucleus (see page 28).

Melting point

All of the elements in Group II are metals. Each of the Group II metals provide two delocalised electrons per atom. As the atomic radius increases, the strength of the metallic bonding decreases. Hence the melting point decreases.

Element	Melting point/K
Be	1553
Mg	924
Ca	1124
Sr	1073
Ba	987
Ra	973

The melting point of magnesium is less than that of calcium. The reason for this anomalous value is that the metallic structure of magnesium is different from that of calcium.

Reactions of Group II elements with water

There is a well-defined trend in the reactions of the Group II metals with water.

Element	Reaction with water	Product
Be	does not react with water	none
Mg	reacts very slowly with cold water but reacts readily with steam	MgO
Ca	all react with cold water	$Ca(OH)_2$
Sr	*vigour of reaction increases* ↓	$Sr(OH)_2$
Ba		$Ba(OH)_2$
Ra		$Ra(OH)_2$

The increase in reactivity down the Group is a consequence of the electronic structure of the elements. The elements react by losing electrons to form **dipositive ions**. Going down the Group, the outer s sub-level electrons are further from the nucleus of the atom and are lost more easily. (This is a similar explanation to the trend in reactivity of the Group I metals with water studied at GCSE.)

The equation for the reaction of magnesium with steam (Figure 1) is:

$$Mg(s) + H_2O(g) \rightarrow MgO(s) + H_2(g)$$

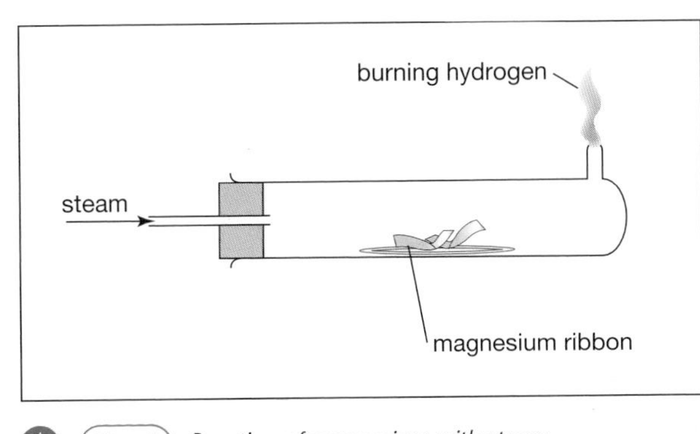

Fig. 1 *Reaction of magnesium with steam*

The reaction of the other Group II metals (M) with water can be represented by a general equation:

$$M(s) + 2H_2O(l) \rightarrow M(OH)_2(aq) + H_2(g)$$

For example:

$$Ca(s) + 2H_2O(l) \rightarrow Ca(OH)_2(s) + H_2(g)$$

$$Ba(s) + 2H_2O(l) \rightarrow Ba(OH)_2(aq) + H_2(g)$$

Q

1 Explain why barium reacts more vigorously with water than calcium.
2 Explain why the melting point of the elements decreases down Group II.
3 Write an equation for the reaction of barium with water. Name the products.

Key Ideas 50 – 51

- **The trends in the physical properties and chemical reactions of the elements depend upon the electronic structure of the atoms.**
- **Atomic radius increases down the Group, as more energy levels are occupied.**
- **Ionisation energy, electronegativity and melting points decrease down the Group.**
- **The reactivity of the elements with water increases down the Group, since the outer electrons are farther from the nucleus and are lost more easily when the atoms form ions.**

3 Group II compounds

These pages consider some compounds of the Group II metals. There are well-defined trends in the properties and reactions of the compounds. The chemistry of beryllium compounds shows some differences from that of other elements in the Group. These atypical properties are the result of the greater polarising power of the beryllium ion.

Group II metal hydroxides

There are two clear trends in the properties of the Group II metal hydroxides. Solubility in water *increases* down the Group; as a consequence, the strength as a base also *increases* since there are more hydroxide ions in solution. It is the hydroxide ions that are responsible for the basic properties of the metal hydroxides.

$Be(OH)_2$	insoluble	amphoteric	↓ *solubility increases and strength as a base increases*
$Mg(OH)_2$	insoluble	basic	
$Ca(OH)_2$	slightly soluble	basic	
$Sr(OH)_2$	soluble	basic	
$Ba(OH)_2$	soluble	basic	
$Ra(OH)_2$	soluble	basic	

$$M(OH)_2 + \text{water} \rightarrow M^{2+}(aq) + 2OH^-(aq)$$

Metal hydroxides neutralise acids to form a salt and water:

$$Mg(OH)_2(s) + 2HCl(aq) \rightarrow MgCl_2(aq) + 2H_2O(l)$$

$$Ba(OH)_2(aq) + 2HNO_3(aq) \rightarrow Ba(NO_3)_2(aq) + 2H_2O(l)$$

Beryllium hydroxide can react *both* as an acid *and* as a base: it is **amphoteric**.

Group II metal sulphates

The trend in solubility is the reverse of that for the Group II metal hydroxides. The solubility of the sulphates decreases down the Group.

$MgSO_4$	soluble	↓ *solubility decreases*
$CaSO_4$	slightly soluble	
$SrSO_4$	insoluble	
$BaSO_4$	insoluble	
$RaSO_4$	insoluble	

The insolubility of barium sulphate is the basis of a **test for sulphate ions in solution**. If a few drops of barium chloride (or barium nitrate) solution are added to a solution containing sulphate ions, a white precipitate of barium sulphate will form.

$$Ba^{2+}(aq) + SO_4^{2-}(aq) \rightarrow \underset{\text{white ppt}}{BaSO_4(s)}$$

The solution being tested is usually **acidified** with dilute hydrochloric acid (if using barium chloride) or dilute nitric acid (if using barium nitrate) to prevent the formation of unwanted precipitates.

Atypical properties of beryllium compounds

Beryllium hydroxide, $Be(OH)_2$, is amphoteric, whereas the other Group II hydroxides are basic. Beryllium hydroxide will react as an acid and as a base.

As a base: $Be(OH)_2(s) + H_2SO_4(aq) \rightarrow BeSO_4(aq) + 2H_2O(l)$

or $Be(OH)_2(s) + 2H^+(aq) \rightarrow Be^{2+}(aq) + 2H_2O(l)$

As an acid: $Be(OH)_2(aq) + 2KOH(aq) \rightarrow 2K^+(aq) + Be(OH)_4^{2-}(aq)$

or $Be(OH)_2(aq) + 2OH^-(aq) \rightarrow Be(OH)_4^{2-}(aq)$

The beryllium ion, Be^{2+}, is very small compared to the other Group II metal ions. The high charge-to-size ratio of the Be^{2+} ion makes the ion highly **polarising**. This results in many simple beryllium compounds having an appreciably **covalent character**.

Beryllium chloride, $BeCl_2$, is soluble in non-polar organic solvents, whereas the other Group II chlorides are ionic and dissolve in water. The fact that beryllium chloride dissolves in non-polar solvents illustrates its covalent character.

A **complex** is formed when a metal ion forms co-ordinate bonds with species that have a lone pair of electrons. The number of co-ordinate bonds formed in the complex is called the **co-ordination number.** (See A2 Chemistry for a more detailed study of complexes.)

Beryllium forms complexes that have a maximum co-ordination number of four, such as $[Be(OH)_4]^{2-}$ and $[BeCl_4]^{2-}$. In each of these species beryllium is co-ordinately bonded to hydroxide and chloride ions.

Q

1. Predict the shape of $[Be(OH)_4]^{2-}$ and $[BeCl_4]^{2-}$.
2. Define the term *amphoteric*, and write equations to show the amphoteric action of beryllium hydroxide.
3. Explain why barium hydroxide is a stronger base than calcium hydroxide.
4. Write equations for the reactions of $Ba(OH)_2$ and $Ca(OH)_2$ with hydrochloric acid, HCl(aq).

Key Ideas 52 – 53

- **The solubility of the metal hydroxides increases down the Group.**
- **As the solubility of the metal hydroxides increases, the strength as a base increases.**
- **The solubility of the metal sulphates decreases down the Group.**
- **The insolubility of barium sulphate forms the basis of the qualitative test for sulphate ions.**
- **Beryllium compounds have atypical properties caused by the high polarising power of the beryllium ion.**

Unit 3 Questions

4

1 The melting points of the elements depend upon their structure and bonding. The melting points of the elements in Period 3 are shown in the graph.

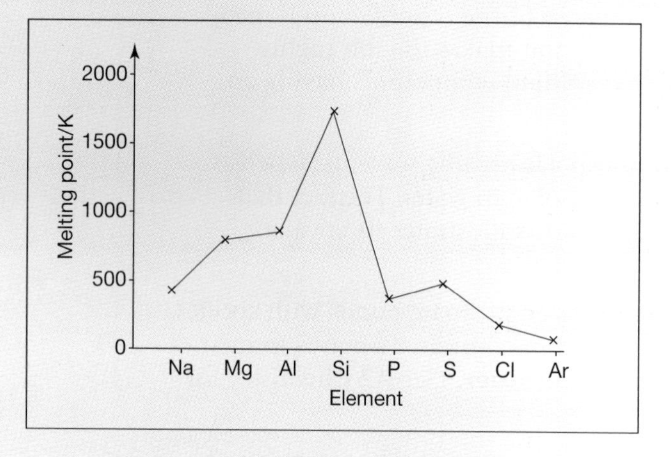

Use your knowledge of the structure and bonding in the elements to explain each of the following:

a) aluminium has a higher melting point than magnesium (2)
b) sulphur has a higher melting point than phosphorous (2)
c) silicon has a very high melting point (2)
d) argon has the lowest melting point in the Period. (2)

2 Describe and explain the trend in electronegativity values:

a) across Period 3 (3)
b) down Group II. (3)

3 Describe and explain the trend in atomic radius:

a) across Period 3 (3)
b) down Group II. (3)

4 Describe, with essential details, how you would test for the presence of the sulphate group in solid sodium sulphate. (4)

5 **a)** Describe and explain the trend in the reactivity of the Group II metals with water. (3)
b) Calcium metal reacts with cold water.
(i) Write an equation for the reaction.
(ii) Account for the formation of a white precipitate during the reaction.
(iii) How could you test for the hydrogen produced? (3)

6) a) State the trend in solubility in water of the Group II metal hydroxides. (1)

b) State the trend in the strengths as bases of the Group II metal hydroxides. (1)

c) Explain how these two trends are connected. (2)

7) Beryllium hydroxide is amphoteric.

a) Explain the meaning of the term *amphoteric*. (1)

b) Write equations to show beryllium hydroxide acting as an amphoteric hydroxide. (2)

8) The Group II chlorides are all ionic except $BeCl_2$. Explain why the bonding in $BeCl_2$ is predominantly covalent. (3)

9) a) Write out the full electronic structures for magnesium and calcium. (2)

b) Explain why these elements are called *s-block elements*. (1)

c) Write equations to represent the first three ionisation energies of calcium. (3)

d) Explain why the third ionisation energy of calcium is much higher than the second ionisation energy. (3)

1 Quantity matters ...

You will be familiar with the need, when carrying out experiments in the laboratory, to measure out an amount of each reagent. You need to know how much to measure and how you are going to measure it. Quantitative chemistry is all about 'How much?' We must therefore consider the **mass** of solids, the **concentration** of solutions, and the **volume** of gases.

The common linking factor between these quantities, which you will learn about at AS level, is the **mole**, which is the chemist's unit for amount of substance.

Measuring amounts

Mass is measured in grams (g) or kilograms (kg). All matter has mass, but in practice at AS level we usually limit mass measurements to solids.

Volume is measured in cubic centimetres (cm^3), cubic decimetres (dm^3) or cubic metres (m^3). Volume is used to measure the amount of a liquid, for example using a measuring cylinder, or the amount of a gas, for example using a gas syringe.

The *concentration* of a solution is measured in grams per dm^3 ($g\,dm^{-3}$) or moles per dm^3 ($mol\ dm^{-3}$). You can read about **moles** on page 58.

You will have to be able to convert units:

- $1\,kg = 1000\,g$
- $1\,dm^3 = 1000\,cm^3 = 1/1000\,m^3$
- $1/1000$ is usually written as 1×10^{-3}

Q

1 Express the following masses in kg:

a) 500 g b) 25 g c) 5 g

2 Express the following volumes in dm^3:

a) $2000\,cm^3$ b) $0.05\,m^3$ c) $2\,cm^3$

Relative atomic mass (A_r)

Atoms are individually too small to weigh. As explained on page 10, a relative atomic mass scale is used, based on the isotope carbon-12. The **relative atomic mass** of an element is the average mass of the atoms, taking into account the natural relative abundance of its isotopes, compared to 1/12th of the mass of an atom of carbon-12.

$$\textbf{relative atomic mass, } A_r = \frac{\text{the average mass of an atom} \times 12}{\text{the mass of one atom of carbon-12}}$$

The use of carbon-12 as the reference standard, whereby one atom of carbon weighs exactly 12.000 **atomic mass units**, was introduced in 1961. This superseded the use of hydrogen as the standard, since carbon-12 is easier to handle.

Table 1 shows the exact relative atomic masses of some elements.

Element	A_r	Value used
aluminium	26.981	27
bromine	79.904	80
calcium	40.080	40
chlorine	35.453	35.5
copper	63.540	63.5
hydrogen	1.0076	1
oxygen	15.999	16

Table 1 *Relative atomic masses*

Points to note in Table 1:

1 Copper and chlorine are the only two non-integer values used.
2 The relative atomic masses are not whole numbers because they are the average (mean) masses of the naturally occurring isotopes, taking into account their relative abundance.
3 A_r values can be calculated from mass spectra data (see page 10).
4 The values that are used in calculations are quoted in the Periodic Table (see Appendix A).

Using the data in Table 1, you can see that one atom of oxygen is sixteen times heavier than one atom of hydrogen, and that one atom of bromine is twice as heavy as one atom of calcium.

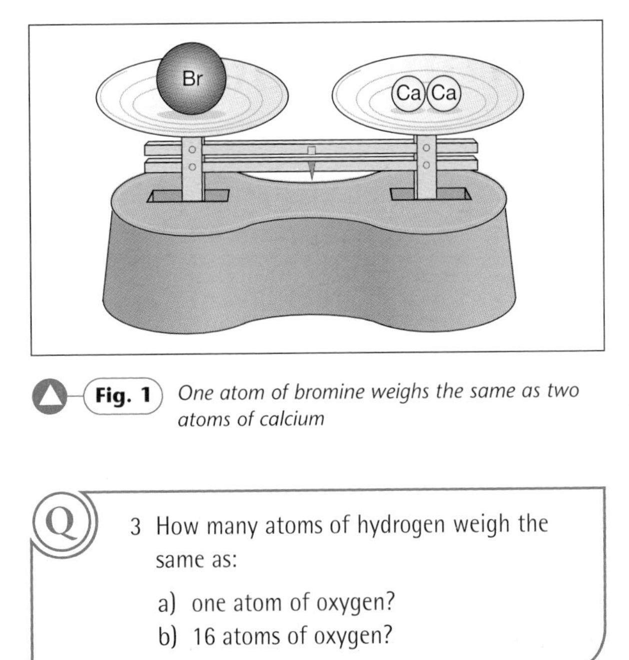

Fig. 1 *One atom of bromine weighs the same as two atoms of calcium*

> **Q** 3 How many atoms of hydrogen weigh the same as:
> a) one atom of oxygen?
> b) 16 atoms of oxygen?

Relative molecular mass (M_r)

Compounds are made up of a specific number of atoms or ions: this number is represented by the formula of the compound. Their masses are again compared to carbon-12.

$$\textbf{relative molecular mass, } M_r = \frac{\text{the average mass of an 'entity'} \times 12}{\text{the mass of one atom of carbon-12}}$$

The term **entity** is used rather than 'molecule' since ionic compounds are made up of ions, not molecules.

To find the relative molecular mass of a compound you must first find the formula and then add up the relative atomic masses of the atoms present, taking into account the number of each type of atom. Some examples are given in Table 2.

Name	Formula	Adding up the A_r values	M_r
Sodium chloride	NaCl	23 + 35.5	58.5
Nitric acid	HNO_3	1 + 14 + (16 × 3)	63
Potassium sulphate	K_2SO_4	(39 × 2) + 32 + (16 × 4)	174
Calcium hydroxide	$Ca(OH)_2$	40 + (16 × 2) + (1 × 2)	74

Table 2 *Calculating relative molecular masses for compounds*

Key Ideas 56 – 57

- **The relative atomic mass scale (A_r) is used to 'weigh' atoms.**
- **The formula of a compound is used to 'count' the atoms present in a molecule.**
- **The relative molecular mass (M_r) of a compound is found by adding together the A_r values of each atom in the formula of one molecule.**

2 Atoms count

How do we know that *one* atom of carbon joins with *two* atoms of oxygen to form *one* molecule of carbon dioxide? We cannot react single atoms together because they are too small to weigh, so we need a more convenient scale of measuring the amount of substance. We use *moles*.

The mole

The use of the mole is a way of scaling up easily from relative atomic masses and relative molecular masses of substances to grams. The mole is a unit just as the gram is a unit; its symbol is **mol**. A mole is defined in terms of carbon-12, which is the reference standard:

> 1 **mole** is the amount of substance that contains as many particles as there are atoms in exactly 12g of carbon-12.

How many atoms are there in 12g of carbon-12? The answer is 6.023×10^{23} atoms. This is a lot of atoms!

Fig. 1 *(Clockwise from top left) 1 mole of carbon, 1 mole of iron, 1 mole of water, 1 mole of calcium carbonate*

One mole of any substance always contains 6.023×10^{23} particles. The particles are not necessarily atoms – they may be molecules, ions or electrons – but the number is always constant. It is called the **Avogadro constant** (or **Avogadro's number**) after the Italian physicist, Amedeo Avogadro (1776–1856), who did some of the fundamental work involved in 'counting' atoms.

The symbol used for the Avogadro constant is ***L***. This is in recognition of the Austrian school teacher, Loschmidt, who originally calculated a value for this constant.

Working with Avogadro

Here is an example of the sort of conversion you need to know and how to set out your answer.

One atom of hydrogen (^1H) weighs 1.673×10^{-24}g. What is the mass of one mole of these atoms?

1 mol of any element contains 6.023×10^{23} atoms. So if each atom weighs 1.673×10^{-24}g, then 1 mol must weigh

$$6.023 \times 10^{23} \times 1.673 \times 10^{-24}\text{g} = 10.0765 \times 10^{-1}\text{g}$$
$$= 1.0076\text{g}$$

The mass of one mole of atoms is called the **molar mass**. We have calculated the molar mass of hydrogen using the expression:

molar mass
= mass of one atom × Avogadro's number

Q

1 If one atom of ^{12}C weighs 1.9925×10^{-23}g, calculate the accurate mass of one mole of ^{12}C.

Measuring out moles

The mass of one mole of hydrogen calculated above is 1.0076 g. On page 56 this number was quoted as the accurate relative atomic mass of hydrogen, so now we have a relationship between moles of an element and the relative atomic mass:

1 mole of an element = the relative atomic mass in grams.

This provides a quick way of working with moles without having to use Avogadro's number.

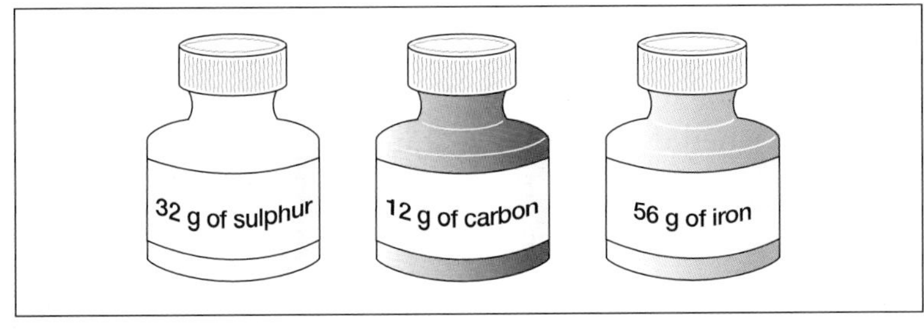

Fig. 2 *How many atoms are there in each of these bottles?*

Each bottle in Figure 2 contains the relative atomic mass of the element in grams. Since this equals one mole, then each bottle contains 6.023×10^{23} atoms. Provided that the relative atomic mass of the element is known, molar quantities can be measured out by weighing.

The relative atomic masses of the elements are shown in the Periodic Table. Some examples of how to convert moles into grams and vice versa are shown on the right.

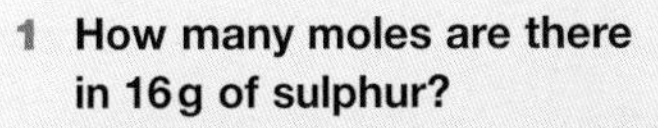

1 How many moles are there in 16 g of sulphur?

The A_r of sulphur is 32. (Look this up in the Periodic Table.)
So 1 mol of sulphur weighs 32 g.
Since 16 is half of 32, 16 g of sulphur
= half a mole
= 0.5 mol.

2 What mass of iron must be weighed out to obtain 3 moles?

The A_r of iron is 56.
So 1 mol of iron contains 56 g.
Therefore 3 mol contains
56×3 g = 168 g.

3 How many moles are there in 0.12 g of carbon?

The A_r of carbon is 12.
So 12 g of carbon is 1 mol.

$$\frac{0.12}{12} = 0.01$$

Therefore 0.12 g of carbon
= 0.01 mol.

Linking mass and moles

In the above examples, which involve elements, the relationship between moles and mass is:

$$\text{number of moles} = \frac{\text{mass in grams}}{\text{relative atomic mass}}$$

$$n = m/A_r$$

or number of moles × relative atomic mass
= mass in grams

You will need a copy of the Periodic Table.

2 What mass of each element contains:

a) 2 mol of lithium? b) 0.5 mol of iron?

3 How many moles are there in:

a) 39 g of potassium? b) 2.4 g of magnesium?

Key Ideas 58 – 59

- **If the A_r of an element is weighed out in grams, then one mole of atoms is present.**
- **One mole of any substance contains 6.023×10^{23} particles. This number is called the Avogadro constant.**
- **The relationship between moles and mass for elements is: number of moles = mass/A_r.**

3 Atoms count (2)

Most chemicals that you will use in the laboratory will be compounds, not elements, so we need to extend the ideas about moles and apply them to compounds.

Compounds and the mole

We have seen that for elements one mole equals the relative atomic mass in grams. It therefore follows that when dealing with compounds we can use:

1 mole of a compound = the relative molecular mass in grams

On page 57, we calculated the relative molecular masses for some compounds. We can use these again to look at moles.

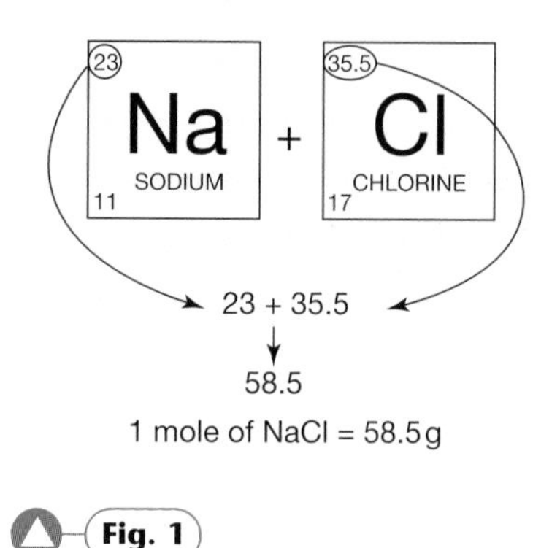

Fig. 1

Name	Formula	M_r	Mass of one mole	Mass of two moles	Mass of ten moles
Sodium chloride	NaCl	58.5	58.5 g	117 g	585 g
Potassium sulphate	K_2SO_4	174	174 g	348 g	1740 g
Calcium hydroxide	$Ca(OH)_2$	74	74 g	148 g	740 g

Table 1 *Relative molecular masses of compounds*

For more complicated conversions you will need to use the expression:

$$\text{number of moles} = \frac{\text{mass in grams}}{\text{relative molecular mass}}$$

$$n = m/M_r$$

or number of moles × relative molecular mass = mass in grams

Here are some examples of the sorts of conversions you will have to perform, and how to set out your answers.

1 What mass of solid must be weighed out to give 0.5 mol of calcium carbonate?

Step 1: First work out the formula for calcium carbonate.
It is $CaCO_3$. This contains 1 atom of calcium (Ca = 40), 1 atom of carbon (C = 12), and 3 atoms of oxygen (O = 16).

Step 2: Now calculate the M_r by adding up the atomic masses.
40 + 12 + (16 × 3) = 100.
This means that 1 mol contains 100 g.

Step 3: Use **number of moles × M_r = mass**:
0.5 mol × 100 = 50 g.

2 How many moles are there in 34.8 g of potassium sulphate?

Step 1: The formula for potassium sulphate is K_2SO_4.

Step 2: Now calculate M_r:
(39 × 2) + 32 + (16 × 4) = 174.

So there are 174 g in 1 mol.

Step 3: Using the mass/Mr = moles,
34.8/174 = 0.2 mol.

Q

1 Use the relative atomic masses on the Periodic Table to calculate the relative molecular mass of:

a) $Ca(OH)_2$ b) NaCl c) $ZnCO_3$ d) CaO e) $MgSO_4$ f) $AlCl_3$

2 Calculate the mass of each of the following:

a) 2 mol of $Ca(OH)_2$ b) 0.25 mol of NaCl c) 0.01 mol of $ZnCO_3$

3 How many moles are there in each of the following?

a) 112 g of CaO b) 12 g of $MgSO_4$ c) 33.375 g of $AlCl_3$

What about gases?

Gases such as oxygen, chlorine and hydrogen are elements, not compounds, but since they exist as **diatomic molecules** it is necessary to use their relative molecular masses in calculations – for example, oxygen exists as O_2, so its $M_r = 32$ (not 16). Other gases such as methane and ammonia are compounds, and their formulae need to be learnt.

Elements		Compounds	
Name	Formula	Name	Formula
Oxygen	O_2	Ammonia	NH_3
Nitrogen	N_2	Carbon dioxide	CO_2
Chlorine	Cl_2	Methane	CH_4
Fluorine	F_2	Sulphur dioxide	SO_2
Hydrogen	H_2	Nitrogen dioxide	NO_2

Table 2 *Formulae of some common gases*

Gases are more often measured by volume. This is dealt with on page 72.

Q

4 Calculate the relative molecular masses for each gas shown in Table 2. You will need a copy of the Periodic Table.

5 How many moles are there in:

a) 7.1 g of chlorine? b) 20 g of hydrogen? c) 3.2 g of sulphur dioxide?

6 What is the mass of each of the following, in grams?

a) 2 mol of fluorine b) 0.4 mol of nitrogen c) 0.05 mol of carbon dioxide

Key Idea 60 – 61

- **The relationship between moles and mass for compounds is: number of moles = mass/M_r.**

4 What's in a formula?

You will know from GCSE that compounds are made up of a specific number of atoms or ions, and that chemists represent these by formulae. At AS level you need to know about two types of formulae: empirical and molecular.

Looking at formulae

The **empirical formula** of a compound gives the *simplest ratio* of the number of atoms of each element present in a molecule.

The **molecular formula** of a compound gives the *actual number* of atoms of each element present in a molecule.

The molecular formula, which is the more commonly used, is a multiple of the empirical formula. For example, phosphorus(V) oxide exists as P_4O_{10} molecules. P_4O_{10} is the *molecular* formula, showing that one molecule of phosphorus(V) oxide contains 4 atoms of phosphorus bonded to 10 atoms of oxygen. This formula can be simplified to P_2O_5 – this is the *empirical* formula of phosphorus(V) oxide, showing that the simplest ratio of phosphorus to oxygen atoms is 2:5.

For inorganic compounds, the empirical and molecular formulae are usually the same. Some examples are shown in Table 1.

Compound	M_r	Empirical formula	Molecular formula
Water	18	H_2O	H_2O
Hydrogen peroxide	34	HO	H_2O_2
Carbon dioxide	44	CO_2	CO_2
Sulphuric acid	98	H_2SO_4	H_2SO_4
Chlorine	71	Cl	Cl_2
Ammonia	17	NH_3	NH_3

Table 1 *Empirical and molecular formulae*

Calculating the formula

Historically, the only method of finding the formula of a compound was by experiment. Here is a page taken from a student's notebook.

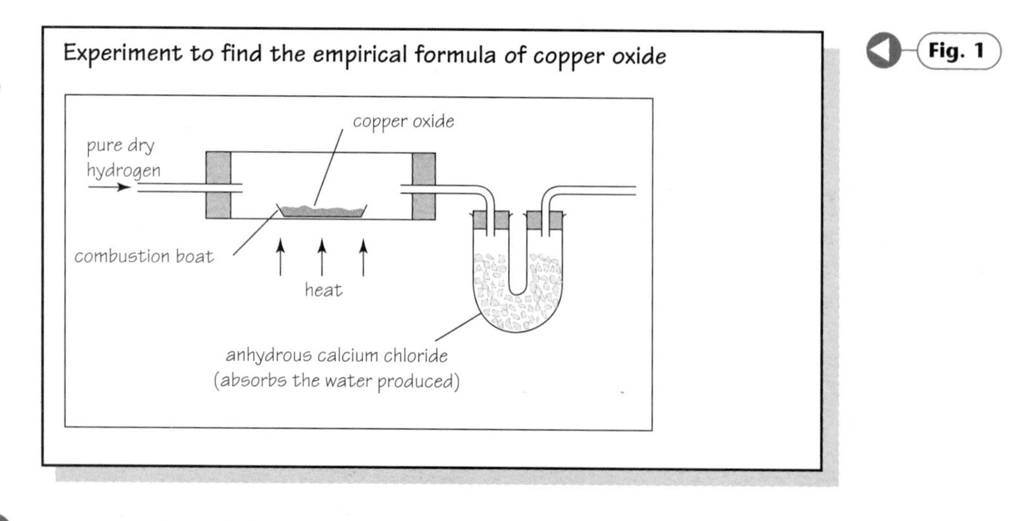

Fig. 1

Fig. 1 Continued

A weighed sample of copper oxide is reduced to copper by hydrogen gas. After cooling the sample is re-weighed.

Results

Mass of boat = 1.72 g
Mass of boat + copper oxide = 2.92 g
Mass of boat + copper = 2.68 g

Calculation

Mass of copper oxide = 2.92 − 1.72 = 1.20 g
Mass of copper = 2.68 − 1.72 = 0.96 g
Mass of oxygen which combines with the copper to make copper oxide = 1.20 − 0.96 = 0.24 g

Use number of moles of an element = mass/A_r, where Cu = 63.5 and O = 16.

So $\frac{0.96}{63.5}$ mol of copper combines with $\frac{0.24}{16}$ mol of oxygen.

So 0.015 mol of copper combines with 0.015 mol of oxygen.

This means that the molar ratio of Cu : O = 0.015 : 0.015.

As these numbers are the same, this gives a ratio of 1 : 1, so 1 mol of Cu combines with 1 mol of O.

The empirical formula of copper oxide is therefore Cu_1O_1, or CuO.

Q

1 An oxide of lead was weighed in a test tube. The oxide was heated and reduced to lead by passing a stream of hydrogen gas over it. After heating to constant mass it was re-weighed. Find the formula for this oxide of lead, given the results below:

Mass of test tube empty = 10.20 g
Mass of test tube + oxide = 17.37 g
Mass of test tube + lead = 16.41 g

Key Ideas 62 – 63

- **Compounds can be represented by empirical or molecular formulae.**
- **Molecular formulae are more useful since these provide the actual numbers of the different types of atoms present.**
- **Empirical formulae can be determined by experiment.**

5 Finding the right formula

Using data from experiments you can calculate the empirical formula or molecular formula for any compound. Here are some worked examples to show you how to calculate formulae and how to set out your answers.

Calculating the empirical formulae

1 Find the empirical formula of an oxide of sulphur, given that 3.2 g of sulphur combines with 3.2 g of oxygen.

Step 1: Write down the reacting masses.
3.2 g of sulphur combines with 3.2 g of oxygen.

Step 2: Convert the masses to moles using **number of moles = mass/A_r.**

$\frac{3.2}{32}$ mol of sulphur combines with $\frac{3.2}{16}$ mol of oxygen

= 0.1 mol = 0.2 mol

Step 3: Express these answers as the molar ratio.
S:O = 0.1:0.2

Step 4: Simplify the ratio by dividing by the smallest number.

$S:O = \frac{0.1}{0.1}:\frac{0.2}{0.1}$

= 1:2

So the empirical formula = $S_1O_2 = SO_2$.

2 14.2 g of an oxide of phosphorus contains 6.2 g of phosphorus. Determine the empirical formula of this compound.

Step 1: First you need to find how much oxygen there is in the oxide.
Mass of oxygen = 14.2 – 6.2 = 8.0 g.

Step 2: Convert to moles.

$\frac{6.2}{31}$ mol phosphorus : $\frac{8.0}{16}$ mol of oxygen

= 0.2 mol = 0.5 mol

Step 3: The molar ratio of P:O = 0.2:0.5.

Step 4: The simplified ratio is $P:O = \frac{0.2}{0.2}:\frac{0.5}{0.2}$ mol = 1:2.5

but the ratio must be in whole numbers, so P:O = 2:5.

So the empirical formula is P_2O_5.

Q

1 5 g of a sulphide of iron contains 2.31 g of sulphur. Calculate the empirical formula for this compound.

2 2.52 g of magnesium is completely burned in oxygen forming 4.20 g of magnesium oxide. Find the formula of the oxide.

This type of calculation is not restricted to compounds that only contain two elements. The molar ratio can be calculated for any number of elements, as illustrated by the following examples.

3 A drum of white powder, compound *X*, has been dumped in the river and may cause a pollution problem. The laboratory analysis of a sample of *X* shows that it contains 2.66 g of calcium, 0.80 g of carbon and 3.22 g of oxygen. Calculate the empirical formula of *X* to decide whether or not this compound is a hazard.

Step 1: Write down the reacting masses.
2.66 g of calcium combines with 0.80 g of carbon and 3.22 g of oxygen.

Step 2: Convert the masses to moles using **number of moles = mass/A_r.**

calcium	:	carbon	:	oxygen
$\frac{2.66}{40}$ mol	:	$\frac{0.80}{12}$ mol	:	$\frac{3.32}{16}$ mol
= 0.067 mol		= 0.067 mol		= 0.201 mol

Step 3: Express these answers as the molar ratio.

Ca:C:O = 0.067:0.067:0.201.

Step 4: Simplify the ratio by dividing by the smallest number.

$$Ca:C:O = \frac{0.067}{0.067}:\frac{0.067}{0.067}:\frac{0.200}{0.067}$$

$$= 1:1:3.$$

So the empirical formula = $Ca_1C_1O_3$ = $CaCO_3$.

You will recognise the empirical formula calculated in the last example as that of calcium carbonate. It is insoluble in water and so is unlikely to cause a pollution problem. In fact if the water is polluted by acid rain, calcium carbonate will help to neutralise it.

Q

3 A compound containing sodium, carbon and oxygen was found on analysis to contain 4.6 g of sodium, 1.2 g of carbon and 4.8 g of oxygen. Find the empirical formula for this compound and name it.

4 A sample of compound **A** weighs 6.8 g. When this sample was analysed 2.0 g was found to be calcium, 1.6 g was sulphur and the rest was oxygen. Find the empirical formula of **A**.

4 Find the empirical formula of hydrated calcium chloride using the following experimental data: 14.7 g of a sample of the salt contains 4.0 g of calcium, 7.1 g of chlorine and 3.6 g of water.

To answer this question it is essential to realise the importance of the word 'hydrated'. This refers to molecules of water which are attached to the compound in the crystal. This water is called 'water of crystallisation' and the number of molecules present are shown in the formula.

Step 1: Write down the relative masses.
4.0 g of calcium combines with 7.1 g of chlorine and 3.6 g of water.

Step 2: Convert to moles.

$\frac{4.0}{40}$ mol of Ca combines with $\frac{7.1}{35.5}$ mol of Cl and $\frac{3.6}{18}$ mol of H_2O

= 0.1 mol = 0.2 mol = 0.2 mol

Step 3: Express these as a molar ratio.
$Ca:Cl:H_2O$ = 0.1:0.2:0.2

Step 4: Simplify the ratio.

$$Ca:Cl:H_2O = \frac{0.1}{0.1}:\frac{0.2}{0.1}:\frac{0.2}{0.1} = 1:2:2$$

So the empirical formula is $CaCl_2.2H_2O$.

Q

5 A hydrated sample of iron(II) sulphate has the formula $FeSO_4.xH_2O$, where x equals the number of molecules of water of crystallisation. Calculate the value for x, given that 3.8 g of anhydrous salt remains when 6.95 g of the hydrated salt is heated to constant mass.

Key Idea 64 – 65

- **The empirical formula of a compound can be calculated from experimental data using moles.**

6 Finding the right formula (2)

The distinction between empirical and molecular formulae is more apparent when studying *organic* molecules. You will have met some examples when you studied the oil industry at GCSE. Crude oil is a mixture of organic molecules called *hydrocarbons*. Most of the calculations on molecular formulae will involve organic molecules.

Molecular formulae

Remember that the empirical formula of a compound gives the *simplest ratio* of the number of the different atoms present, whereas the molecular formula gives the *actual number* of each different atom present in a molecule. We can illustrate these ideas by looking at the formula for the hydrocarbon, ethane.

It is usually written as C_2H_6: this is the *molecular formula*. One molecule of ethane contains 2 atoms of carbon and 6 atoms of hydrogen.

If this formula is simplified, the result is CH_3: this is the *empirical formula*, showing the simplest ratio of C:H in ethane as 1:3, that is 1 atom of carbon to 3 atoms of hydrogen.

If we add up the masses of the atoms present in the empirical formula we arrive at the empirical formula mass: $(12 \times 1) + (1 \times 3) = 15$.

The molecular formula, which is more commonly used, is a multiple of the empirical formula. For example, for ethane the molecular formula is $(CH_3)_n$, where n=2. The value of n is found from the relative molecular mass (M_r). The M_r for ethane is 30. This is twice the empirical formula mass, so the molecular formula is twice the empirical formula.

Calculating molecular formulae

To calculate the molecular formula of a compound it is necessary first to calculate the empirical formula of the compound as shown previously. Then compare the relative molecular mass with the empirical formula mass to find the value for n, and hence the molecular formula.

The example below will illustrate this.

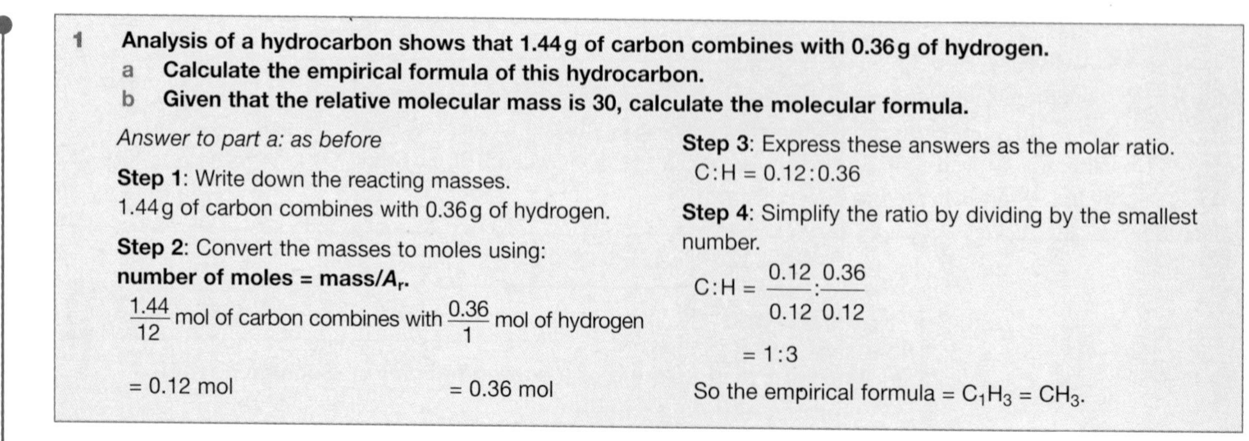

1 Analysis of a hydrocarbon shows that 1.44 g of carbon combines with 0.36 g of hydrogen.

a Calculate the empirical formula of this hydrocarbon.

b Given that the relative molecular mass is 30, calculate the molecular formula.

Answer to part a: as before

Step 1: Write down the reacting masses.
1.44 g of carbon combines with 0.36 g of hydrogen.

Step 2: Convert the masses to moles using:
number of moles = mass/A_r.

$\frac{1.44}{12}$ mol of carbon combines with $\frac{0.36}{1}$ mol of hydrogen

= 0.12 mol = 0.36 mol

Step 3: Express these answers as the molar ratio.
C:H = 0.12:0.36

Step 4: Simplify the ratio by dividing by the smallest number.

$$C:H = \frac{0.12}{0.12}:\frac{0.36}{0.12}$$

$$= 1:3$$

So the empirical formula = C_1H_3 = CH_3.

Answer to part b: new calculation

First you need to calculate the empirical formula mass.

For CH_3 this is $12 + (1 \times 3) = 15$

but the molecular mass given in the question is 30.

Since $\frac{30}{15} = 2$, the molecular formula is twice the empirical formula, that is $(CH_3)_2$.

So the answer is given as C_2H_6.

For organic compounds the results of analysis of the compound are often given in terms of the **percentage composition by mass**. In the calculation it is assumed that there are 100 g of the compound, so the percentages can be converted into grams and the calculation continues as before. This is illustrated in the next example.

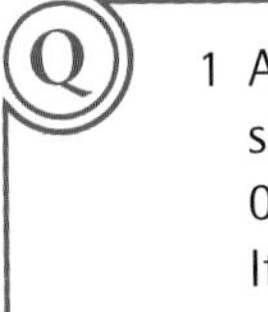

1 Analysis of a sample of compound **X** shows that it contains 1.2 g of carbon, 0.2 g of hydrogen and 1.6 g of oxygen. If the relative molecular mass of **X** is 60, calculate the molecular formula.

2 a Compound A contains 12.78% carbon, 2.13% hydrogen and 85.20% bromine by mass. Calculate the empirical formula of A.

b Calculate the molecular formula of A, given that its $M_r = 188$.

Answer to part a

Step 1:
12.78 g of carbon combines with 2.13 g of hydrogen and 85.20 g of bromine.

Step 2:

$$\frac{12.78}{12} \text{ mol of C} : \frac{2.13}{1} \text{ mol of H} : \frac{85.20}{80} \text{ mol of Br}$$

= 1.065 mol = 2.13 mol = 1.065 mol

Step 3:
The molar ratio of C:H:Br = 1.065:2.13:1.065.

Step 4: The simplified ratio is

$$\text{C:H:Br} = \frac{1.065}{1.065}:\frac{2.13}{1.065}:\frac{1.065}{1.065} = 1:2:1.$$

So the empirical formula = CH_2Br.

Answer to part b

First you need to calculate the empirical formula mass.

For CH_2Br this is $12 + (1 \times 2) + 80 = 94$.

But the molecular mass given in the question is 188.

Since $\frac{188}{94} = 2$, the molecular formula is twice the empirical formula, that is $(CH_2Br)_2$.

So the answer is given as $C_2H_4Br_2$.

If the M_r is not given in this type of question it will be necessary to calculate its value. This type of calculation is dealt with on page 76.

Q

2 Calcium bromide contains 20% by mass of calcium. Find its empirical formula.

3 A hydrocarbon contains 85.7% by mass of carbon. Calculate the empirical formula of the hydrocarbon. Given that the relative molecular mass is 56, calculate the molecular formula.

Key Ideas 66 – 67

- **The empirical formula of a compound can be calculated from percentage composition by mass data.**
- **The molecular formula can be calculated from the empirical formula, provided that the M_r is known.**

7 Balancing act ...

At GCSE you used chemical equations to summarise reactions. The equation shows the formulae of the reactants and products. When the equation is balanced it shows the *relative amounts* of each substance involved. These amounts are moles of substance – that is, the **molar ratio**. At AS level we use several different types of equations to illustrate chemical reactions, and we use equations to perform calculations with moles of substance.

Different types of chemical equations

We can illustrate the different types of equations with specific examples.

Molecular equations

$$Ca(s) + 2H_2O(l) \rightarrow Ca(OH)_2(aq) + H_2(g)$$

This is a **molecular equation**, showing the reaction between calcium metal and water.

The equation shows that one mole of calcium reacts with two moles of water to form one mole of calcium hydroxide and one mole of hydrogen gas.

Ionic equations

$$Ba^{2+}(aq) + SO_4^{2-}(aq) \rightarrow BaSO_4(s)$$

This is an **ionic equation**, showing the formation of barium sulphate from its ions.

Here one mole of barium ions in solution reacts with one mole of sulphate ions in solution to form one mole of barium sulphate, which is precipitated as a solid.

Ion-electron equations

$$Li(g) \rightarrow Li^+(g) + e^-$$

This is an **ion-electron equation**, showing the ionisation of a lithium atom.

The equation shows one mole of gaseous lithium atoms losing one mole of electrons to form one mole of lithium ions. This equation is sometimes called a **half-equation**. At GCSE you may have used this type of equation to show the reactions which occur at the electrodes during electrolysis.

In each of these equations state symbols are used.

The **state symbols** are **(g)** gas, **(l)** liquid, **(s)** solid and **(aq)** aqueous (i.e. dissolved in water). These should be used wherever possible.

Which type of equation to use

The type of equation used depends on what is being represented.

An ionic equation is preferable to a molecular equation in some instances. It actually simplifies the equation, showing only the ions which are reacting. Look at the neutralisation reaction between hydrochloric acid and sodium hydroxide:

$$HCl(aq) + NaOH(aq) \rightarrow NaCl(aq) + H_2O(l)$$

The molecular equation shown above can be broken down to show the ions present:

$$H^+(aq) + Cl^-(aq) + Na^+(aq) + OH^-(aq) \rightarrow H_2O(l) + Na^+(aq) + Cl^-(aq)$$

The individual ions separate in solution and behave independently of one another. Notice that the sodium ions and the chloride ions appear on both sides of the equation. This is because they do not actually take part in the reaction. They are called **spectator ions** and can be left out of the equation. This leaves the ionic equation:

$$H^+(aq) + OH^-(aq) \rightarrow H_2O(l)$$

This is a very useful equation: it represents the ionic equation for the neutralisation of any acid by any alkali. Ionic equations are used extensively at AS level.

Balancing equations

Whichever type of equation is used, the equation must be **balanced**.

- Balancing is necessary in order to represent a reaction in which matter cannot be created or destroyed.
- The equation must contain the same number of each type of atom on each side of the equation.
- If the equation contains charged particles (ions and/or electrons), the overall charge on each side of the equation must be the same.

Q

1 Consider the formation of barium sulphate from a solution of barium nitrate and sodium sulphate.

a) Write the molecular equation.
b) Write the ionic equation.

2 Balance the following equations. Remember that the formulae are not altered.

a) $Mg(s) + HCl(aq) \rightarrow MgCl_2(aq) + H_2(g)$
b) $CH_4(g) + O_2(g) \rightarrow CO_2(g) + H_2O(l)$
c) $Fe_2O_3(s) + CO(g) \rightarrow Fe(s) + CO_2(g)$
d) $Mg^{2+}(aq) + OH^-(aq) \rightarrow Mg(OH)_2(s)$
e) $Cl_2(g) + e^- \rightarrow Cl^-(aq)$

3 Write balanced equations for the following reactions. Use the table of ions in Appendix B to help you to write the formulae. In each case choose which type of equation best represents the reaction, and include state symbols.

a) The formation of magnesium oxide when magnesium is burned in oxygen.
b) The neutralisation reaction when sodium hydroxide reacts with sulphuric acid.
c) The formation of a calcium ion (Ca^{2+}) from a calcium atom.
d) The precipitation of silver chloride when silver nitrate solution reacts with copper(II) chloride solution.

Key Ideas 68 – 69

- **Balanced equations are used to represent chemical reactions using the formulae of the reactants and products.**
- **Balanced equations show the rearrangement of atoms to make new substances.**
- **The state symbols are (g) gas, (l) liquid, (s) solid and (aq) aqueous.**
- **Ionic equations show only the reacting ions and simplify equations containing ionic compounds.**

8 Counting the cost

In a recent year Hydro Agri (UK) Ltd produced 650,000 tonnes of ammonium nitrate. This is one of the world's leading fertilisers. The final stage in the process involves reacting ammonia with nitric acid:

$$NH_3(l) + HNO_3(aq) \rightarrow NH_4NO_3(s)$$

The industrial chemists need to know exactly how much ammonia and nitric acid are needed to make the required amount of ammonium nitrate. Wastage is costly, so they carry out calculations to find out how much they need.

Reacting masses

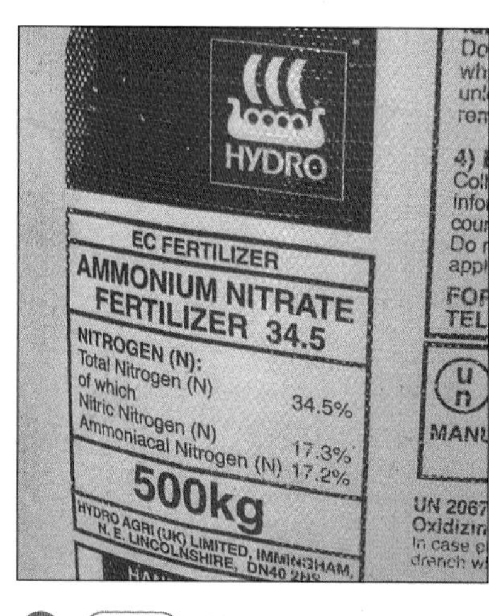

Fig. 1 *Ammonium nitrate fertiliser*

These calculations link your knowledge of moles and balanced equations.

The equation below shows that ammonia and nitric acid react in a ratio of 1 : 1. This is a molar ratio – 1 mol of ammonia reacts with 1 mol of nitric acid:

$$NH_3 + HNO_3 \rightarrow NH_4NO_3$$
$$1\,mol + 1\,mol \rightarrow 1\,mol$$

Since 1 mole equals the relative molecular mass (M_r) in grams, then calculating the M_r for each compound gives:

$$17\,g \text{ of } NH_3 + 63\,g \text{ of } HNO_3 \rightarrow 80\,g \text{ of } NH_4NO_3$$

Notice that the masses on the left-hand side of the equation equal the mass on the right-hand side (this is in accordance with the **Law of Conservation of Mass**).

Industry will work in larger mass units than grams – kilograms or tonnes. These can be applied to the equation as long as the same mass units are used throughout:

$$17\text{ kg of } NH_3 + 63\text{ kg of } HNO_3 \rightarrow 80\text{ kg of } NH_4NO_3$$

Calculating reacting masses

Here are some examples to show you how to do these calculations and how to set out your answer.

1 What mass of ammonium nitrate fertiliser can be made from 18.9 tonnes of nitric acid?

Step 1: Write out the balanced equation. This gives the molar ratio of the compounds you are dealing with, in this case ammonium nitrate and nitric acid. The amount of ammonia is not used in this calculation.

$$NH_3 + HNO_3 \rightarrow NH_4NO_3$$
$$1\,mol \rightarrow 1\,mol$$

Step 2: Convert the mass of nitric acid given in the question first to grams, then to moles, using **number of moles = mass/M_r**, where mass = 18.9g and M_r = 63.

$$\frac{18.9}{63} \text{ mol} = 0.3 \text{ mol}$$

Step 3: Find how many moles of ammonium nitrate are made.

The molar ratio is 1:1, so 0.3 mol of nitric acid makes 0.3 mol of ammonium nitrate.

Step 4: Calculate the mass of ammonuim nitrate produced **using number of moles × M_r = mass,** where number of moles = 0.3 and M_r = 80.

The mass of ammonium nitrate produced is $0.3 \times 80 = 24$g.

If 18.9g of nitric acid produces 24g of ammonium nitrate, then 18.9 tonnes will produce 24 tonnes of ammonium nitrate.

2 A farmer requires 33kg of ammonium sulphate fertiliser. How much ammonia is needed to make this quantity of fertiliser?

Step 1: First write the balanced equation.

$$2NH_3 + H_2SO_4 \rightarrow (NH_4)_2SO_4$$
$$2\text{ mol} \rightarrow 1\text{ mol}$$

Step 2: Calculate the number of moles of ammonium sulphate required using **number of moles = mass/M_r**, where mass = 33g and M_r = 132.

$$\frac{33}{132} \text{ mol} = 0.25 \text{ mol.}$$

Step 3: From the equation:

1 mol of ammonium sulphate is made from 2 mol of ammonia, i.e. the molar ratio is 1:2, so 0.25 mol of ammonium sulphate is made from (0.25×2) mol of ammonia = 0.5 mol.

Step 4: Using **number of moles × M_r = mass,**

the mass of ammonia required = $0.5 \times 17 = 8.5$g.

If 33g of fertiliser is made form 8.5g of ammonia, then 33kg is made from 8.5kg of ammonia.

What about ions?

The mass of ions is the same as the constituent atoms since they differ only in the number of electrons, which have no significant mass.

For example, consider the ions present in ammonium sulphate:

$$(NH_4)_2SO_4 \rightarrow 2NH_4^+ + SO_4^{2-}$$
$$1\text{ mol} \rightarrow 2\text{ mol} + 1\text{ mol}$$

Converting moles to mass gives:

$$132\text{g} \rightarrow 36\text{g} + 96\text{g}$$

The M_r of the sulphate ion = $32 \times (16 \times 4) = 96$.

The M_r of the ammonium ion = $14 + (1 \times 4)$ = 18, so 2 mol weighs 36g.

Q

1 Ammonia can be converted industrially by oxidation into nitric acid. The overall reaction can be represented by:

$$NH_3(g) + 2O_2(g) \rightarrow HNO_3(aq) + H_2O(l)$$

a) Calculate the mass of nitric acid which can be obtained from 34 tonnes of ammonia.

2 Quicklime (calcium oxide) is made industrially by roasting limestone (calcium carbonate) in a limekiln.

$$CaCO_3(s) \rightarrow CaO(s) + CO_2(g)$$

Calculate the mass of limestone needed to produce 16.8kg of quicklime.

Key Ideas 70 – 71

- **The balanced equation shows the molar ratio of the reactants and products.**
- **The molar ratio is used to calculate reacting masses.**
- **Molar quantities can be applied to ions as well as atoms and molecules.**

9 It's a gas ...

So far we have looked at moles and molar ratios in terms of mass. Gases also react in molar amounts, but it is much more convenient to consider the *volume* of a gas rather than its mass.

Reacting volumes

We are now going to look at how volumes of gases can be calculated. A good starting point is to look at the **Haber Process**, which you studied at GCSE. The Haber Process is important industrially in producing ammonia. It is the first stage in fertiliser manufacture. It is useful to look again at the Haber Process because the reaction involves gases. The equation for this process is shown below:

$3H_2(g) + N_2(g) \rightarrow 2NH_3(g)$

The molar ratio of the reacting gases in this process is 3:1 H_2:N_2. The gases are mixed in this ratio before they are passed over the iron catalyst.

The reaction conditions for this process vary depending on the plant, but typically the temperature is between 400 °C and 450 °C, and the pressure is 200 atmospheres (see page 116).

Molar volumes

For gaseous reactions it is important to state the temperature and pressure, because the volume of a gas changes if either the temperature or pressure is altered. If different gases are to be compared, the comparison must be made under the same conditions of temperature and pressure:

The standard temperature used is 273 K (0°C). **K** stands for **kelvin**. The standard pressure is 100 kPa (1 atmosphere). **kPa** stands for **kilopascal**.

In 1811 Avogadro proposed his hypothesis that 'Equal volumes of all gases at the same temperature and pressure contain an equal number of molecules'. At **standard temperature and pressure (stp)**, a volume of 22.4 dm^3 of any gas contains 6.023×10^{23} molecules – that is, Avogadro's number of molecules (see page 58). Since one mole of any substance contains 6.023×10^{23} particles, it follows that:

1 mole of any gas occupies 22.4 dm^3 at standard temperature and pressure.

The volume 22.4 dm^3 is known as the **molar volume** and provides us with a method of linking moles of any gas to volumes or masses. For example, oxygen (O_2) has a relative molecular mass of 32, so one mole of oxygen weighs 32 g. One mole of oxygen at stp has a volume of 22.4 dm^3, so 32 g of oxygen occupies a volume of 22.4 dm^3.

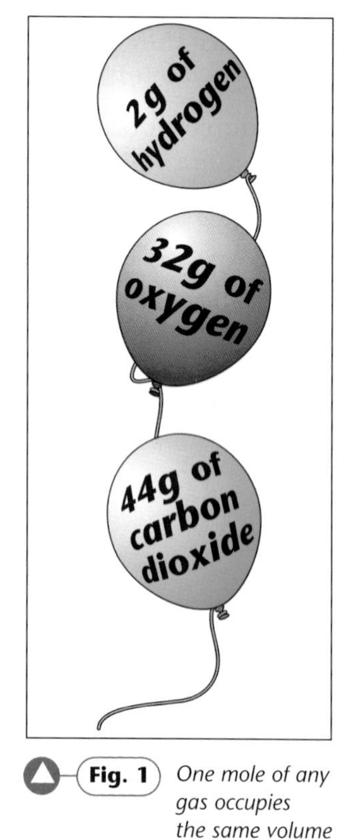

Fig. 1 *One mole of any gas occupies the same volume*

Calculating reacting volumes

1 A volume of 134.4 dm^3 of pure hydrogen is reacted with excess nitrogen in the presence of an iron catalyst. The reaction is carried out at stp. Assuming the reaction goes to completion, what volume of ammonia is produced?

Step 1: Write the balanced equation.

$3H_2(g) + N_2(g) \rightarrow 2NH_3(g)$
3 mol → 2 mol

Step 2: Remember that 1 mol of any gas at stp = 22.4 dm^3.

So in 134.4 dm^3 of hydrogen there are

$$\frac{134.4}{22.4} = 6\,\text{mol}.$$

Step 3: Since the molar ratio for $H_2:NH_3$ is 3:2 (from the equation), then 6 mol of H_2 makes 4 mol of NH_3.

Step 4: Convert the number of moles of ammonia back to a volume.

4 mol of $NH_3 = 4 \times 22.4\,\text{dm}^3$
$= 89.6\,\text{dm}^3$.

Q

1 Propane burns in oxygen according to the equation below:

$C_3H_8(g) + 5O_2(g) \rightarrow 3CO_2(g) + 4H_2O(g)$

What is the minimum volume of oxygen gas needed for the complete combustion of 25 dm^3 of propane at stp?

This type of calculation is not just restricted to gaseous reactions. It may involve the production of a gas from a solid, as the example below shows.

2 How many grams of magnesium will react with excess sulphuric acid to produce 896 cm^3 of hydrogen at stp?

Step 1: Write the balanced equation to find the molar ratio.

$Mg(s) + H_2SO_4(aq) \rightarrow MgSO_4(aq) + H_2(g)$
1 mol 1 mol

Step 2: Convert the units of volume to dm^3.

$$896\,\text{cm}^3 \text{ of hydrogen} = \frac{896}{1000}\,\text{dm}^3 = 0.896\,\text{dm}^3$$

Step 3: Find how many moles of hydrogen are made.

22.4 dm^3 of hydrogen = 1 mol

$$\text{so } \frac{0.896}{22.4} = 0.04\text{ mol}.$$

Step 4: Since the molar ratio is 1:1 (from the equation), then

0.04 mol of hydrogen is made from 0.04 mol of magnesium.

Step 5: Magnesium is a solid, so convert to mass using **moles** × $\boldsymbol{A_r}$ = **mass**, where $\boldsymbol{A_r}$ = 24.

$0.04 \times 24 = 0.96\,\text{g}$

Q

2 A volume of 112 cm^3 of carbon dioxide is produced when limestone is reacted with an excess of dilute hydrochloric acid. Given that the equation for the reaction is

$CaCO_3(s) + 2HCl(aq) \rightarrow CaCl_2(aq) + H_2O(l) + CO_2(g)$

calculate the mass of limestone required, in grams, if the reaction is carried out at stp.

Key Idea 72 – 73

- **The molar gas volume at stp is 22.4 dm^3. This means that one mole of any gas at 273 K and 100 kPa occupies a volume of 22.4 dm^3.**

10 It's an ideal gas ...

A new problem arises with gas calculations if the volumes of gases are *not* at standard temperature and pressure. Because the volume of a gas varies depending on the temperature and pressure, any changes in these must be taken into account. The problem is solved by using the ideal gas equation.

The ideal gas equation

A lot of experiments were carried out on gases as long ago as 1662, when Robert Boyle measured the volumes of gases at different pressures. He observed that at constant temperature the volume of a gas decreases if the pressure is increased. This can be expressed mathematically as:

$V \propto 1/p$ where V = volume and p = pressure

That is, volume is inversely proportional to pressure.

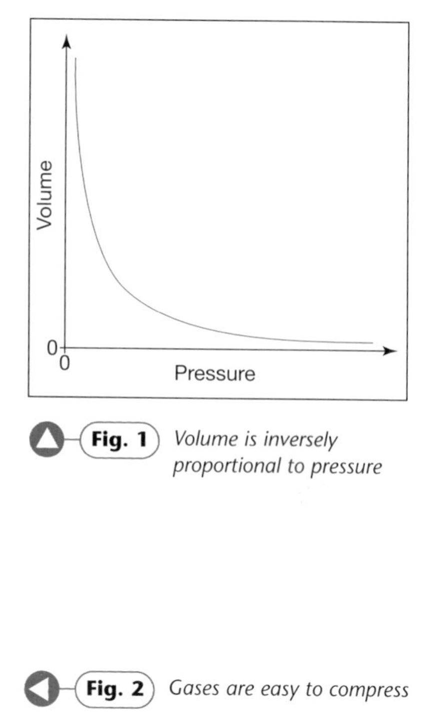

Fig. 1 *Volume is inversely proportional to pressure*

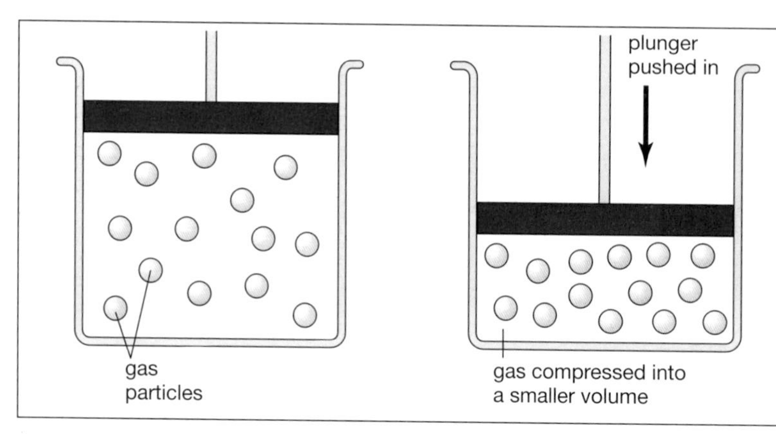

Fig. 2 *Gases are easy to compress*

Gases are easy to compress because the particles are so widely spaced (see page 36). If a gas is heated it expands. The relationship between the volume of a gas and temperature was investigated by Jacques Charles. His law states that 'At constant pressure the volume of a given amount of gas is directly proportional to its absolute temperature'. That is:

$V \propto T$ where T = absolute temperature

The absolute temperature scale was proposed by Lord Kelvin. Absolute zero of −273 °C was redesignated 0 K (zero kelvin), and this is the temperature scale used in all gas calculations.

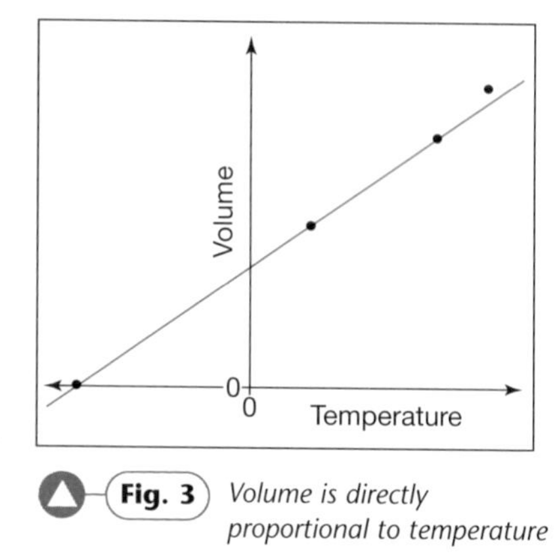

Fig. 3 *Volume is directly proportional to temperature*

Another way of expressing Avogadro's Law (see page 58) is to say that 'At constant temperature and pressure the volume of any gas is directly proportional to the number of moles of the gas'. That is:

$V \propto n$ where n = number of moles

If these three equations are combined we get:

$V \propto \frac{nT}{p}$ or $V = \frac{RnT}{p}$ where R = the gas constant

Rearranging this formula gives $pV = nRT$, which is the usual way of writing the ideal gas equation.

- p = pressure measured in Pa (pascals)
- V = volume measured in m^3
- n = number of moles = $\frac{m}{M_r}$ where m = mass in grams
- R = the gas constant = $8.31\,J\,K^{-1}\,mol^{-1}$ (joules per kelvin per mole)
- T = temperature measured in K (kelvin).

Using the ideal gas equation

Here are some examples of calculations and how to set out your answers. Don't forget always to quote the units.

1 What volume is occupied by 0.25 mol of a gas at a pressure of 200 kPa and a temperature of 27 °C?

Step 1: First convert the pressure and temperature to the correct units:

$200\,kPa = (200 \times 1000) = 2 \times 10^5\,Pa$

$27\,°C = (27 + 273) = 300\,K$

Step 2: Next rearrange the equation so that the unknown is on its own:

$V = \frac{nRT}{p}$ $n = 0.25$, $R = 8.31$, $T = 300$ and $p = 2 \times 10^5$

Step 3: Now substitute the numbers into the equation:

$$V = \frac{0.25 \times 8.31 \times 300}{2 \times 10^5}$$

$$= \frac{0.25 \times 8.31 \times 300 \times 10^{-5}}{2}$$

$$= 3.12 \times 10^{-3}\,m^3$$

2 At 571 K a 0.6 g sample of helium occupies a glass bulb of volume $7.0\,dm^3$. Calculate the pressure exerted by the helium under these conditions.

Step 1: First convert the mass of the sample to moles using $\boldsymbol{n = m/A_r}$, where A_r for helium = 4.

$$n = \frac{0.6}{4} = 0.15\,mol$$

Step 2: Now convert the volume into the correct units $= 7.0 \times 10^{-3}\,m^3$.

Step 3: Rearrange the equation and substitute the numbers:

$$p = \frac{nRT}{V} = \frac{0.15 \times 8.31 \times 571}{7.0 \times 10^{-3}}$$

$$= \frac{0.15 \times 8.31 \times 571 \times 10^3}{7}$$

$$= 1.02 \times 10^5\,Pa$$

Q

1 What volume is occupied by 0.1 mol of a gas at 35 °C and 150 kPa pressure?

2 Calculate the pressure exerted by 71 g of chlorine gas in a volume of $50\,dm^3$ at 298 K.

3 At what temperature will 4.4 g of carbon dioxide occupy a volume of $20\,dm^3$ at standard pressure?

Key Ideas 74 – 75

- **The volume of a gas alters depending on the temperature and pressure.**
- **The ideal gas equation can be used to calculate the volume of a gas under different conditions of temperature and pressure.**

11 Masses of molecules ...

The relative molecular mass of a compound can be calculated in three different ways. You have already met two of them.

1 If the formula of the compound is known, the addition of the A_r values for the atoms present gives the M_r value (see page 57).

2 From mass spectrum data: the peak with the largest *m*/*z* value gives the M_r (see page 10).

The third method uses the ideal gas equation.

Calculating relative molecular masses from the ideal gas equation

$$pV = nRT$$

p = pressure measured in Pa (pascals)

V = volume measured in m^3

n = number of moles = $\frac{m}{M_r}$ where m = mass in grams

R = the gas constant = 8.31 J K^{-1} mol^{-1} (joules per kelvin per mole)

T = temperature measured in K (kelvin)

By rearranging the equation, we get:

$$n = \frac{pV}{RT} \text{ where } n = \text{number of moles}$$

but $n = \frac{m}{M_r}$

Putting these two equations together gives:

$$\frac{m}{M_r} = \frac{pV}{RT}$$

and rearranging we get:

$$M_r = \frac{mRT}{pV}$$

This expression can be used to calculate the relative molecular mass of a gas. Here is a typical examination question.

0.71 g of a gas when contained in a vessel of volume 0.821 dm^3 exerted a pressure of 50.65 kPa at a temperature of 227 °C. Use these data to calculate the M_r of the gas.

Step 1: Convert the values to the correct units.

Volume in m^3 = 0.821×10^{-3}.

Pressure in Pa = 50.65×10^3.

Temperature in K = (227 + 273) = 500.

Step 2: Now substitute these values into the equation.

$$M_r = \frac{0.71 \times 8.31 \times 500}{50.65 \times 10^3 \times 0.821 \times 10^{-3}} = \frac{0.71 \times 8.31 \times 500}{50.65 \times 0.821}$$

$$= 70.94.$$

Always check your final answer. Think about it: gases are *small* molecules – they rarely have M_r values above 100.

Q

1 Find the relative molecular mass of gas **A** given the following data: 12.02 g of **A** occupies a volume of 0.003 m^3 at 37 °C and 172 kPa pressure.

2 0.636 g of a gas occupy a volume of 154 cm^3 at 150 kPa pressure and a temperature of 310 K. Calculate the relative molecular mass of the gas.

3 A glass bulb of volume 1 dm^3 is filled with 1.60 g of a gas **B** to a pressure of 100 kPa. Calculate the relative molecular mass of gas **B** if the temperature inside the bulb is 301 K.

Finding M_r by experiment

The easiest and most accurate way to find the relative molecular mass of a gas is to use a mass spectrometer. From the spectrum produced, the *m/z* value for the molecular ion peak equals the value of M_r.

This is how industry will determine relative molecular masses but in school or college we must revert to a cheaper method. A simple experiment is illustrated below.

Fig. 1 *A page taken from a student's notebook*

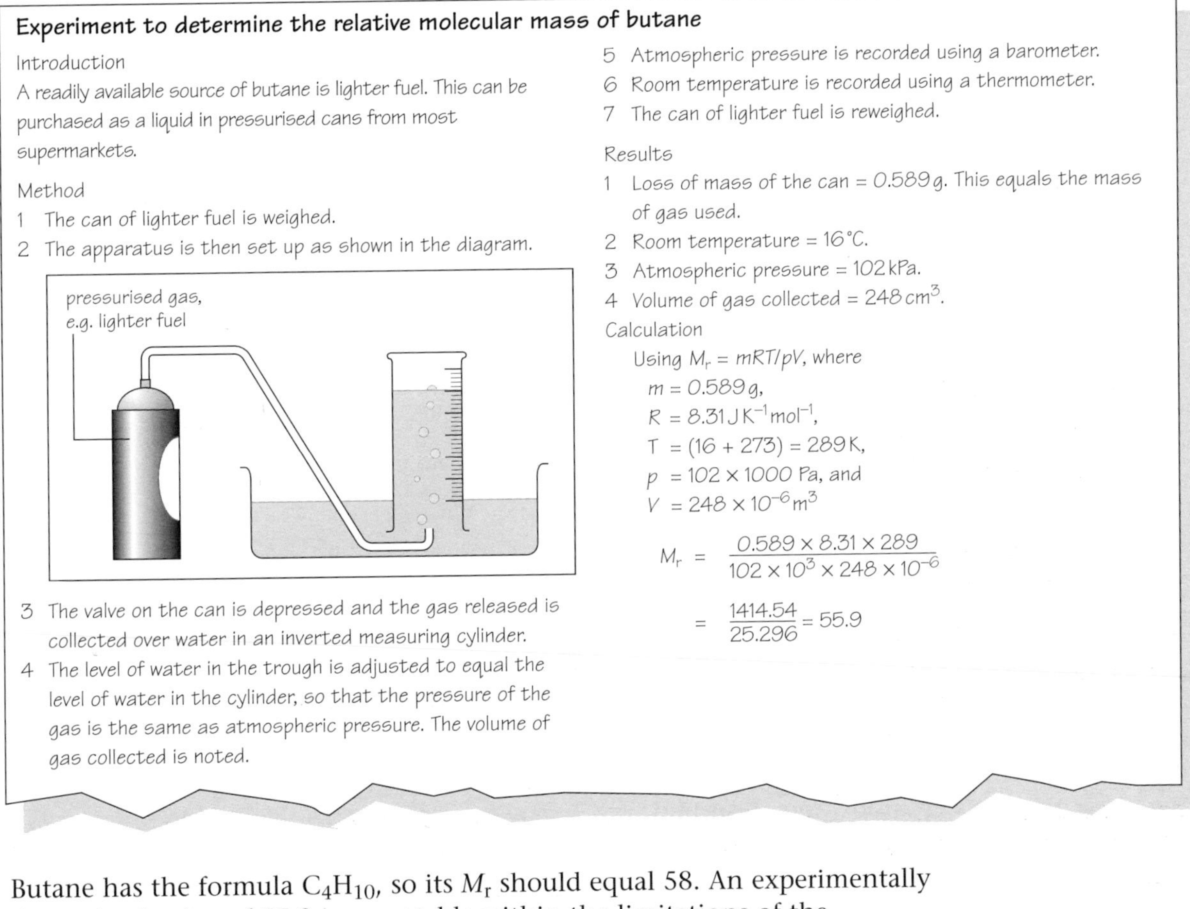

Experiment to determine the relative molecular mass of butane

Introduction

A readily available source of butane is lighter fuel. This can be purchased as a liquid in pressurised cans from most supermarkets.

Method

1. The can of lighter fuel is weighed.
2. The apparatus is then set up as shown in the diagram.

3. The valve on the can is depressed and the gas released is collected over water in an inverted measuring cylinder.
4. The level of water in the trough is adjusted to equal the level of water in the cylinder, so that the pressure of the gas is the same as atmospheric pressure. The volume of gas collected is noted.
5. Atmospheric pressure is recorded using a barometer.
6. Room temperature is recorded using a thermometer.
7. The can of lighter fuel is reweighed.

Results

1. Loss of mass of the can = 0.589 g. This equals the mass of gas used.
2. Room temperature = 16 °C.
3. Atmospheric pressure = 102 kPa.
4. Volume of gas collected = 248 cm^3.

Calculation

Using $M_r = mRT/pV$, where

$m = 0.589\,g$,

$R = 8.31\,J\,K^{-1}\,mol^{-1}$,

$T = (16 + 273) = 289\,K$,

$p = 102 \times 1000\,Pa$, and

$V = 248 \times 10^{-6}\,m^3$

$$M_r = \frac{0.589 \times 8.31 \times 289}{102 \times 10^3 \times 248 \times 10^{-6}}$$

$$= \frac{1414.54}{25.296} = 55.9$$

Butane has the formula C_4H_{10}, so its M_r should equal 58. An experimentally determined value of 55.9 is acceptable within the limitations of the experiment, and bearing in mind that the lighter fuel may not be pure butane.

Q

4 The empirical formula of compound Z is CH_2. At 100 °C, 0.24 g of Z occupies 134 cm^3 at a pressure of 99 kPa. Calculate the relative molecular mass of Z and use this value to find its molecular formula.

Key Idea 76 – 77

- **The ideal gas equation can be used to calculate the M_r of a gas.**

12 The solution is molar ...

If you stir sugar into your coffee you are making an aqueous solution with a particular concentration. To make an aqueous solution, a soluble solid is dissolved in water. The solid is called the *solute* and the water is the *solvent.* (These are terms you will have used at GCSE.) If a known amount of solute is dissolved in water to produce a solution of known volume, then the *concentration* of that solution can be calculated.

Molarity

If you study the label on a bottle of mineral water you will notice that the water contains a number of different solutes.

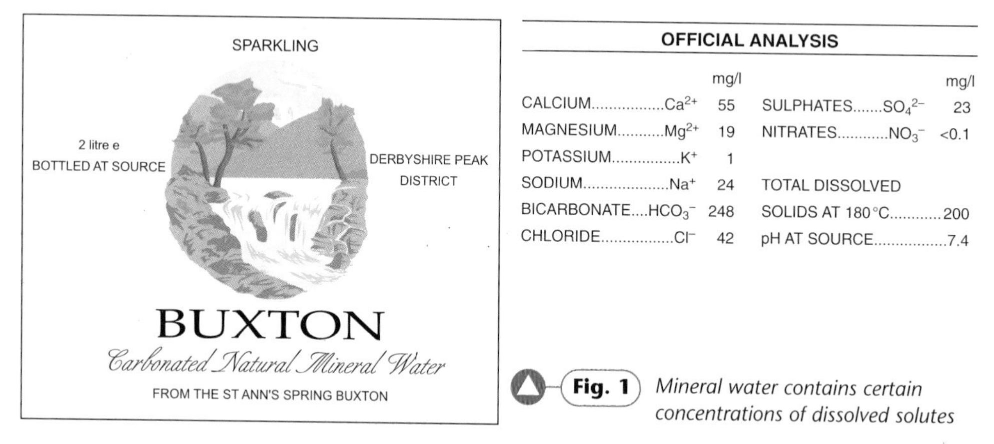

OFFICIAL ANALYSIS

	mg/l		mg/l
CALCIUM................Ca^{2+}	55	SULPHATES.......SO_4^{2-}	23
MAGNESIUM...........Mg^{2+}	19	NITRATES............NO_3^-	<0.1
POTASSIUM...............K^+	1		
SODIUM...................Na^+	24	TOTAL DISSOLVED	
BICARBONATE....HCO_3^-	248	SOLIDS AT 180 °C............	200
CHLORIDE................Cl^-	42	pH AT SOURCE.................	7.4

Fig. 1 Mineral water contains certain concentrations of dissolved solutes

The label shows the concentration of each of the solutes present expressed as mg/l. This means that the concentrations were measured in milligrams per litre. At AS level we measure the amount of solute in moles (mol) and the volume of a solution in cubic decimetres (dm^3), so the concentration is measured in mol dm^{-3}. The number of moles per cubic decimetre of a solution is called its **molarity** (M).

It is important to know the concentration of solutions used in the laboratory. The concentration will affect not only how fast the reaction occurs but how much product is obtained. There is also a safety aspect. Concentrated acids (about 20 M) need to be handled with extreme caution: they are very corrosive. In other situations the concentration of a solution also needs to be known accurately, for example formula milk for babies, alcohol production, and administration of medicinal drugs. In each of these cases the concentration must not exceed a certain value or it might be dangerous to the consumer.

Making molar solutions

A one molar solution contains one mole of solute dissolved in water to make one cubic decimetre of solution. This is written as 1M or 1 mol dm^{-3}. If the solution is made up very accurately then it is recorded as 1.000 M or 1.000 mol dm^{-3}. A solution whose molarity is known exactly is called a **standard solution**.

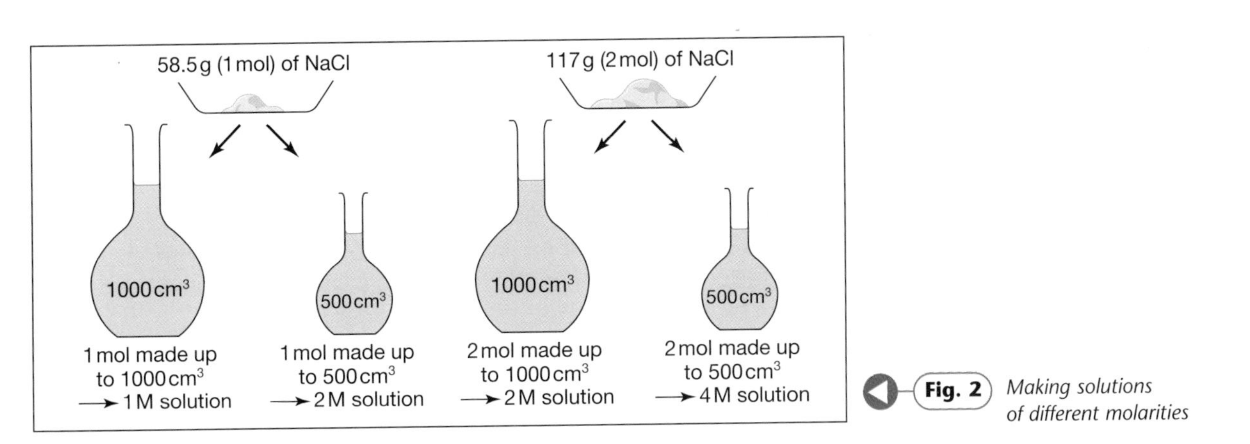

Fig. 2 *Making solutions of different molarities*

Suppose that you are asked to make a 1 M solution of sodium chloride.

1 First you would need to measure out one mole of salt. Remember that one mole of a compound equals its relative molecular mass in grams, so to obtain one mole of sodium chloride you would weigh out 58.5 g (the M_r value for sodium chloride is 58.5).

2 Next, this mass of solid needs to be dissolved in water. The final volume of the sodium chloride solution should be 1 dm^3. The best way to make up a solution to the correct volume is to use a volumetric flask.

Notice that one mole of solute is *not* added to 1 dm^3 of solvent to make a 1 M solution, but one mole of solute is *made up to* 1 dm^3 of solution. Can you see the difference? One mole of solute in 1 dm^3 of solvent would make *more* than 1 dm^3 of solution.

The next task is to make a 2 M solution of sodium chloride. If 1 dm^3 of solution is required then two moles of solid, that is 117 g of sodium chloride, need to be dissolved in water and made up to 1 dm^3. An alternative is to dissolve one mole in half the volume of solution, that is 58.5 g of sodium chloride in 500 cm^3 of solution.

The important point here is that the molarity of the solution depends on the amount of solid and the amount of solution, both of which can be varied. These relationships are shown in the following expression:

$$\boldsymbol{M} = \frac{\boldsymbol{n} \times \mathbf{1000}}{\boldsymbol{V}}$$

Rearranging gives $n = \dfrac{VM}{1000}$

where M = molarity in mol dm^{-3},
V = volume in cm^3,
n = number of moles.

The next section shows you how to use these expressions in calculations.

Q

1 Calculate the relative molecular mass for the following compounds:

a) HNO_3 b) NaOH c) $CaCl_2$

2 Use the values found in question 1 to decide how much of each of the solids you would need to weigh out to produce:

a) 1000 cm^3 of 1 M HNO_3
b) 1000 cm^3 of 2 M NaOH
c) 500 cm^3 of 2 M $CaCl_2$
d) 100 cm^3 of 5 M NaOH

Key Ideas 78 – 79

- **Chemists often carry out reactions in aqueous solution.**
- **The concentration of a solution is measured in moles per cubic decimetre (mol dm^{-3}).**
- **A molar solution (M) contains 1 mol dm^{-3}.**

13 This is technical ...

The previous section introduced an expression which can be used to calculate the molarity of a solution. Here we are going to look at some examples of calculations that a technician working in a chemistry laboratory would have to perform daily.

Calculating the molarity of a solution

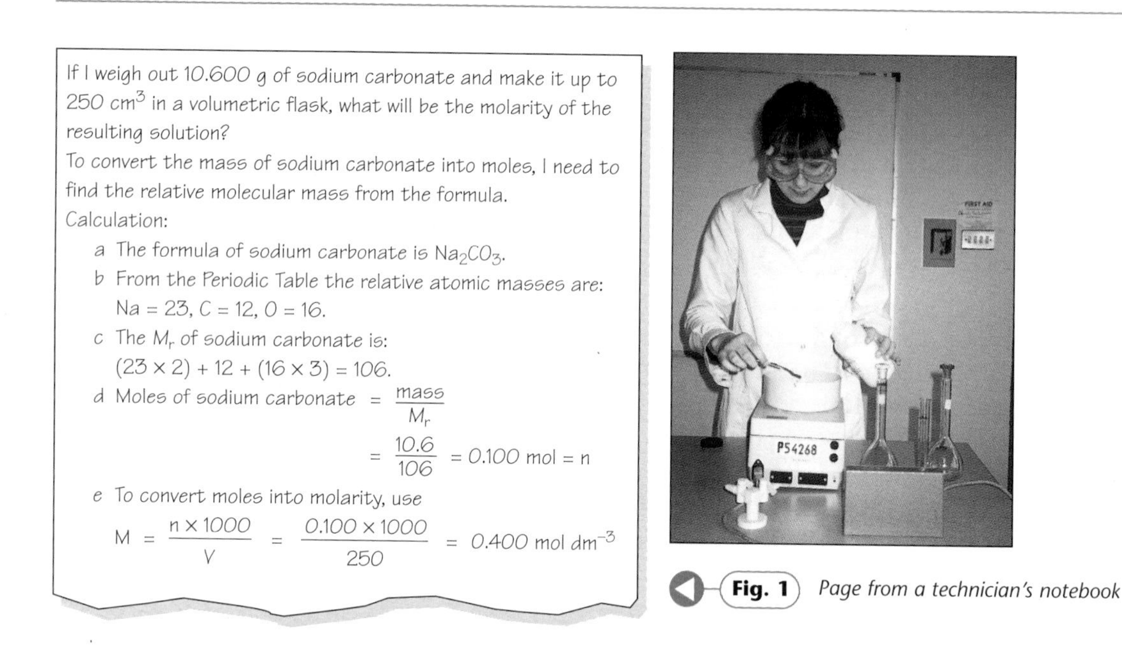

If I weigh out 10.600 g of sodium carbonate and make it up to 250 cm^3 in a volumetric flask, what will be the molarity of the resulting solution?
To convert the mass of sodium carbonate into moles, I need to find the relative molecular mass from the formula.
Calculation:

a The formula of sodium carbonate is Na_2CO_3.
b From the Periodic Table the relative atomic masses are: Na = 23, C = 12, O = 16.
c The M_r of sodium carbonate is: $(23 \times 2) + 12 + (16 \times 3) = 106$.
d Moles of sodium carbonate $= \frac{\text{mass}}{M_r} = \frac{10.6}{106} = 0.100 \text{ mol} = n$
e To convert moles into molarity, use
$$M = \frac{n \times 1000}{V} = \frac{0.100 \times 1000}{250} = 0.400 \text{ mol dm}^{-3}$$

Fig. 1 *Page from a technician's notebook*

Here is another example of this type of calculation.

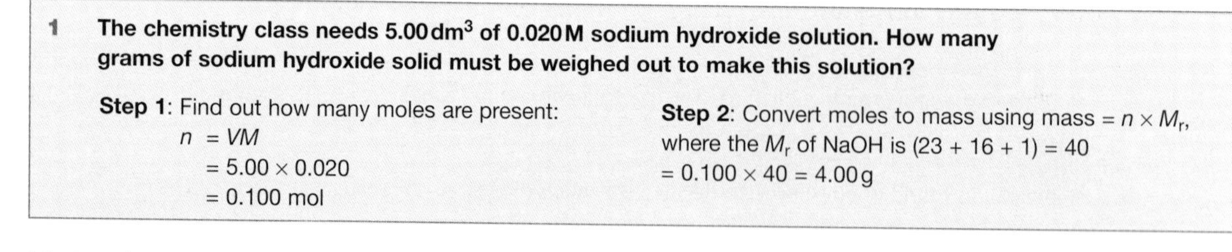

1 The chemistry class needs 5.00 dm^3 of 0.020 M sodium hydroxide solution. How many grams of sodium hydroxide solid must be weighed out to make this solution?

Step 1: Find out how many moles are present:
$n = VM$
$= 5.00 \times 0.020$
$= 0.100$ mol

Step 2: Convert moles to mass using mass $= n \times M_r$, where the M_r of NaOH is $(23 + 16 + 1) = 40$
$= 0.100 \times 40 = 4.00$ g

Notice that this answer is quoted to three significant figures. It would be necessary to weigh out the solid sample to this degree of accuracy if you needed the solution to have the exact molarity of 0.020 M.

Alternatively you could do this calculation in one step.

Notice that there are two expressions above which both involve moles:

$n = VM$ and mass $= n \times M_r$

Putting these two expressions together, we get:

$$\textbf{mass} = \boldsymbol{VM \times M_r}$$

Substituting the values from the question gives:

mass of sodium hydroxide $= 5.00 \times 0.020 \times 40 = 4.00$ g

Q

1 Calculate the mass of each of the following compounds required to make up the following solutions. (Hint: you first need to calculate the relative molecular mass of each compound.)

Compound	Molarity (mol dm^{-3})	Volume (cm^3)
a) KOH	1	1000
b) NaCl	0.1	500
c) $AgNO_3$	0.5	250
d) $NaNO_3$	0.2	100

Diluting solutions

Often solutions are purchased which need diluting to the required concentration. Here is an example.

2 A stock solution of hydrochloric acid contains 10 mol dm^{-3}. What volume of the stock solution should be dissolved in water to make 6 dm^3 of acid of concentration 0.1 mol dm^{-3}?

There are several different ways to approach this calculation. The method using moles is outlined here.

Step 1: Calculate how many moles are required in the diluted solution, using $n = VM$:

$n = 6 \times 0.1 = 0.6$ mol

Step 2: But the stock solution contains 10 mol in every dm^3, so the volume which contains 0.6 mol is given by:

$0.6 = V \times 10$

so $V = \frac{0.6}{10}$

$= 0.06\,dm^3$ or $60\,cm^3$

If 60 cm^3 of 10 M acid is made up to 6 dm^3, the resulting solution will be 0.1 M.

Q

2 Sulphamic acid (NH_2SO_3H) is used to clean out coffee machines. A standard solution of this acid was prepared by dissolving 5.210 g in water and making up the volume to exactly 250 cm^3 in a volumetric flask.

a) Calculate the number of moles of acid used.

b) Calculate the molarity of the acid solution.

3 Washing soda is hydrated sodium carbonate ($Na_2CO_3.10H_2O$). If 2.86 g of the solid is dissolved in water and made up to 100 cm^3, what is the concentration of the resulting solution in mol dm^{-3}?

4 Solutions of laboratory reagents are usually stored in 500 cm^3 bottles. How many bottles of 1 M copper(II) sulphate solution can be made from a stock sample of 1.595 kg of anhydrous copper(II) sulphate solid?

Key Ideas 80 – 81

- **The concentration of a solution is measured in moles per cubic decimetre (mol dm^{-3}).**
- **A molar solution (M) contains 1 mol dm^{-3}.**
- **Useful expressions for performing molarity calculations are $n = VM$ and mass = $n \times M_r$.**
- **Putting these two expressions together gives: mass = $VM \times M_r$, where volume is measured in dm^3.**

14 Let's experiment ...

How would you find the concentration of a solution by experiment? The answer is to carry out a **titration**. This involves the use of accurate volumetric apparatus. At AS level you will perform acid/alkali titrations. If you know the exact molarity of the alkali, you can find the exact molarity of the acid.

Volumetric analysis

When an acid reacts with an alkali, a **neutralisation** reaction occurs. For example:

$$HCl(aq) + NaOH(aq) \rightarrow NaCl(aq) + H_2O(l)$$

The **neutral point** or **end-point** of this reaction can be detected using an indicator.

The volume of acid required for the complete neutralisation of a specific volume of alkali of known concentration can be found by performing a titration experiment. The alkali is known as a **standard solution**, because its concentration is known exactly.

Fig. 1 *Students carrying out titrations*

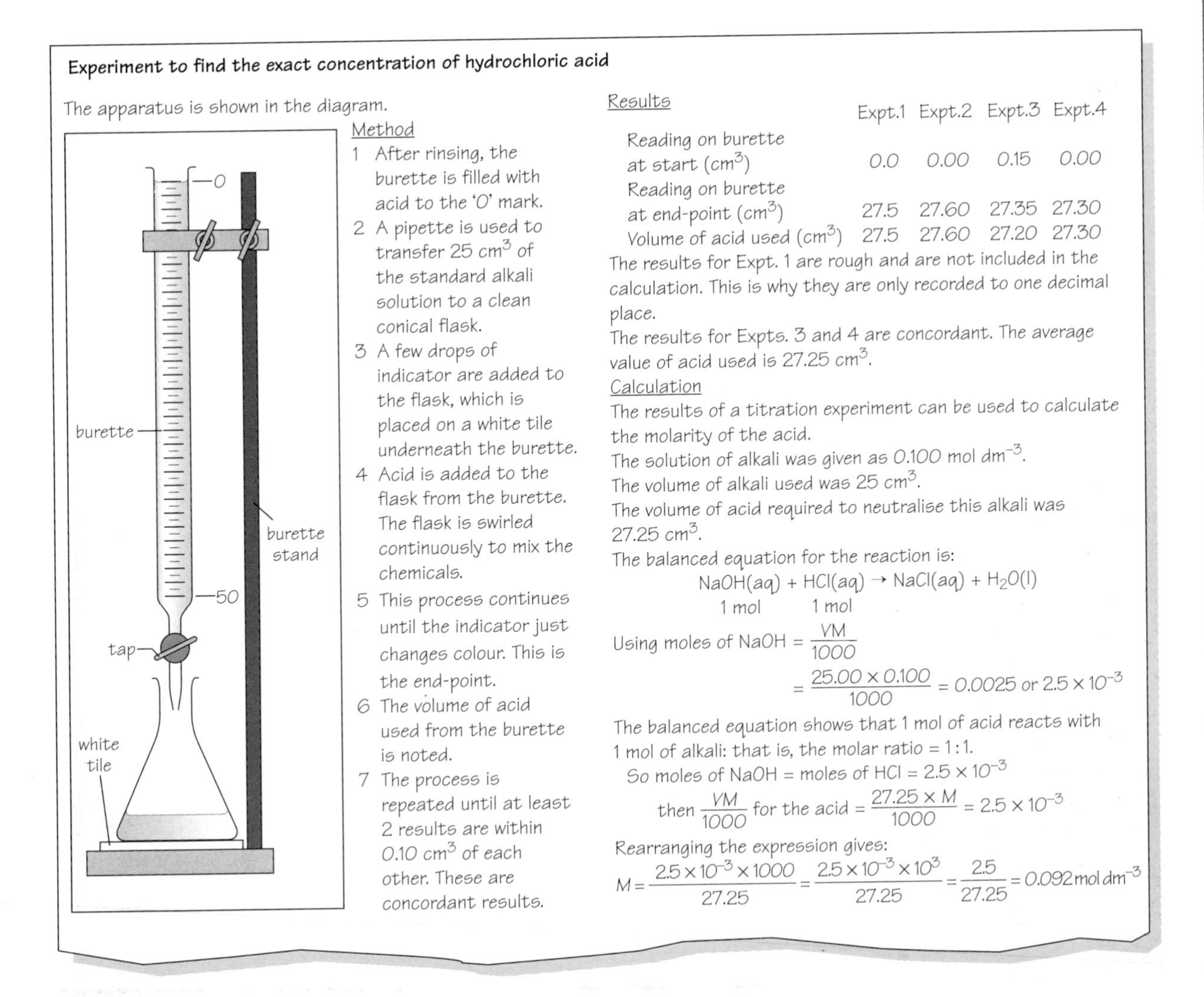

Experiment to find the exact concentration of hydrochloric acid

The apparatus is shown in the diagram.

Method

1 After rinsing, the burette is filled with acid to the 'O' mark.
2 A pipette is used to transfer 25 cm^3 of the standard alkali solution to a clean conical flask.
3 A few drops of indicator are added to the flask, which is placed on a white tile underneath the burette.
4 Acid is added to the flask from the burette. The flask is swirled continuously to mix the chemicals.
5 This process continues until the indicator just changes colour. This is the end-point.
6 The volume of acid used from the burette is noted.
7 The process is repeated until at least 2 results are within 0.10 cm^3 of each other. These are concordant results.

Results

	Expt.1	Expt.2	Expt.3	Expt.4
Reading on burette at start (cm^3)	0.0	0.00	0.15	0.00
Reading on burette at end-point (cm^3)	27.5	27.60	27.35	27.30
Volume of acid used (cm^3)	27.5	27.60	27.20	27.30

The results for Expt. 1 are rough and are not included in the calculation. This is why they are only recorded to one decimal place.
The results for Expts. 3 and 4 are concordant. The average value of acid used is 27.25 cm^3.

Calculation

The results of a titration experiment can be used to calculate the molarity of the acid.
The solution of alkali was given as 0.100 mol dm^{-3}.
The volume of alkali used was 25 cm^3.
The volume of acid required to neutralise this alkali was 27.25 cm^3.
The balanced equation for the reaction is:

$$NaOH(aq) + HCl(aq) \rightarrow NaCl(aq) + H_2O(l)$$

1 mol 1 mol

Using moles of NaOH $= \frac{VM}{1000}$

$$= \frac{25.00 \times 0.100}{1000} = 0.0025 \text{ or } 2.5 \times 10^{-3}$$

The balanced equation shows that 1 mol of acid reacts with 1 mol of alkali: that is, the molar ratio = 1 : 1.
So moles of NaOH = moles of HCl = 2.5×10^{-3}

then $\frac{VM}{1000}$ for the acid $= \frac{27.25 \times M}{1000} = 2.5 \times 10^{-3}$

Rearranging the expression gives:

$$M = \frac{2.5 \times 10^{-3} \times 1000}{27.25} = \frac{2.5 \times 10^{-3} \times 10^3}{27.25} = \frac{2.5}{27.25} = 0.092 \text{ mol dm}^{-3}$$

Q

1 25.00 cm^3 of sodium hydroxide solution of concentration 0.20 mol dm^{-3} neutralises exactly 20.00 cm^3 of dilute hydrochloric acid. Calculate the concentration of the acid.

2 What volume, in cm^3, of 0.15 M potassium hydroxide solution is required to neutralise 32 cm^3 of 0.20 M nitric acid?

3 Baking powder is sodium hydrogen carbonate. It reacts with hydrochloric acid, in solution, according to the equation below:

$$NaHCO_3(aq) + HCl(aq) \rightarrow NaCl(aq) + CO_2(g) + H_2O(l)$$

What is the molarity of the acid if 28.5 cm^3 neutralises 25.0 cm^3 of 0.10 M $NaHCO_3$?

Key Ideas 82 – 83

- **A titration experiment can be performed to find the concentration of a solution.**
- **The concentration of the solution is calculated using molar ratios.**

15 The final solution

Sometimes you will be asked to perform titration calculations for a reaction where the molar ratio is *not* 1:1. In such cases the equation will be given. Some worked examples of more difficult calculations are shown in this final section to Module 1.

Volumetric calculations

Here is an example of a calculation in which the molar ratio is not 1:1.

1 Two moles of nitric acid are needed to neutralise one mole of sodium carbonate solution. What volume of 0.100 M nitric acid is required to neutralise 25.00 cm³ of 0.048 M sodium carbonate solution?

$Na_2CO_3(aq) + 2HNO_3(aq) \rightarrow 2NaNO_3(aq) + CO_2(g) + H_2O(l)$

1 mol   2 mol

moles of $Na_2CO_3 = \frac{25.00 \times 0.048}{1000} = 1.2 \times 10^{-3}$

From the equation, 1 mol of Na_2CO_3 will neutralise 2 mol of acid.

So moles of acid needed $= 1.2 \times 10^{-3} \times 2$
$= 2.4 \times 10^{-3}$

Moles of acid $= \frac{VM}{1000}$

Rearranging:

$V = \frac{\text{moles} \times 1000}{M}$

$= \frac{2.4 \times 10^{-3} \times 1000}{0.100}$

$= 24.00 \text{ cm}^3$

Q

1 Sulphuric acid reacts with sodium hydroxide solution according to the equation shown below:

$H_2SO_4(aq) + 2NaOH(aq) \rightarrow Na_2SO_4(aq) + 2H_2O(l)$

a) 30 cm³ of the acid is needed to neutralise 25 cm³ of 0.20 M NaOH. Calculate the molarity of the acid.

b) What is the molarity of the alkali if 26.5 cm³ are neutralised by 25 cm³ of 0.025 M H_2SO_4?

Some calculations deal with the mass of solids as well as the concentration of solutions. This type of question is illustrated in the next example.

2 A 5 litre bottle of concentrated hydrochloric acid (10 M) has smashed on delivery. The technician has decided to neutralise the spillage with slaked lime (calcium hydroxide). Given that the equation for the reaction is $Ca(OH)_2(s) + 2HCl(aq) \rightarrow CaCl_2(aq) + 2H_2O(l)$ calculate the mass, in kg, of slaked lime needed to completely neutralise the acid.

Step 1: Find the number of moles of acid.
5 litres = 5 dm³.
$n = VM = 5 \times 10$
$= 50 \text{ mol}$

Step 2: From the equation,
1 mol of $Ca(OH)_2$ reacts with 2 mol of acid

So 25 mol of $Ca(OH)_2$ is needed for neutralisation

Step 3: Using **mass =** $n \times M_r$,
the mass of $Ca(OH)_2$ required $= 25 \times 74$
$= 1850 \text{ g} = 1.85 \text{ kg}$

2 The equation for the reaction between limestone (calcium carbonate) and hydrochloric acid is:

$$CaCO_3(s) + 2HCl(aq) \rightarrow CaCl_2(aq) + CO_2(g) + H_2O(l)$$

With what volume of 2.0 M HCl would 5.0 g of limestone react?

3 In an experiment, 2.601 g of anhydrous sodium carbonate were dissolved in water and the solution made up to 250 cm^3 in a volumetric flask.

a) Calculate the molarity of the sodium carbonate solution.

25.0 cm^3 of this sodium carbonate solution required 26.35 cm^3 of dilute hydrochloric acid for complete neutralisation.

b) Calculate the concentration of the acid in this experiment, given that the equation for the reaction is:

$$Na_2CO_3(aq) + 2HCl(aq) \rightarrow 2NaCl(aq) + CO_2(g) + H_2O(l)$$

Another practical use for titrations is to find the percentage purity of a compound. In hard water areas limescale (calcium carbonate) builds up inside kettles and filter coffee machines. This can be removed using a descaler. The active ingredient in a descaler is sulphamic acid (NH_2SO_3H). A student determined the percentage purity by mass of the sulphamic acid in the descaler using the following method.

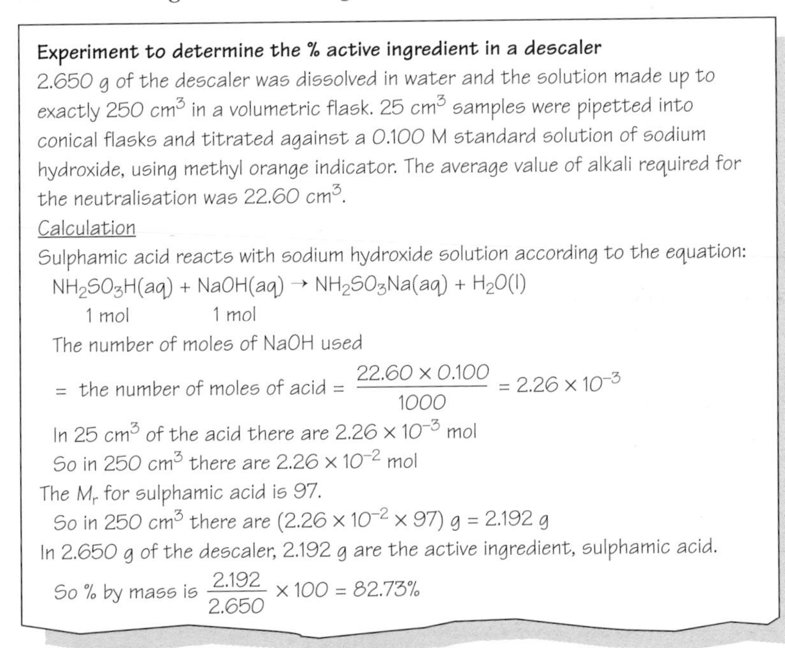

Experiment to determine the % active ingredient in a descaler

2.650 g of the descaler was dissolved in water and the solution made up to exactly 250 cm^3 in a volumetric flask. 25 cm^3 samples were pipetted into conical flasks and titrated against a 0.100 M standard solution of sodium hydroxide, using methyl orange indicator. The average value of alkali required for the neutralisation was 22.60 cm^3.

Calculation

Sulphamic acid reacts with sodium hydroxide solution according to the equation:

$NH_2SO_3H(aq) + NaOH(aq) \rightarrow NH_2SO_3Na(aq) + H_2O(l)$

1 mol 1 mol

The number of moles of NaOH used

$= \text{the number of moles of acid} = \frac{22.60 \times 0.100}{1000} = 2.26 \times 10^{-3}$

In 25 cm^3 of the acid there are 2.26×10^{-3} mol

So in 250 cm^3 there are 2.26×10^{-2} mol

The M_r for sulphamic acid is 97.

So in 250 cm^3 there are $(2.26 \times 10^{-2} \times 97)$ g = 2.192 g

In 2.650 g of the descaler, 2.192 g are the active ingredient, sulphamic acid.

So % by mass is $\frac{2.192}{2.650} \times 100 = 82.73\%$

4 Washing soda is hydrated sodium carbonate. Calculate the percentage by mass of sodium carbonate in washing soda using the following results. 4.00 g of washing soda were dissolved in water and made up to 250 cm^3 of solution. In a titration, 25.0 cm^3 of this solution required 30.0 cm^3 of 0.10 M HCl for complete neutralisation.

A summary of some of the calculations that involve moles is given in Appendix C.

Key Ideas 84 – 85

- **A titration experiment can be performed to find the concentration of a solution or the percentage purity of a solid sample.**
- **The concentration of the solution is calculated using molar ratios.**

Unit 4 Questions

16

1. a) Define the term *relative atomic mass*. (2)
   b) Why is ^{12}C referred to when defining this term? (1)
   c) The mass of one atom of ^{12}C is 1.9925×10^{-23} g. Use this value to calculate the Avogadro constant. (2)

2. Copy and complete the following table using a Periodic Table to look up the relative atomic masses. (8)

Chemical name	Formula	State at room temperature and pressure	Relative molecular mass
Water			18
	NH_3	gas	
Methane			16
Sulphuric acid		liquid	
	$CuSO_4.5H_2O$	solid	
Calcium carbonate			
	C_4H_{10}		

3. During fermentation using yeast, 1.8 g of glucose ($C_6H_{12}O_6$) can be converted into 0.92 g of ethanol (C_2H_6O) and 448 cm^3 of carbon dioxide, measured at stp.

   a) Calculate the number of moles of glucose used in this experiment. (1)
   b) Calculate the number of moles of ethanol and carbon dioxide produced in this experiment. (3)
   c) Hence write a balanced equation showing the fermentation reaction. (1)

4. Calculate the mass of helium needed to fill a weather balloon which has a volume of 120 dm^3 at standard pressure and a temperature of 18 °C. (4)

5. a) Analysis of a sample of compound **X** showed that it contained 1.12 g of carbon, 0.187 g of hydrogen and 1.49 g of oxygen. Calculate the empirical formula of compound **X**. (3)
   b) A sample of compound X was vaporised and it was found that 0.126 g of the compound occupied 37.8 cm^3 at a pressure of 172 kPa and a temperature of 373 K. Find the relative molecular mass of compound **X** using this information. (2)
   c) Use your answers to parts **a** and **b** to deduce the molecular formula of compound **X**. (2)

6. Compound **Y** (M_r = 58) has this percentage composition by mass: C = 62.04%, H = 10.41%, O = 27.55%. Calculate the molecular formula of compound **Y**. (5)

7. An organic compound **Z** has the following percentage composition by mass:

   carbon 52.2% hydrogen 13.0% oxygen 34.8%

   In an experiment it was found that 0.023 g of the vapour of the compound **Z** occupied a volume of 11.2 cm^3 at stp.

   a) Determine the molecular formula of the compound **Z**. (6)

   Compound **Z** burns completely in oxygen to produce carbon dioxide and water.

   b) Write an equation for this reaction. (1)
   c) Calculate the volume of oxygen required to completely burn two moles of compound **Z** at 398 K and 120 kPa. (3)

8. **a)** Write an equation to show the reaction between magnesium metal and sulphuric acid. (1)
   b) What mass of magnesium will react completely with 100 cm^3 of 1 M H_2SO_4? (2)
   c) Calculate the volume of hydrogen gas produced, in cm^3, when 4.8 g of magnesium reacts with excess acid at stp. (2)

9. Hydrochloric acid reacts with sodium carbonate according to the following equation:

   $$Na_2CO_3(aq) + 2HCl(aq) \rightarrow 2NaCl(aq) + CO_2(g) + H_2O(l)$$

   a) An ionic equation for this reaction is:

   $$CO_3^{2-}(aq) + H^+(aq) \rightarrow CO_2(g) + H_2O(l)$$

   Give two reasons why this equation is not balanced. (2)

   b) 5.202 g of anhydrous sodium carbonate were dissolved in water and the solution made up to exactly 500 cm^3. Calculate the concentration in mol dm^{-3} of the resulting solution. (3)
   c) 25 cm^3 of this solution were titrated against dilute hydrochloric acid. The amount of acid needed for complete neutralisation was 24.50 cm^3. Calculate the molarity of the acid. (3)
   d) One of the products of this reaction is sodium chloride solution. If this solution is evaporated to dryness, what mass of sodium chloride will be made from the reaction between 100 cm^3 of the hydrochloric acid with excess sodium carbonate? (2)

Module 1 Questions

1

(1) a) Explain the trends in atomic radius and in electronegativity down Group II of the Periodic Table. (6)

b) Discuss two ways in which the chemistry of beryllium is not typical of Group II compounds. (5)

(2) a) Discuss the structure and bonding in the elements of Period 3 (Na to Ar). Describe how the structure and bonding influence the melting points of these elements. (12)

b) Explain why Period 3 only contains 8 elements given that the third energy level can hold up to 18 electrons. (2)

(3) a) Explain the meaning of the term *first ionisation energy* of an element. (3)

b) Sketch a graph to show how the ionisation energies for the successive removal of all the electrons from a sodium atom vary with the number of electrons removed. Explain how the graph provides evidence for the existence of electron energy levels. (7)

c) Give the electronic structures of a sodium atom, and a magnesium atom, and explain why these metals are regarded as s-block elements. (3)

d) State and explain how you would expect the second ionisation energy of sodium to compare with:

(i) the first ionisation energy of neon;

(ii) the second ionisation energy of magnesium. (5)

(4) Explain how the physical properties of sodium chloride, graphite and iodine are related to their structures and bonding. The properties you should discuss are melting points and the ability, or lack of it, to conduct electricity. (12)

(5) Ammonium nitrite decomposes completely at 100 °C according to the equation below:

$$NH_4NO_2(s) \rightarrow N_2(g) + 2H_2O(g)$$

If 0.01 mol of ammonium nitrite is heated to 100 °C, at standard pressure, what would be the total volume of gas produced? (4)

(6) a) Draw a simple diagram to show the positions and numbers of the protons, neutrons and electrons in an atom of ^{7}Li. Describe, in terms of charge and mass, the properties of these sub-atomic particles. (7)

b) What are *isotopes*? Illustrate your answer by reference to chlorine and explain why the relative atomic mass of chlorine (35.5) is not a whole number. (5)

c) Isotopes can be separated in a mass spectrometer and a spectrum produced. Explain how a mass spectrometer is able to do this. In your answer explain the following processes: (i) ionisation; (ii) acceleration; (iii) deflection; (iv) detection. (14)

d) The mass spectrum of a compound has a molecular ion peak at m/z = 168. Analysis of this compound shows that it contains 42.9% carbon, 2.4% hydrogen and 16.7% nitrogen by mass. The remainder is oxygen. Calculate the molecular formula of this compound. (6)

7 **a)** Explain the shapes of the following molecules: (i) CH_4; (ii) BCl_3; (iii) NH_3. (6)

b) Predict the shapes of the following ions: (i) NH_2^- and (ii) IF_4^+. Sketch diagrams to illustrate your answer, including bond angles. (5)

c) Explain why the bond angle in NH_3 is greater than that in H_2O. (4)

8 Hydrated sodium carbonate ($Na_2CO_3.xH_2O$) is sold commercially as washing soda. A student carried out an experiment to determine the value of x, which represents the number of moles of water of crystallisation present in one mole of hydrated sodium carbonate. The following method was used:

> 2.995 g of hydrated sodium carbonate were dissolved in water and the solution made up to exactly 250 cm^3 in a volumetric flask. 25 cm^3 samples of this solution were measured by pipette and titrated against a standard solution of hydrochloric acid. The concentration of the acid was 0.113 M.

The average value of acid required to completely neutralise the sodium carbonate solution was 21.20 cm^3.

Given that the equation for this reaction is:

$$Na_2CO_3(aq) + 2HCl(aq) \rightarrow 2NaCl(aq) + CO_2(g) + H_2O(l)$$

a) Calculate the mass of sodium carbonate in the sample using the titration results. (5)

b) Using your answer to part **a** plus the original mass of the hydrated solid, calculate the value for x. (5)

9 **a)** Describe the structure of, and the bonding in, a water molecule. (4)

b) A block of ice at –10 °C is heated up until the temperature reaches 105 °C. Explain the energy changes associated with the changes in state which will occur between these temperatures. In your answer you should refer to the movement of water particles and the forces acting between them. (8)

10 **a)** How does the atomic radius of the atoms change across Period 3 (Na to Ar)? Explain your answer. (4)

b) How does this change affect the electronegativities of these elements? Explain your answer. (3)

c) Given the following electronegativity values, predict the predominant type of bonding in: (i) CsCl; (ii) $BeCl_2$; (iii) NCl_3. Briefly explain the reasons behind your answers. (5)

Element	Cs	Be	N	Cl
Electronegativity	0.7	1.5	3.0	3.0

Module 1 Questions

2

11 A road tanker has crashed and 73 kg of pure hydrochloric acid are spilling onto the motorway. Two methods are suggested to neutralise the acid.

The first suggestion is to hose the acid down with 5 M sodium hydroxide solution.

a) Write an equation for the reaction between hydrochloric acid and sodium hydroxide. (1)

b) How many moles of acid are there in 73 kg? (2)

c) What volume of sodium hydroxide, in dm^3, is needed to neutralise this amount of acid? (2)

The second suggestion is to scatter powdered slaked lime over the spillage. Slaked lime reacts with hydrochloric acid according to the equation below:

$$Ca(OH)_2(s) + 2HCl(aq) \rightarrow CaCl_2(aq) + 2H_2O(l)$$

d) Calculate the mass, in kg, of slaked lime required to neutralise 73 kg of acid. (4)

e) Which method of neutralising the acid is preferable? Explain your answer. (2)

Slaked lime is manufactured by roasting limestone in a kiln and then adding water. The equations for these reactions are:

$$CaCO_3(s) \rightarrow CaO(s) + CO_2(g)$$

$$CaO(s) + H_2O(l) \rightarrow Ca(OH)_2(s)$$

f) Calculate how much limestone, in kg, is needed to make 5 kg of slaked lime. (4)

12 **a)** The melting points of the elements in Period 3 are given below:

Element	Na	Mg	Al	Si	P	S	Cl
M.p./°C	98	651	660	1410	44	114	–101

With reference to these values, explain the following.

(i) The melting point of magnesium is higher than that of sodium.
(ii) Silicon has a very high melting point.
(iii) The melting point of chlorine is lower than that of sulphur. (9)

b) Analysis of a compound of sulphur shows that 1.78 g of the sulphur present is combined with 0.89 g of oxygen and 3.98 g of chlorine.

(i) Calculate the empirical formula of this compound.

(ii) What type of bonding would you expect to be present in this compound? Explain your answer.

(iii) The compound reacts readily with water to produce a mixture of sulphur dioxide and hydrogen chloride. Write an equation for this reaction. (7)

13 a) Distinguish between the *empirical* formula and the *molecular* formula of a compound. (3)

b) The percentage composition by mass of a hydrocarbon **X** was found to be C = 85.72%, H = 14.28%. At stp 0.207 g of **X** occupies a volume of 84 cm^3. Use these data to find the molecular formula of **X**. (7)

14 a) The graph in Figure 1 shows the trend in boiling points of the hydrides of Group 7 with increasing relative molecular mass.

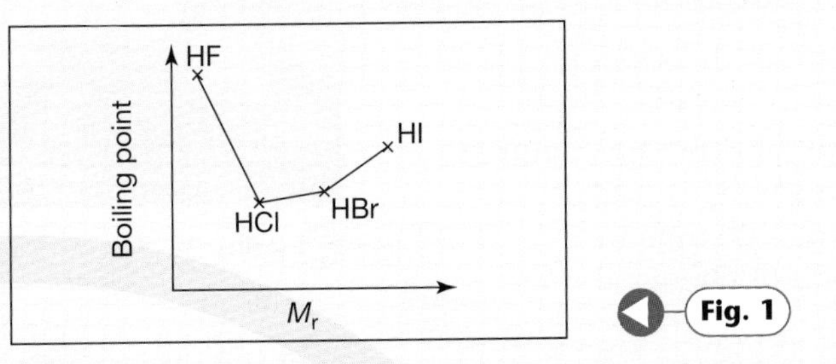

Fig. 1

(i) Explain the trend in increasing boiling points from HCl to HI.

(ii) Account for the anomalously high boiling point of HF. (6)

b) How would the boiling point of F_2 compare with that of HF? Explain your answer. (3)

15 a) Figure 2 shows a mass spectrometer.

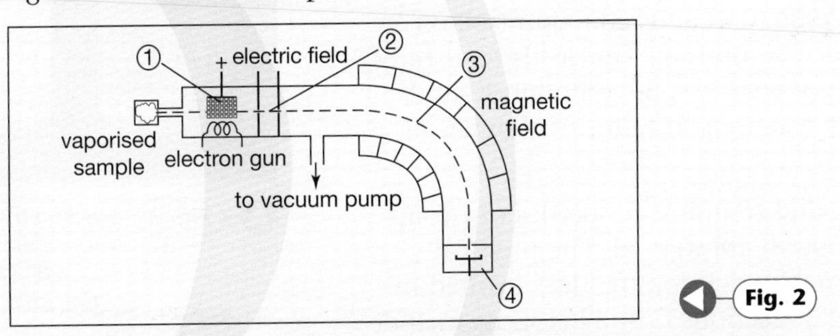

Fig. 2

Name and explain briefly the processes which occur at stages 1–4. (12)

b) The data obtained from a mass spectrum of magnesium are shown below.

m/z	24	25	26
Relative intensity	1	0.127	0.139

(i) Magnesium has three isotopes. What evidence of this is given in the data?

(ii) What are the similarities and differences between these three isotopes?

(iii) Define the term *relative atomic mass.*

(iv) Use these data to calculate the accurate relative atomic mass of magnesium. (12)

Introduction to Module 2

In GCSE Science you may have studied some aspects of metal extractions, rates of reaction, energy changes in reactions, and reversible reactions. In Module 2 these ideas are developed further, together with a detailed study of oxidation and reduction.

In Unit 5 the ideas of energy change studied at GCSE lead to a study of enthalpy changes (heat energy changes at constant pressure) in chemical reactions, and of methods of determining enthalpy changes experimentally using calorimeters. The unit includes calculations of enthalpy changes using energy cycles or bond energies.

In Unit 6 the study of factors that affect the rate of chemical reactions (kinetics) is built on GCSE understanding. This includes a study of how the energy is distributed among the particles in a gas. This leads to a deeper understanding of the effect of a change in temperature and catalysts on the rate of chemical reactions.

In Unit 7 the ideas of reversible reactions are developed into a study of dynamic chemical equilibria reactions. The factors that affect the position of chemical equilibria are studied. Together with the ideas in Units 5 and 6, this leads to an understanding of the conditions selected in industrial processes.

In Unit 8 the ideas of oxidation and reduction (redox) are studied in terms of electron transfer. This leads to a study of the oxidation states of elements in compounds, and consideration of redox reactions in terms of changes in oxidation states.

In Unit 9 the physical and chemical properties of Group VII elements (the halogens) are studied. The physical properties are considered by developing ideas learned in Module 1. The chemical reactions of the halogens include their reactions as oxidising agents and the reactions of the halides as reducing agents, so this study builds on the oxidation states studied in Unit 8. A more detailed study of the chemical reactions of chlorine is included.

In Unit 10 the extraction methods of the metals iron, aluminium and titanium are studied. This work builds on the extraction methods studied at GCSE. Some of the environmental and economic factors associated with recycling used metals are considered.

Fig. 1 *The Thermite reaction is violently exothermic.*

Fig. 2 *Kinetics is the study of rates of reaction; here, marble chips react with different concentrations of hydrochloric acid.*

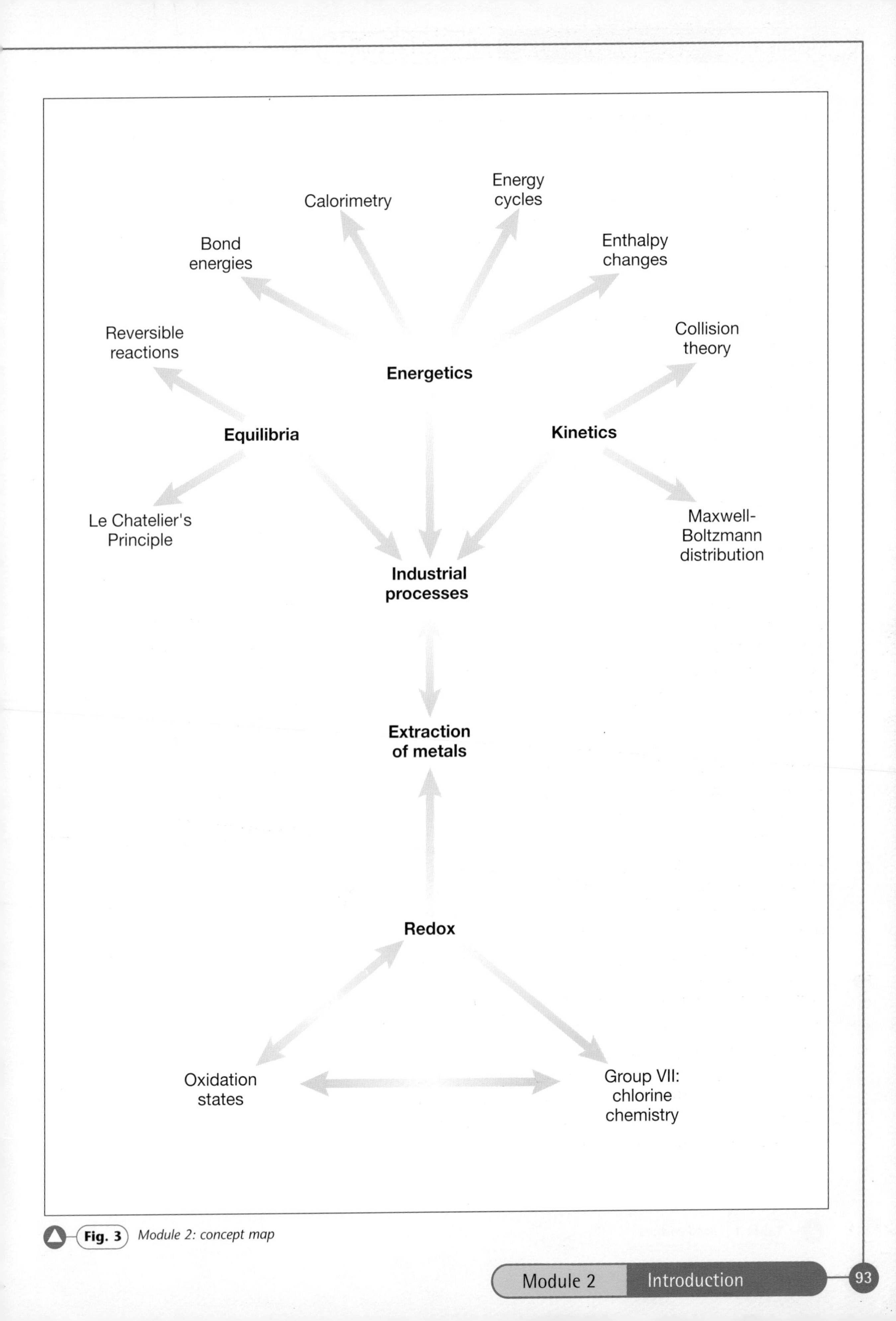

Fig. 3 *Module 2: concept map*

1 The driving force ...

Energetics is the study of energy changes that occur during chemical reactions. From studying GCSE science you will know that usually whenever chemical reactions occur, energy is transferred to or from the surroundings. An *exothermic* reaction transfers heat *to* the surroundings; an *endothermic* reaction takes in heat *from* the surroundings.

Breaking and making chemical bonds

When chemical reactions occur, bonds are broken and new bonds are formed. Energy is needed to *break* bonds, and energy is released when bonds are *formed*.

In a reaction, if more energy is released when the new bonds in the product are formed than is needed to break the bonds in the reactants, then the difference in energy is *released* to the surroundings. This is an **exothermic** reaction.

$$CH_4(g) + 2O_2(g) \rightarrow CO_2(g) + 2H_2O(l)$$

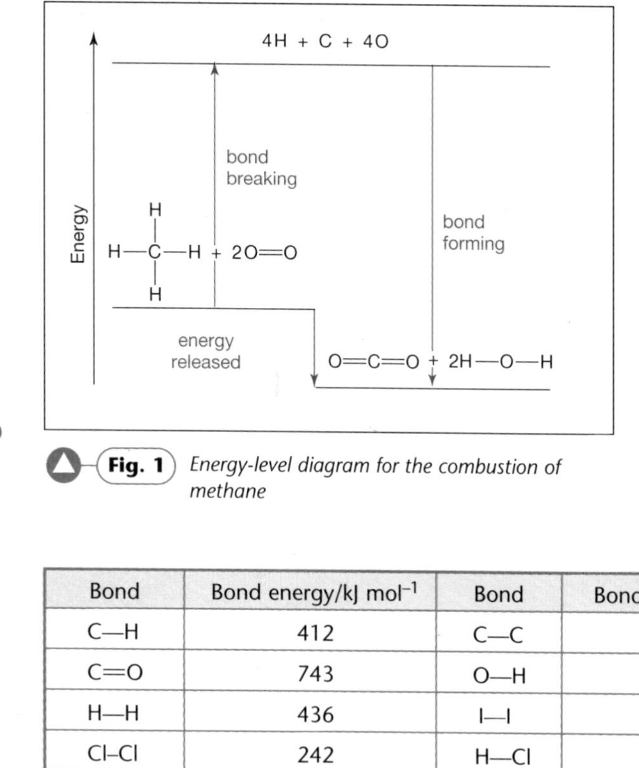

Fig. 1 Energy-level diagram for the combustion of methane

In a reaction, if more energy is needed to break the bonds in the reactants than is released when the new bonds are formed in the products, then the difference in energy is *absorbed* from the surroundings. This is an **endothermic** reaction.

$$2HI(g) \rightarrow H_2(g) + I_2(g)$$

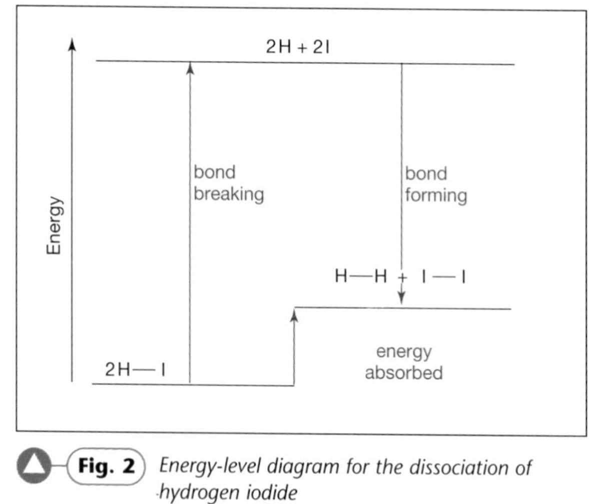

Fig. 2 Energy-level diagram for the dissociation of hydrogen iodide

The energy change in a reaction can be calculated using **bond energies**. Table 1 gives the values of some bond energies.

Bond	Bond energy/kJ mol^{-1}	Bond	Bond energy/kJ mol^{-1}	Bond	Bond energy/kJ mol^{-1}
C—H	412	C—C	348	C—O	360
C=O	743	O—H	463	O=O	496
H—H	436	I—I	151	H—I	299
Cl-Cl	242	H—Cl	431	O—O	146

Table 1 Bond energies

1 What is the energy change during the combustion of methane? (See Figure 1.)

$CH_4(g) + 2O_2(g)$			$CO_2(g) + 2H_2O(l)$			Overall energy change, ΔH:
bonds broken:			bonds formed:			bonds broken – bonds formed
4 C—H	4 × 412 =	1648	2 C=O	2 × 743 =	1486	2640 – 3338 = −698 kJ mol^{-1}
2 O=O	2 × 496 =	992	4 O—H	4 × 463 =	1852	
	Total	2640		Total	3338	

Since more energy is released forming the new bonds than is required in breaking the original bonds, the reaction is exothermic and releases 698 kJ mol^{-1}.

2 What is the energy change during the dissociation of hydrogen iodide? (See Figure 2.)

2HI(g)			$H_2(g) + I_2(g)$			Overall energy change, ΔH:
bonds broken:			bonds formed:			bonds broken – bonds formed
2 H—I	2 × 299 = 598		H—H		436	598 – 587 = +11 kJ mol^{-1}
			I—I		151	
	Total	598		Total	587	

Since less energy is released forming the new bonds than is required in breaking the original bonds, the reaction is endothermic and *absorbs* 11 kJ mol^{-1}.

Mean bond enthalpies/bond dissociation energies

The energy needed to break a *particular* covalent bond (or the energy released when the bond is formed) is called the **bond dissociation energy**. For example:

$$Cl—Cl(g) \rightarrow Cl(g) + Cl(g) \quad +242\ kJ\ mol^{-1}$$

The Cl—Cl bond is only found in Cl_2. However, many other types of bonds, such as C—C, C—H and O—H, are found in many different compounds. For these bonds we use mean bond enthalpy values in calculations.

The **mean bond enthalpy** is the mean (average) value of the bond dissociation energy of a particular type of bond in a range of different compounds. For example, there are millions of compounds that have C—H bonds. The mean bond enthalpy value of the C—H bond is an average value that takes into account the different C—H values in a range of different compounds.

Mean bond enthalpies can be used to calculate the enthalpy change in a reaction. Since mean bond enthalpies use average values, answers for enthalpy changes obtained using them are usually slightly different from experimentally determined values.

Q

1 Explain why 436 kJ mol^{-1} represents the bond dissociation energy for the H—H bond rather than a mean bond enthalpy.

2 Use the bond energies in Table 1 to calculate the energy change for this reaction:

$$H_2(g) + Cl_2(g) \rightarrow 2HCl(g)$$

Key Ideas 94 – 95

- **Energy is needed to break bonds, and energy is released when bonds are formed. The balance of energy needed and energy released results in an overall energy change in a reaction.**

2 Enthalpy changes – no pressure

Most reactions are carried out in open vessels and the pressure remains constant. The heat energy change at constant pressure is called an **enthalpy change, ΔH**. The units of enthalpy change are $kJ\,mol^{-1}$.

Standard enthalpy changes

Standard enthalpy changes, $\Delta H^{\ominus}$, refer to heat energy changes under standard conditions of temperature and pressure.

The standard conditions for enthalpy changes are a temperature of 298 K and a pressure of 100 kPa.

- For an *exothermic* reaction, ΔH is *negative*.
- For an *endothermic* reaction, ΔH is *positive*.

Two specific enthalpy changes are studied at AS level: standard enthalpy of formation and standard enthalpy of combustion.

Standard enthalpy of formation

The **standard enthalpy of formation** of a compound, $\Delta H_f^{\ominus}$, is the enthalpy change which occurs when one mole of a compound is formed from its elements in their standard states at 298 K and 100 kPa.

Enthalpies of formation can be exothermic or endothermic.

The enthalpy changes for the reactions shown in the equations in Table 1 are standard enthalpies of formation. State symbols are essential: in these equations the elements are shown in their standard states.

Reaction	Standard enthalpy of formation /kJ mol^{-1}
$C(gr) + 2H_2(g) \rightarrow CH_4(g)$	–74.9
$C(gr) + O_2(g) \rightarrow CO_2(g)$	–394
$2C(gr) + 2H_2(g) \rightarrow C_2H_4(g)$	+52.3
$C(g) + 2H_2(g) + \frac{1}{2}O_2(g) \rightarrow CH_3OH(l)$	–239

C(gr) indicates carbon in the form of graphite

Table 1 *Standard enthalpies of formation*

By definition, the standard enthalpy of formation of an *element* in its standard state is zero.

Q

1 For each of the following compounds, write an equation for the reaction in which the enthalpy change is the standard enthalpy of formation:

a) $Na_2CO_3(s)$ b) $CH_3CH_2OH(l)$ c) $NH_4NO_3(s)$

Standard enthalpy of combustion

The **standard enthalpy of combustion** of a substance (element or compound), $\Delta H_c^{\ominus}$, is the enthalpy change which occurs when one mole of the substance undergoes *complete* combustion in excess oxygen at 298 K and 100 kPa. Combustion reactions are always exothermic.

The enthalpy changes for the reactions shown in Table 2 are standard enthalpies of combustion.

Note that the same equation can represent more than one type of reaction. For example, the following equation represents both the combustion of carbon *and* the formation of carbon dioxide:

$$C(gr) + O_2(g) \rightarrow CO_2(g) \qquad \Delta H^{\ominus} = -394\ \text{kJ mol}^{-1}$$

Similarly, the following equation represents both the combustion of hydrogen *and* the formation of water:

$$H_2(g) + \tfrac{1}{2}O_2(g) \rightarrow H_2O(l) \quad \Delta H^{\ominus} = -286\ \text{kJ mol}^{-1}$$

Reaction	Standard enthalpy of combustion /kJ mol^{-1}
$C(gr) + O_2(g) \rightarrow CO_2(g)$	–394
$CH_4(g) + 2O_2(g) \rightarrow CO_2(g) + 2H_2O(g)$	–890

Table 2 *Standard enthalpies of combustion*

Q

2 For each of the following compounds, write an equation for the reaction in which the enthalpy change is the standard enthalpy of combustion of the compound.

a) $C_3H_8(g)$ b) $CH_3OH(l)$ c) $C_3H_6(g)$

Hess's Law

The **First Law of Thermodynamics** states that energy cannot be made or destroyed, but can only be changed from one form of energy into a different form of energy. An alternative version of the first law is that the energy content of a closed system is constant.

Hess's Law is a restatement of the first law of thermodynamics:

> In a reaction, the total enthalpy change is independent of the route.

Suppose that a reactant **A** can be changed into a product **B** by two different methods. In the first, **A** is converted directly into product **B**. In the second, **A** is converted into product **C**, which is then converted into product **B** (see Figure 1). Then the *total* energy change for the two routes must be the same.

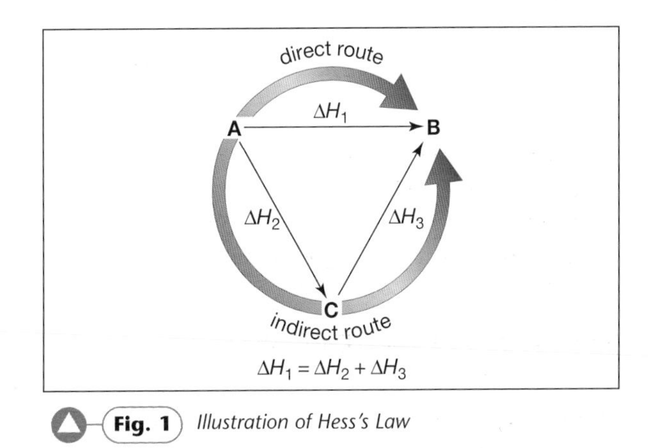

Fig. 1 *Illustration of Hess's Law*

Hess's Law is very useful as it can be used to calculate enthalpy changes that cannot be determined experimentally.

Key Ideas 96 – 97

- **The heat energy change at constant pressure is called an enthalpy change, ΔH.**
- **The conditions for standard enthalpy changes, $\Delta H^{\ominus}$, are a temperature of 298 K and a pressure of 100 kPa.**
- **Two important enthalpy changes are enthalpies of formation and combustion.**
- **Hess's Law states that the enthalpy change in a reaction is independent of the route.**

3 Using energy cycles ...

Some enthalpy changes, for example enthalpies of combustion, can be determined by experiment. Many other enthalpy changes cannot be determined directly by experimental methods. Hess's Law allows us to calculate enthalpy changes that cannot be determined directly by experiment.

Using enthalpies of combustion

The enthalpy of formation of a compound can be calculated using the enthalpies of combustion. This involves constructing an **energy cycle** using enthalpies of combustion. Hess's Law is then used to calculate the enthalpy of formation from the cycle.

Substance	Enthalpy of combustion/kJ mol^{-1}
C(gr)	–394
H_2(g)	–286
CH_4(g)	–890
C_2H_6(g)	–1560
C_3H_8(g)	–2220

Table 1 *Enthalpies of combustion*

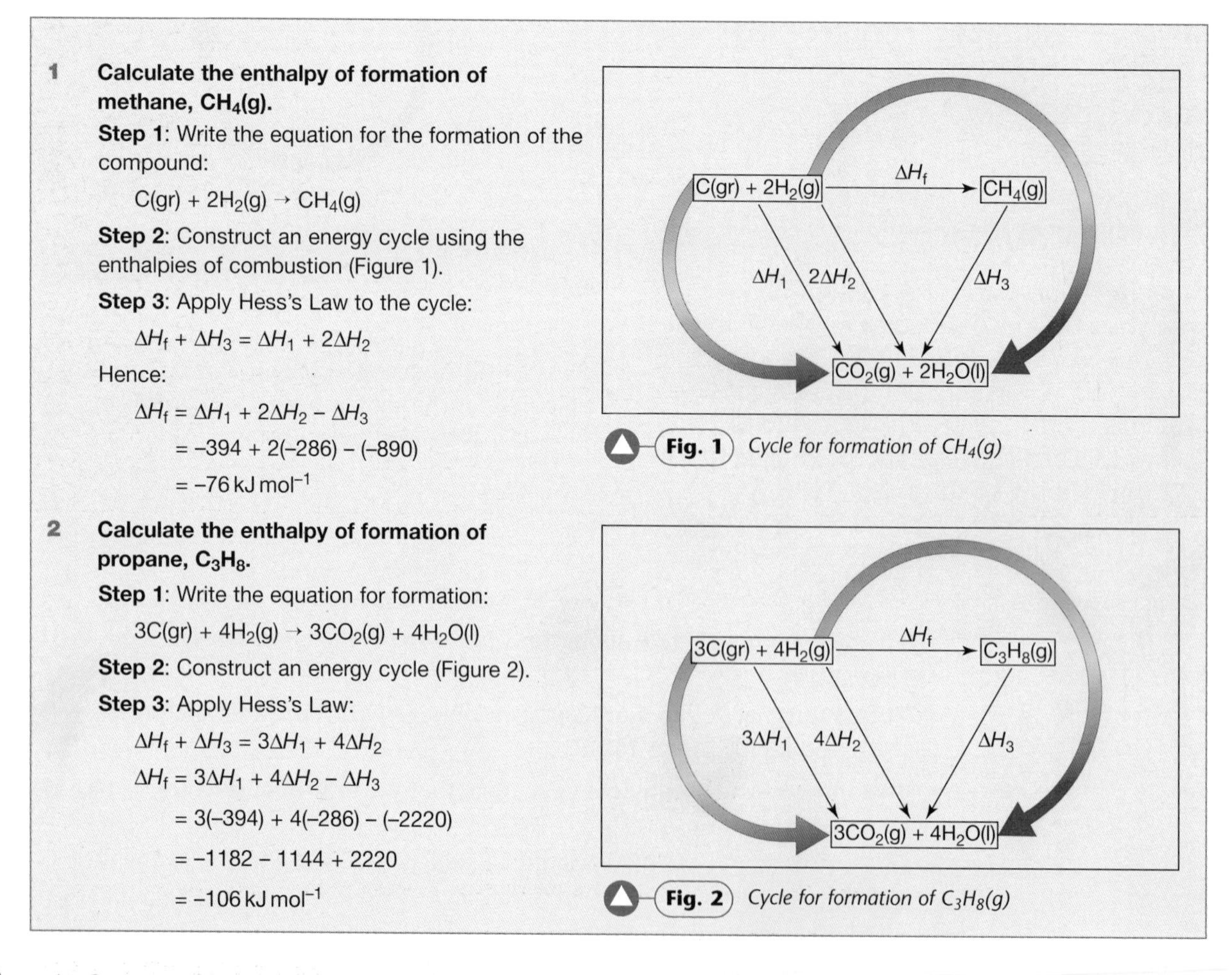

1 Calculate the enthalpy of formation of methane, CH_4(g).

Step 1: Write the equation for the formation of the compound:

$C(gr) + 2H_2(g) \rightarrow CH_4(g)$

Step 2: Construct an energy cycle using the enthalpies of combustion (Figure 1).

Step 3: Apply Hess's Law to the cycle:

$\Delta H_f + \Delta H_3 = \Delta H_1 + 2\Delta H_2$

Hence:

$\Delta H_f = \Delta H_1 + 2\Delta H_2 - \Delta H_3$

$= -394 + 2(-286) - (-890)$

$= -76\ \text{kJ mol}^{-1}$

Fig. 1 *Cycle for formation of CH_4(g)*

2 Calculate the enthalpy of formation of propane, C_3H_8.

Step 1: Write the equation for formation:

$3C(gr) + 4H_2(g) \rightarrow 3CO_2(g) + 4H_2O(l)$

Step 2: Construct an energy cycle (Figure 2).

Step 3: Apply Hess's Law:

$\Delta H_f + \Delta H_3 = 3\Delta H_1 + 4\Delta H_2$

$\Delta H_f = 3\Delta H_1 + 4\Delta H_2 - \Delta H_3$

$= 3(-394) + 4(-286) - (-2220)$

$= -1182 - 1144 + 2220$

$= -106\ \text{kJ mol}^{-1}$

Fig. 2 *Cycle for formation of C_3H_8(g)*

Q 1 Calculate the standard enthalpy of formation of ethane, $C_2H_6(g)$, using the standard enthalpies of combustion data provided in Table 1.

Using enthalpies of formation

The enthalpy of a reaction, ΔH_R, can be calculated from the enthalpies of formation of all the reactants and products. This involves drawing a cycle using enthalpies of formation, and then using Hess's Law to calculate the enthalpy of reaction.

Enthalpies of formation **Table 2**

Compound	Enthalpies of formation (kJ mol^{-1})
$Na_2CO_3(s)$	–1131
$NaHCO_3(s)$	–948
$CO_2(g)$	–394
$H_2O(l)$	–286
$Li_2CO_3(s)$	–1216
$Li_2O(s)$	– 596

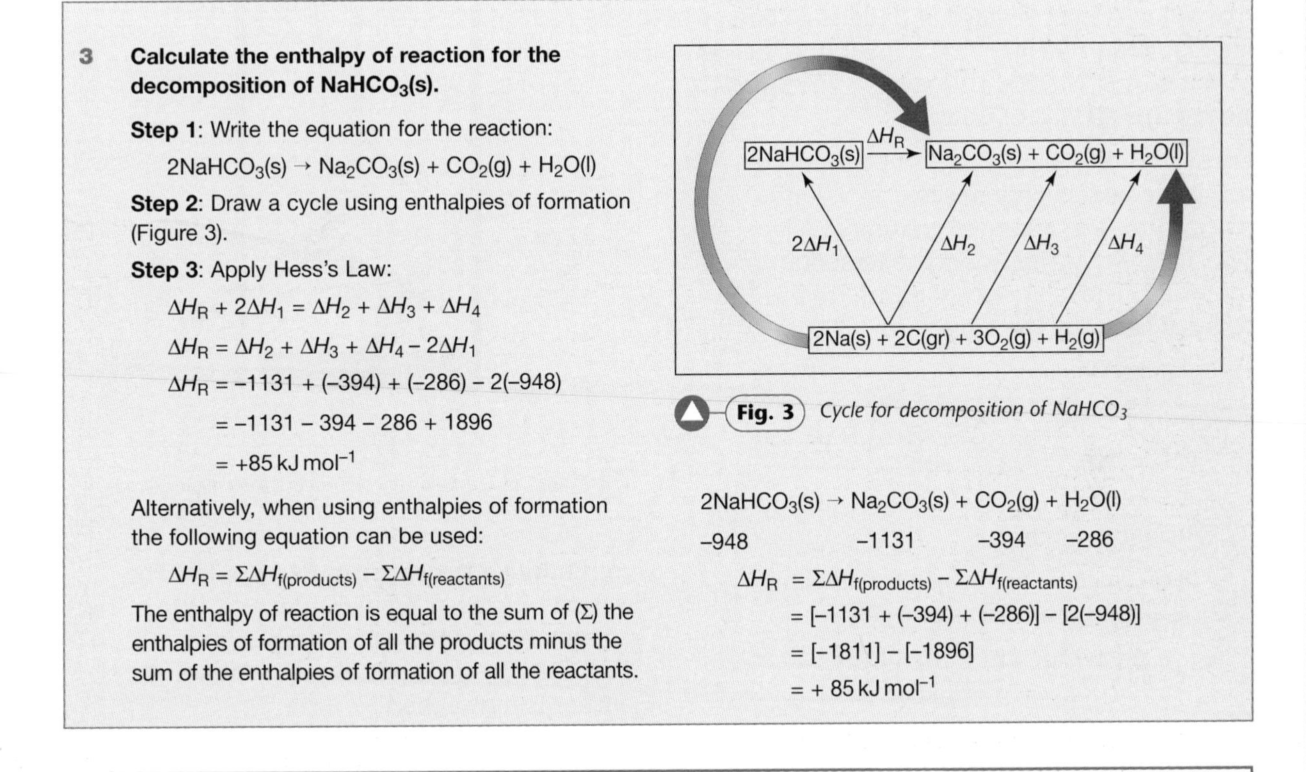

3 Calculate the enthalpy of reaction for the decomposition of $NaHCO_3(s)$.

Step 1: Write the equation for the reaction:

$2NaHCO_3(s) \rightarrow Na_2CO_3(s) + CO_2(g) + H_2O(l)$

Step 2: Draw a cycle using enthalpies of formation (Figure 3).

Step 3: Apply Hess's Law:

$\Delta H_R + 2\Delta H_1 = \Delta H_2 + \Delta H_3 + \Delta H_4$

$\Delta H_R = \Delta H_2 + \Delta H_3 + \Delta H_4 - 2\Delta H_1$

$\Delta H_R = -1131 + (-394) + (-286) - 2(-948)$

$= -1131 - 394 - 286 + 1896$

$= +85$ kJ mol^{-1}

Fig. 3 *Cycle for decomposition of $NaHCO_3$*

Alternatively, when using enthalpies of formation the following equation can be used:

$\Delta H_R = \Sigma\Delta H_{f(products)} - \Sigma\Delta H_{f(reactants)}$

The enthalpy of reaction is equal to the sum of (Σ) the enthalpies of formation of all the products minus the sum of the enthalpies of formation of all the reactants.

$2NaHCO_3(s) \rightarrow Na_2CO_3(s) + CO_2(g) + H_2O(l)$

–948   –1131   –394   –286

$\Delta H_R = \Sigma\Delta H_{f(products)} - \Sigma\Delta H_{f(reactants)}$

$= [-1131 + (-394) + (-286)] - [2(-948)]$

$= [-1811] - [-1896]$

$= +85$ kJ mol^{-1}

Q 2 Use the enthalpy of formation data in Table 2 to calculate the enthalpy change in the following reaction:

$Li_2CO_3(s) \rightarrow Li_2O(s) + CO_2(g)$

Key Idea 98 – 99

- **Using Hess's Law and energy cycles, the enthalpy change of a reaction can be calculated using enthalpies of formation and enthalpies of combustion.**

4 Measuring heat energy changes

At GCSE you may have performed some simple experiments such as burning a peanut or an alcohol to find the amount of heat released in a combustion reaction.

Calorimetry

Calorimetry (from calorie – heat unit) is used to find the heat energy released in a reaction. The principle of calorimetry relies on the fact that when water is heated, 4.2 joules of energy are required to increase the temperature of 1 g of water by 1 °C (1 K). This value is the **specific heat capacity**, ***c***, of water.

Enthalpies of combustion can be found by using the heat energy released in a combustion reaction to heat a known volume of water. This heat energy increases the temperature of the water. The heat energy transferred to the water is given by the equation:

$$q = mc\Delta T$$

where

q is the heat energy change (joules)

m is the mass of water (grams – the density of water is $1\,g\,cm^{-3}$)

c is the specific heat capacity of water (joules per gram per degree)

ΔT is the temperature rise of the water.

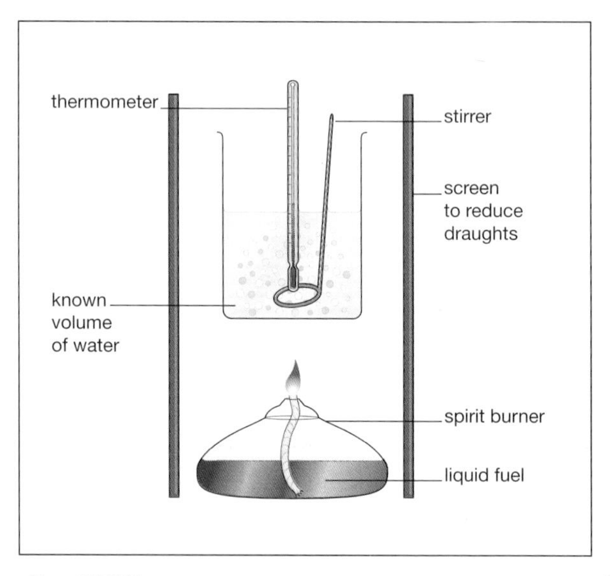

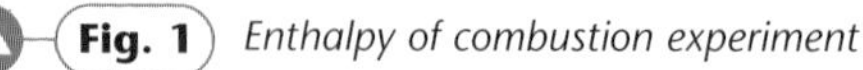
Fig. 1 *Enthalpy of combustion experiment*

Q

1 Calculate the amount of heat energy needed to increase the temperature of 150 g of water from 20 °C to 35 °C.

Enthalpies of combustion can be found experimentally by burning a known mass of a compound or element, and using the heat energy released to increase the temperature of a known mass of water.

One method involves weighing a spirit burner containing a liquid fuel. The spirit burner is lit and placed below a beaker (or calorimeter) containing a known volume of water. The water is stirred at regular intervals. When the temperature of the water has risen by at least 10 °C, the flame is put out and the spirit burner is re-weighed.

Using the mass of fuel burned and the temperature rise of the water, the amount of heat energy released per gram of fuel can be calculated. If the relative molecular mass of the fuel is known, then the enthalpy of combustion, in $kJ\,mol^{-1}$, can be calculated. The main source of error in these experiments is heat loss to the surroundings.

A student followed the method described above to determine the enthalpy of combustion of methanol, CH_3OH. The results of the student's experiment were as follows.

Initial mass of spirit lamp plus methanol	52.76 g	Initial temperature of water	21.0 °C
Final mass of spirit lamp plus methanol	52.45 g	Final temperature of water	36.5 °C
Mass of methanol burned	0.31 g	Temperature rise	15.5 °C
Volume of water in beaker	100 cm^3		

Heat energy released by the combustion of methanol:

100 cm^3 of water has a mass of 100 g since the density of water is 1 g cm^{-3}

$q = mc\Delta T = 100 \times 4.2 \times 15.5 = 6510$ J

This was for 0.31 g of methanol.

The relative molecular mass of methanol, CH_3OH, is 32 (12 + 3 + 16 + 1).

Therefore the heat energy released by the combustion of one mole of methanol is:

$$q = \frac{6510 \times 32}{0.31} = 672\,000 \text{ joules} = 672 \text{ kJ}$$

The enthalpy of combustion of methanol is −672 kJ mol^{-1}.

Bomb calorimeter

A more accurate method of finding enthalpies of combustion is by using a bomb calorimeter. This is based on the same principle as the simple experiment involving spirit burners, but it is more accurate because the heat loss to the surroundings is reduced in effect to zero.

In a bomb calorimeter a sample of a compound is electrically ignited and the heat energy released by combustion heats the water in the calorimeter. The calculation of the enthalpy of combustion follows the same method as in the simple experiment.

Q

2 The combustion of 0.15 g of ethanol, CH_3CH_2OH, in a spirit burner increased the temperature of 75 cm^3 of water by 12.5 °C. Calculate the enthalpy of combustion of ethanol, in kJ mol^{-1}.

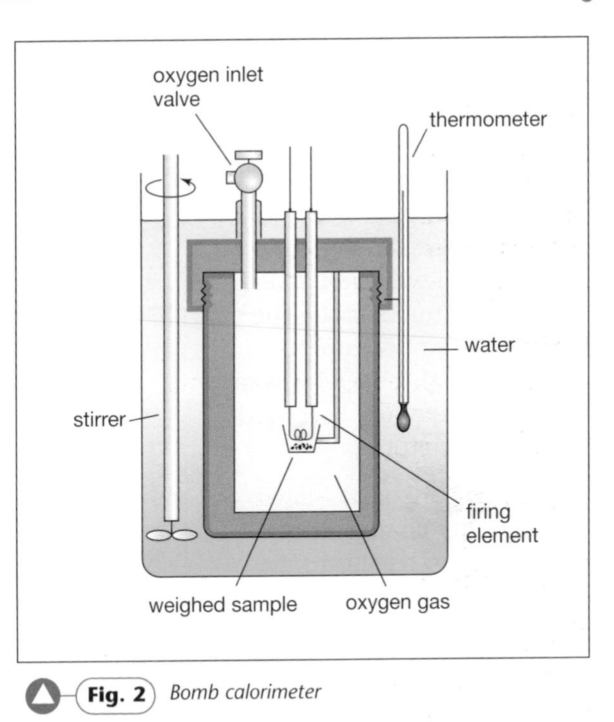

Fig. 2 *Bomb calorimeter*

Key Ideas 100 – 101

- **Enthalpies of combustion can be found experimentally by burning a compound and measuring the temperature rise in a known volume of water that is heated by the combustion of a known mass of the compound.**
- **It takes 4.2 joules of heat energy to increase the temperature of one gram of water by one degree. This value is known as the specific heat capacity of water.**

5 Reactions in solution ...

Calorimetry is used to find the enthalpy changes for a reaction taking place in solution. In an exothermic reaction, the heat energy released increases the temperature of the water in the solution. In an endothermic reaction, the heat energy absorbed comes from the water in the solution and the temperature of the solution falls.

Enthalpy of reactions in solution

A student was asked to calculate the enthalpy change of neutralisation.

$HCl(aq) + NaOH(aq) \rightarrow NaCl(aq) + H_2O(l)$

Ionically,

$H^+(aq) + OH^-(aq) \rightarrow H_2O(l)$

The student placed $50\,cm^3$ of 1 M HCl in a polystyrene cup and recorded the temperature of the acid. $50\,cm^3$ of 2 M NaOH (at the same temperature) was then added, and the mixture stirred. The maximum temperature rise was 6.75 °C (the NaOH was in excess to make sure that all of the HCl reacted).

The student then calculated the enthalpy change in the neutralisation reaction.

When the two solutions were mixed, the total volume of the solution was $100\,cm^3$.

The density of water is 1 g per cm^3, therefore 100 cm^3 of solution contains 100 g of water.

The specific heat capacity of water is $4.2\,J\,g^{-1}\,K^{-1}$.

The heat energy released in the reaction is calculated using the equation:

$q = mc\Delta T = 100 \times 4.2 \times 6.75 = 2835\,J$

This was for the reaction of $50\,cm^3$ of 1 M HCl and $50\,cm^3$ of 1 M NaOH. The number of moles in solution is (molarity × volume)/1000 (MV/1000).

Moles of acid used:

$$= \frac{1 \times 50}{1000} = 0.05 \text{ moles}$$

Hence the heat energy, q, released in the neutralisation of 1 mole of acid:

$$= \frac{2835}{0.05} = 56700\,J = 56.7\,kJ$$

The reaction is exothermic and is carried out at constant pressure, so the enthalpy of neutralisation, ΔH, is $-56.7\,kJ\,mol^{-1}$.

Q

1 In an experiment to find the enthalpy neutralisation of a weak acid, HX, $30\,cm^3$ of a 1 M solution of HX were mixed with $40\,cm^3$ of 1 M KOH (excess) in a polystyrene cup. The temperature rise in the reaction was 5.0 °C. Calculate the enthalpy of neutralisation for the weak acid.

A student carried out an experiment to find the enthalpy of reaction when $NaHCO_3$ reacts with dilute hydrochloric acid.

$NaHCO_3(s) + HCl(aq) \rightarrow NaCl(aq) + CO_2(g) + H_2O(l)$

The student added 3.71 g of $NaHCO_3$ to 30 cm^3 of 1 M hydrochloric acid (excess) in a polystyrene cup. The temperature of the acid fell by 8.5 °C.

30 cm^3 of solution contains 30 cm^3 of water, and hence 30 g of water. The heat energy absorbed in the reaction is found from:

$q = mc\Delta T = 30 \times 4.2 \times 8.5 = 1071$ J

This is for 3.71 g of $NaHCO_3$. The relative molecular mass of $NaHCO_3$ is 84. Therefore for 1 mole of $NaHCO_3$:

$q = \frac{1071}{3.71} \times 84 = 24249 \text{ J} = 24.3 \text{ kJ}$

The reaction is endothermic, so the enthalpy change for the reaction is +24.3 kJ mol^{-1}.

Q

2 A student added 3.41 g of Na_2CO_3 to 40 cm^3 of dilute hydrochloric acid (in excess). The temperature of the acid rose by 7.3 °C. Calculate the enthalpy change in kJ mol^{-1} for the reaction.

$Na_2CO_3(s) + 2HCl(aq) \rightarrow 2NaCl(aq) + CO_2(g) + H_2O(l)$

Cooling curves

In these experiments, the major source of error is heat loss. The polystyrene cup in a beaker provides some insulation. A lid on the top of the polystyrene cup also reduces heat loss.

One experimental technique that compensates for heat loss involves using cooling curves. In the first experiment, the student would record the temperature of the acid every minute for three minutes. At the fourth minute the sodium hydroxide solution would be added and the mixture thoroughly stirred. The temperature of the mixture would then be recorded at the fifth minute, and then every minute for about ten minutes.

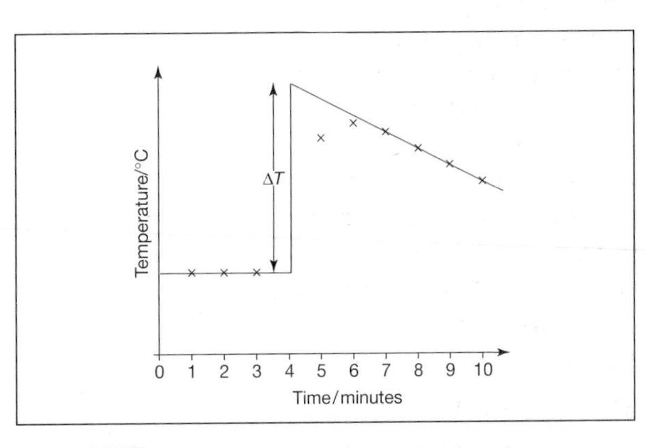

Fig. 1 *Cooling curve for an exothermic reaction*

A graph of temperature against time is then drawn. The points before the addition are joined together and then extrapolated (extended) to the fourth minute. The points after the addition are joined together and extrapolated back to the fourth minute. The theoretical temperature rise at the fourth minute is then found from the graph.

Key Ideas 102 – 103

- **The heat energy change in a reaction in solution can be found experimentally using calorimetry.**
- **In an exothermic reaction the heat energy increases the temperature of the water in the solution. In an endothermic reaction the heat energy is absorbed from the water in the solution and the temperature of the water decreases.**

Unit 5 Questions

1. State the *first law of thermodynamics*. (1)

2. **a)** Define the term *enthalpy change*. (2)
   b) State *two* essential conditions used in standard enthalpy changes. (2)

3. **a)** Define the term *mean bond enthalpy*. (2)
   b) Use the mean bond enthalpy data to calculate the enthalpy change in the following reaction.

   $H_2(g) + Br_2(g) \rightarrow 2HBr(g)$ (3)

   H—H 436 kJ mol^{-1} Br—Br 193 kJ mol^{-1} H—Br 366 kJ mol^{-1}

   c) In the reaction between hydrogen and bromine, state which bond breaks first and give an explanation for your answer. (2)
   d) Explain why *two* of the mean bond enthalpy values quoted in part **b)** should be referred to as *bond dissociation energies*. (1)

4. Use the mean bond enthalpy data to calculate the enthalpy of combustion of ethanol, CH_3CH_2OH. (5)

```
    H   H
    |   |
H — C — C — O — H + 3O=O → 2O=C=O + 3H—O—H
    |   |
    H   H
```

   Mean bond enthalpies (kJ mol^{-1}):
   C—H 412 C—C 348 O=O 496 C=O 743 C—O 360 O—H 463

5. State *Hess's Law* and explain why this law is useful in determining enthalpy values. (3)

6. **a)** Define the term *standard enthalpy of combustion*. (2)
   b) Write an equation for which the enthalpy change is the standard enthalpy of formation of butane, C_4H_{10}. (2)
   c) Use the following enthalpy of combustion data to calculate the standard enthalpy of formation of butane. (3)
   Standard enthalpies of combustion:
   $C_4H_{10}(g)$ –2877 kJ mol^{-1} C(s) –394 kJ mol^{-1} $H_2(g)$ –286 kJ mol^{-1}

7. **a)** Define the term *standard enthalpy of formation*. (2)
   b) Use the enthalpy of formation data to calculate the enthalpy change in the following reaction:

   $2KHCO_3(s) \rightarrow K_2CO_3(s) + CO_2(g) + H_2O(l)$ (3)

   Standard enthalpies of formation:
   $KHCO_3(s)$ –959 kJ mol^{-1} $K_2CO_3(s)$ –1146 kJ mol^{-1} $CO_2(g)$ –394 kJ mol^{-1}
   $H_2O(l)$ –286 kJ mol^{-1}

8) In an experiment to find the enthalpy change in a reaction, 25 cm^3 of 0.5 M copper sulphate solution was placed in a polystyrene cup and the temperature recorded. Excess zinc powder was added and the mixture stirred. The increase in temperature was 21°C. The equation for the reaction is:

$$CuSO_4(aq) + Zn(s) \rightarrow ZnSO_4(aq) + Cu(s)$$

a) Why was excess zinc used in the reaction? (1)
b) Calculate the heat energy (in joules) released in the reaction. (2)
c) Calculate the number of moles of copper sulphate used in the reaction. (2)
d) Hence calculate the molar enthalpy change for the reaction. (2)
e) Give the main source of error in this experiment and suggest *two* methods of reducing this error. (2)

9) When hydrogen gas undergoes complete combustion, the following enthalpy values are obtained:

$$2H_2(g) + O_2(g) \rightarrow 2H_2O(g) \quad \Delta H = -242\,kJ\,mol^{-1}$$

$$2H_2(g) + O_2(g) \rightarrow 2H_2O(l) \quad \Delta H = -286\,kJ\,mol^{-1}$$

a) Explain why these two reactions have different ΔH values. (2)
b) Use the enthalpy values to calculate the enthalpy change for the following process:

$$H_2O(l) \rightarrow H_2O(g)$$ (3)

10) In an experiment to determine the enthalpy of neutralisation of an acid, HX, a student placed 50 cm^3 of 1 M NaOH in a polystyrene cup and recorded the temperature every minute for three minutes. At the fourth minute the student added 50 cm^3 of a 1M solution of HX. The student stirred the mixture and recorded the temperature at the fifth minute and every minute for the next five minutes. The results are shown in the table.

$$HX(aq) + NaOH(aq) \rightarrow NaX + H_2O(l)$$

Time	1	2	3	4	5	6	7	8	9	10
Temperature/°C	19.0	19.0	19.0	–	25.1	24.5	23.9	23.5	23.1	22.7

a) Explain why the mixture was stirred at regular intervals. (1)
b) Plot a cooling curve from the data in the table. (3)
c) From your cooling curve, calculate the theoretical temperature rise at the fourth minute. (1)
d) Calculate the heat energy released in the reaction. (3)
e) Calculate the moles of acid, HX, neutralised in the reaction. (2)
f) Hence calculate the enthalpy of neutralisation per mole of acid. (2)

1 How fast will it go?

At GCSE you will have studied the factors that alter the rate of a chemical reaction. These factors include temperature, concentration and the use of catalysts. At AS level the study of rates of reaction is called **kinetics**. In this unit the ideas studied at GCSE are developed further.

Collision theory

For a reaction to occur, particles have to collide with each other. Only a very small percentage of collisions result in a reaction, however. This is because there is an energy barrier to a reaction (Figure 1). Only those particles with enough energy to overcome the energy barrier will react when they collide. The minimum energy that a particle must have to overcome the energy barrier and react is called the **activation energy, E_a**. The size of this activation energy is different for different reactions.

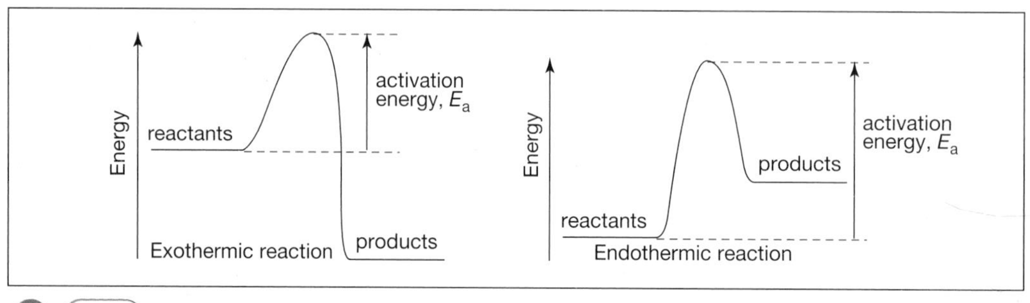

Fig. 1 *Reaction profiles for an exothermic and an endothermic reaction*

1 Explain why few collisions between reactant particles result in a reaction.

Increasing the frequency of collisions

If the *frequency* of collisions is increased, the *rate of reaction* will increase. However, the *percentage* of successful collision will remain the same.

An increase in the frequency of collisions can be achieved by increasing the concentration of the reactants, by increasing the pressure of a gaseous reactant, or by increasing the surface area of a solid reactant.

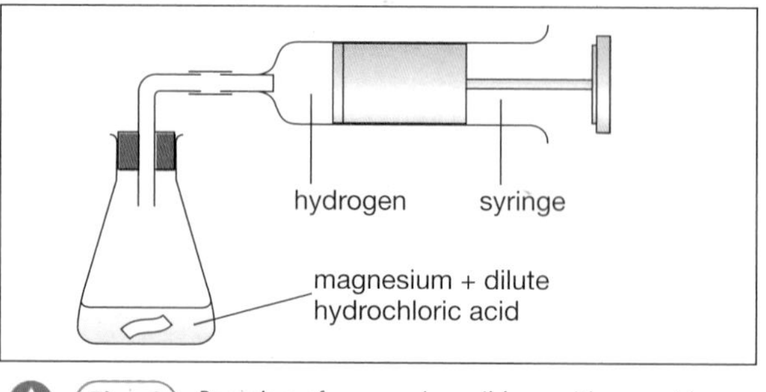

Fig. 2 *Reaction of magnesium ribbon with an acid*

Increasing the concentration of a solution

If the concentration of a solution is increased, there are more reactant particles per unit volume. This increases the probability of reactant particles colliding with each other.

$$Mg(s) + 2HCl(aq) \rightarrow MgCl_2(aq) + H_2(g)$$

If the reaction between magnesium ribbon and excess dilute hydrochloric is carried out twice using the same mass of magnesium ribbon but different concentrations of the acid, then two different reaction rates will result (Figure 3).

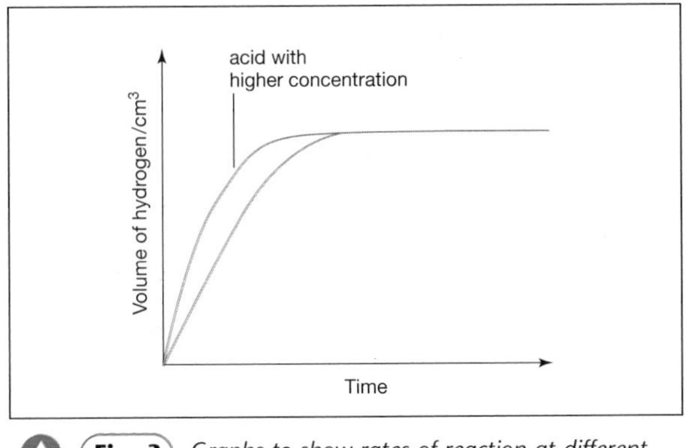

Fig. 3 *Graphs to show rates of reaction at different concentrations*

The steeper curve is for the reaction with the greater acid concentration. The two reactions produce the same volume of hydrogen gas, because the same mass of magnesium was used.

Increasing the pressure of a gaseous reactant

If the pressure of a gas is increased, the particles in the gas are pushed closer together. This increases the concentration of the gas, and hence increases the rate of collisions.

Increasing the surface area of a solid reactant

If one of the reactants is a solid, then the surface area of the solid is an important factor in determining the rate. Only the particles on the *surface* of the solid will be able to undergo collisions with the particles in a solution or a gas. If the solid is powdered then there is a greater surface area available for a reaction, compared with an equal mass of 'unpowdered' solid.

In an experiment, 1 g of marble chips is added to excess dilute hydrochloric acid in a flask.

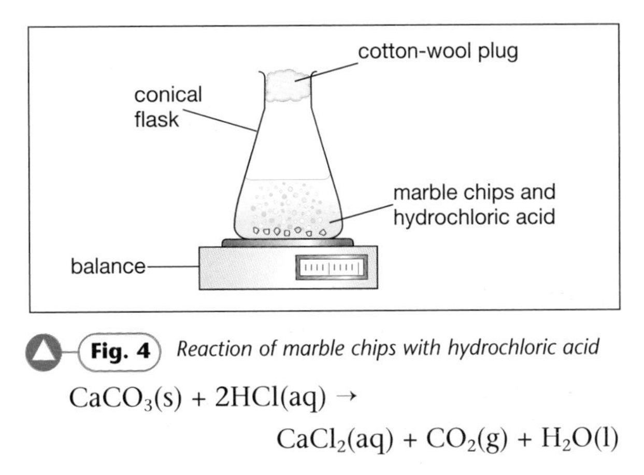

Fig. 4 *Reaction of marble chips with hydrochloric acid*

$$CaCO_3(s) + 2HCl(aq) \rightarrow CaCl_2(aq) + CO_2(g) + H_2O(l)$$

The carbon dioxide formed in the reaction escapes, and the mass of the conical flask and its contents decreases.

In a second experiment, 1 g of powdered marble chips is added to excess hydrochloric acid in a conical flask.

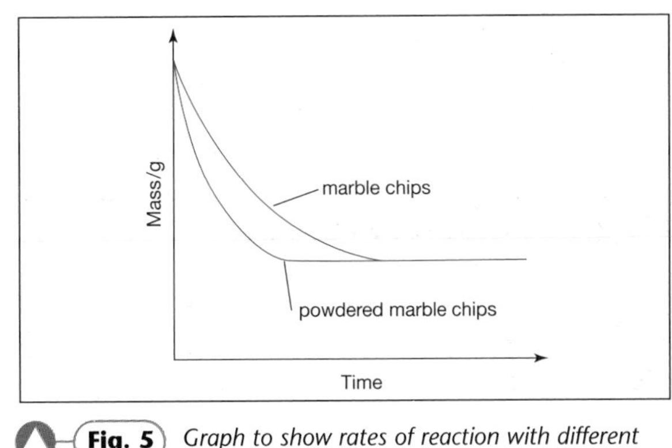

Fig. 5 *Graph to show rates of reaction with different sizes of solid particles*

The steeper curve is for the reaction of powdered marble chips.

Q

2 Explain, in terms of the particles present, why an increase in concentration causes an increase in the rate of a reaction.

Key Ideas 106 – 107

- **A reaction can only occur when collisions take place between particles with sufficient energy.**
- **The minimum energy that a particle must have for a collision to result in a reaction is called the activation energy.**

2 Energy distribution curves

In a gas, the particles are moving with rapid random motion. When the particles collide, energy is transferred between the particles. These random collisions result in the particles having *different energies*. An understanding of the distribution of energy among the particles in a reaction leads to a better understanding of the effect of temperature and catalysts on the rate of reactions.

Maxwell-Boltzmann energy distribution curves

The particles in a gas undergo random collisions in which energy is transferred between the colliding particles. As a result there will be particles with differing energies. **Maxwell-Boltzmann energy distribution curves** show the *distribution* of the energies of the particles in a gas.

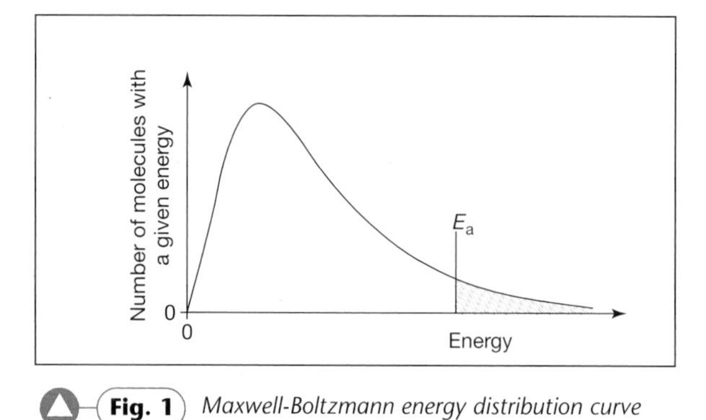

Fig. 1 *Maxwell-Boltzmann energy distribution curve*

The main points to note about the curves are:

1 There are no particles with zero energy.
2 The curve does not touch the *x*-axis at the higher end: this is because there will always be some particles with very high energies.
3 The *area* under the curve is equal to the total number of particles in the system.
4 The *peak* of the curve indicates the most probable energy.

The **activation energy, E_a**, for a given reaction can be marked on the distribution curve. Only those particles with energy equal to or greater than the activation energy can react when a collision occurs.

Although Maxwell-Boltzmann distribution curves are for the particles in a gas, the same distributions can be used for the particles in a liquid or for the particles in a solution.

Effects of a temperature change

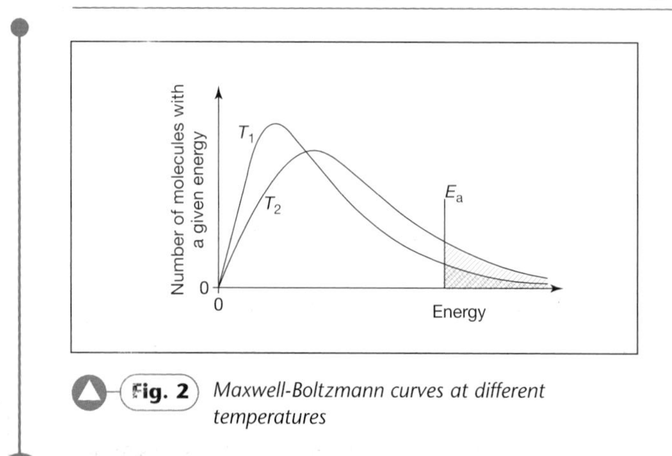

Fig. 2 *Maxwell-Boltzmann curves at different temperatures*

Figure 2 shows Maxwell-Boltzmann distribution curves for a fixed mass of gas at two temperatures T_1 and T_2, where T_2 is about 10 °C higher than T_1. The total *area* under the curve remains the same, since there is no change in the number of particles present.

A small increase in temperature causes significant changes to the distribution of energies. At the higher temperature:

1 the peak is at a higher energy
2 the peak is lower

3 the peak is broader
4 there is a large increase in the number of particles with higher energies.

It is this final change that results in a large increase in rate, even with a relatively small increase in temperature. A small increase in temperature greatly increases in the number of particles with energy greater than the activation energy. This is shown by the shaded areas on the energy distribution curves

Q 1 What does the area under a Maxwell-Boltzmann curve represent?

Effect of a catalyst

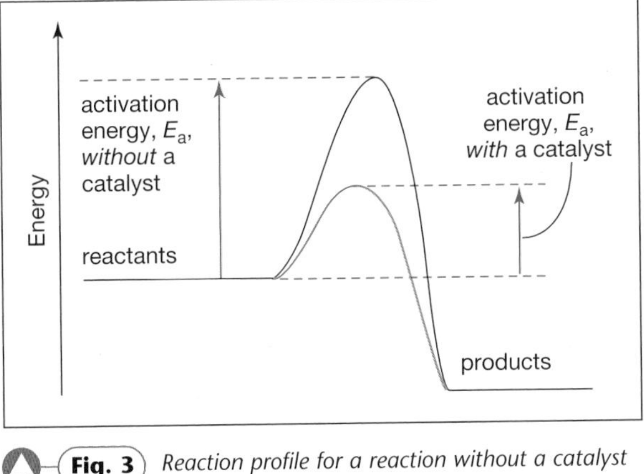

Fig. 3 *Reaction profile for a reaction without a catalyst and with a catalyst*

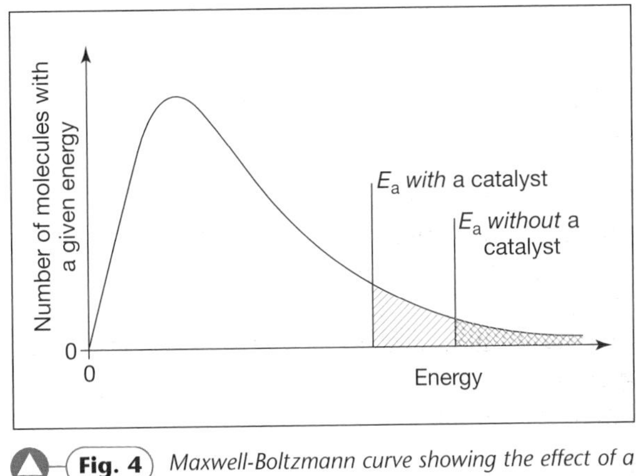

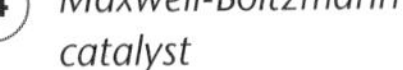

Fig. 4 *Maxwell-Boltzmann curve showing the effect of a catalyst*

A **catalyst** is a substance that alters the rate of a chemical reaction but is itself unchanged at the end of the reaction. A catalyst works by providing an **alternative reaction pathway** that has a lower activation energy (Figure 3).

A catalyst does not *alter* the Maxwell-Boltzmann distribution. Because a catalyst provides a reaction route of lower activation energy, however, a greater proportion of particles will have energy greater than the activation energy. This is shown by the shaded areas on the energy distribution curves (see Figure 4).

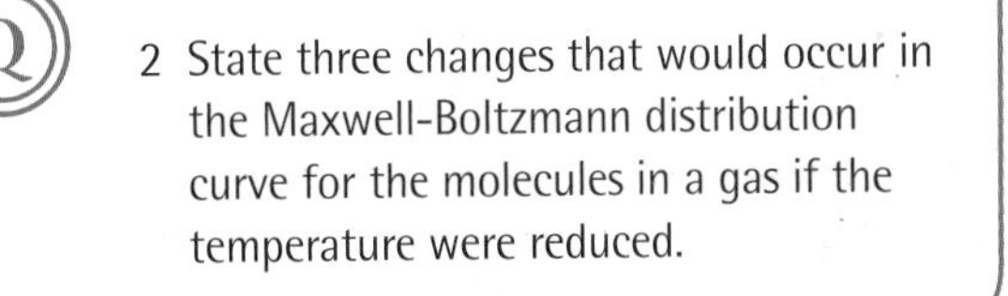

Q 2 State three changes that would occur in the Maxwell-Boltzmann distribution curve for the molecules in a gas if the temperature were reduced.

Key Ideas 108 – 109

- **Maxwell-Boltzmann curves show the distribution of the energies of the molecules in a gas.**
- **A small increase in temperature produces significant changes in the shape of a distribution curve.**
- **A small increase in temperature produces a large increase in the rate of reaction, since there is a large increase in the number of particles with higher energies.**
- **A catalyst provides an alternative reaction pathway with a lower activation energy. As a result, more particles have energy greater than the activation energy.**

Unit 6 Questions

1. a) Explain why very few collisions between particles result in a reaction. (2)
   b) Give *three* methods of increasing the rate of collisions between particles. (3)
   c) Explain what is meant by the term *activation energy*. (2)

2. a) Explain what is meant by the term *catalyst*. (2)
   b) Explain how a catalyst can affect the rate of a chemical reaction. (2)
   c) Draw an energy profile for a endothermic reaction including a catalysed reaction. (2)
   d) Draw a sketch of a Maxwell-Boltzmann distribution curve for the particles in a gas, and use the sketch to show how a catalyst affects the rate of a chemical reaction. (4)

3. Figure 1 shows the Maxwell-Boltzmann energy distribution curves for the particles in a gas at two different temperatures T_1 and T_2.

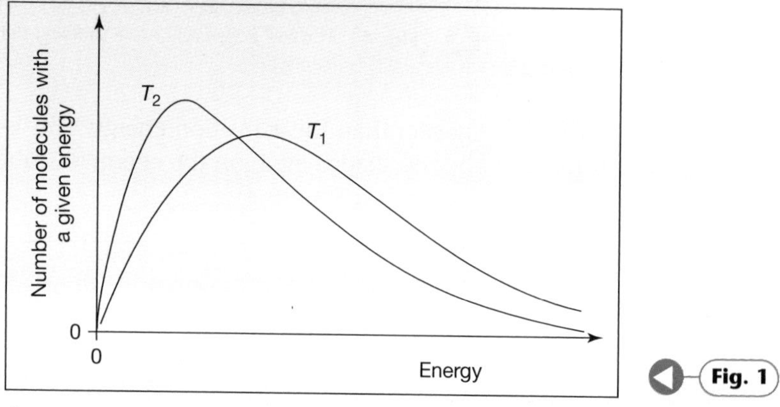

Fig. 1

   a) State which of the two temperatures T_1 or T_2 is the higher temperature. (1)
   b) What does the area under the curve represent? (2)
   c) Explain why the curves do not touch the *x*-axis at high temperatures. (2)
   d) Use the curves to explain why the rate of reaction increases with an increase in temperature. (2)

4. A piece of magnesium ribbon with a mass of 0.1 g was reacted with excess 0.1 M hydrochloric acid. The gas formed during the reaction was collected in a syringe. The graph in Figure 2 shows the volume of gas formed during the reaction.

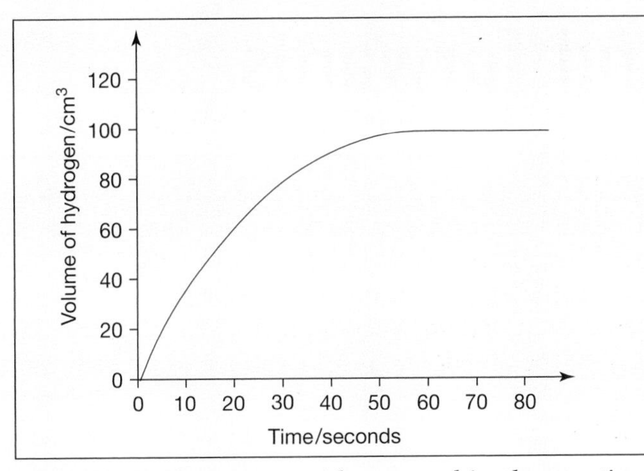

Fig. 2

a) Explain why excess acid was used in the reaction. (1)

b) From the graph, find the time taken for half of the magnesium ribbon to react. (1)

c) Copy out the graph. Using the same axes, sketch curves for the following experiments:

(i) excess 0.1 M hydrochloric acid with 0.1 g of powdered magnesium

(ii) excess 0.2 M hydrochloric acid with 0.1 g of magnesium ribbon

(iii) excess 0.1 M hydrochloric acid with 0.05 g of magnesium ribbon. (6)

5 The curves shown in Figure 3 show the volume of carbon dioxide formed by reacting excess marble chips (calcium carbonate) with dilute hydrochloric acid.

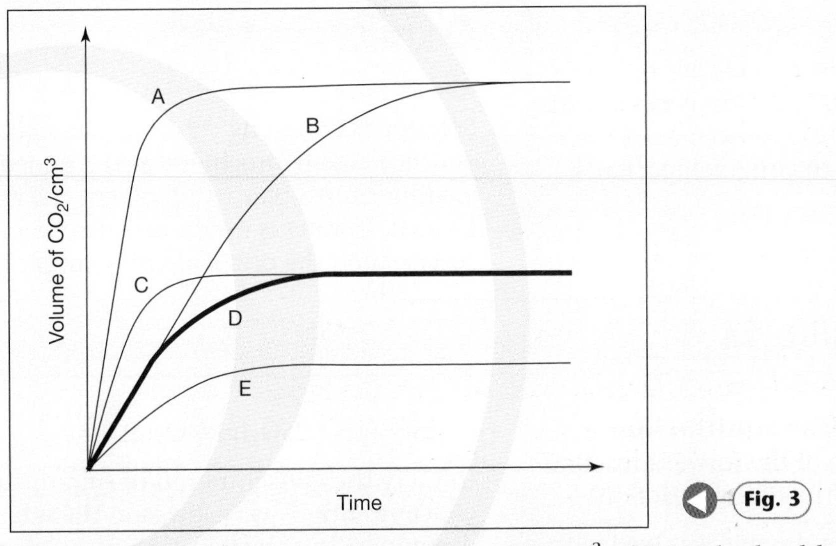

Fig. 3

The curve labelled **D** is for the reaction of 50 cm^3 of 0.1 M hydrochloric acid and excess marble chips.

In the following questions, state which curve, **A–E**, would be obtained by:

a) reacting 100 cm^3 of 0.1 M hydrochloric acid with excess marble chips (1)

b) reacting 50 cm^3 of 0.2 M hydrochloric acid with excess marble chips (1)

c) reacting 50 cm^3 of 0.1 M hydrochloric acid with excess powdered marble chips (1)

d) reacting 50 cm^3 of 0.1 M hydrochloric acid with excess marble chips in a larger vessel. (1)

1 Backwards and forwards

At GCSE you may have studied the idea of reversible reactions. A *reversible reaction* is one in which the products can re-form the original reactants. At AS level, the concept of *dynamic chemical equilibrium* is studied. This includes a study of the factors that affect the position of a chemical equilibrium.

Reversible reactions

Many reactions continue until all of the reactants are used up, and the reaction then stops. The reaction is said to go to **completion**.

A typical type of reaction that goes to completion is the reaction of a metal with an acid. For example, when excess hydrochloric acid is added to magnesium ribbon, the reaction continues until *all* of the magnesium ribbon has reacted. Once all the magnesium ribbon has reacted, the reaction stops.

$$Mg(s) + 2HCl(aq) \rightarrow MgCl_2(aq) + H_2(g)$$

Many reactions, however do *not* go to completion because the reaction is **reversible** – the products of the reaction can react with each other to re-form the original reactants.

$$\underset{\text{pink}}{[Co(H_2O)_6]^{2+}} \rightleftharpoons \underset{\text{blue}}{[CoCl_4]^{2-}}$$

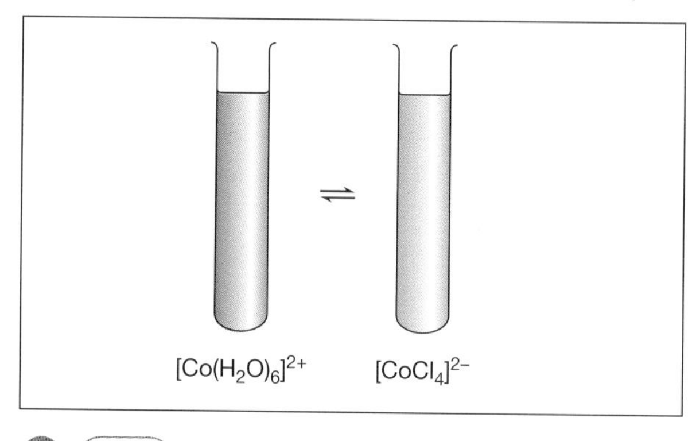

Fig. 1 *Reversible cobalt reaction*

Solutions containing cobalt ions are pink, but if concentrated hydrochloric acid is added the solutions turns blue as a different cobalt ion is formed. If water is now added, the reaction reverses and the original pink solution re-forms.

Chemical equilibria

In a reversible reaction an **equilibrium** is reached when the rate of the forward reaction is equal to the rate of the reverse reaction.

For example, if sulphur dioxide is mixed with oxygen then they react together to form sulphur trioxide:

$$2SO_2(g) + O_2(g) \rightarrow 2SO_3(g)$$

As the reaction continues, the concentration of the sulphur dioxide and oxygen decreases, and the rate of the reaction therefore becomes slower.

The sulphur trioxide formed decomposes to reform sulphur dioxide and oxygen:

$$2SO_3(g) \rightarrow 2SO_2(g) + O_2(g)$$

Initially the concentration of sulphur trioxide concentration is very low, and the rate of the decomposition reaction is slow. As the concentration of the sulphur trioxide increases, however, the rate of the decomposition reaction increases also.

Eventually the rate of the forward reaction and the rate of the backward reaction will be the same (Figure 2), and the concentrations of the reactants will stay the same, even though the reactions continue to occur. (Note that this does *not* mean that the concentrations of the three reactants are all the same!)

At this point a **dynamic equilibrium** has been reached:

$$2SO_2(g) + O_2(g) \rightleftharpoons 2SO_3(g)$$

In a dynamic equilibrium, the reactions continue to occur but there is no change in the overall concentrations of the reactants.

Q 1 Explain the terms *dynamic* and *equilibrium* when applied to a chemical equilibrium.

Chemical equilibria can be either *homogeneous* or *heterogeneous*. In a **homogeneous equilibrium**, all of the reactants are in the same state (phase). Here are two examples:

$$CH_3CH_2OH(aq) + CH_3COOH(aq) \rightleftharpoons CH_3COOCH_2CH_3(aq) + H_2O(l)$$

$$N_2(g) + 3H_2(g) \rightleftharpoons 2NH_3(g)$$

In the first example, all of the reactants are aqueous; in the second, they are all gaseous.

In a **heterogeneous equilibrium**, the reactants are in different states (phases).

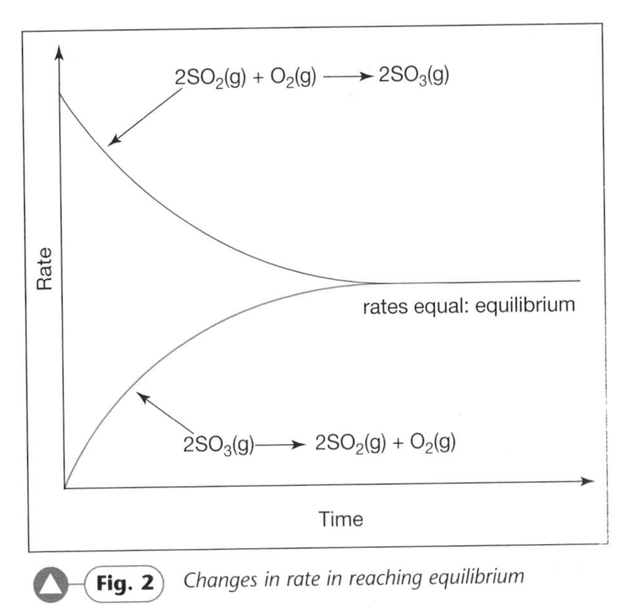

Fig. 2 *Changes in rate in reaching equilibrium*

Here are two examples in which some of the reactants are solids and some are gases.

$$3Fe(s) + 4H_2O(g) \rightleftharpoons Fe_3O_4(s) + 4H_2(g)$$

$$CaCO_3(s) \rightleftharpoons CaO(s) + CO_2(g)$$

Characteristics of a dynamic equilibrium

1. A dynamic equilibrium occurs in a *closed system*. This means that no more reactants are added and no reactants are removed.
2. Reactions continue to occur: the reaction is *dynamic*.
3. The concentrations of reactants and products remain constant: the reaction is in *equilibrium* (even though it may *look* as though it has stopped).
4. The equilibrium can be established from either side of the reaction. For example, if sulphur dioxide and oxygen are mixed then an equilibrium containing the two reactants and sulphur trioxide will result. If sulphur trioxide is allowed to decompose then an equilibrium will result containing all three reactants.

Q 2 Classify these reactions as homogeneous or heterogeneous equilibria:

a) $H_2(g) + I_2(g) \rightleftharpoons 2HI(g)$
b) $Na_2CO_3(s) \rightleftharpoons Na_2O(s) + CO_2(g)$
c) $N_2O_4(g) \rightleftharpoons 2NO_2(g)$
d) $H_2O(l) \rightleftharpoons H_2O(g)$

Key Ideas 112 – 113

- **Many chemical reactions are reversible.**
- **A chemical equilibrium is reached when the forward and reverse rates in a chemical reaction are equal.**
- **In a dynamic chemical equilibrium, reactions continue to occur but the overall concentrations of the reactants remain constant.**

2 The balancing act ...

The *rate* at which an equilibrium is reached depends upon the concentration of the reactants, the pressure in reactions involving gases, the temperature, and whether or not a catalyst is used. These factors can also alter the *position* of a chemical equilibrium.

Le Chatelier's Principle

Le Chatelier's Principle states that if one or more of the factors that affect the position of a chemical equilibrium are altered, then the equilibrium shifts to reduce (oppose) the effect of the change.

Three factors that can affect the position of a chemical equilibrium are temperature, concentration, and pressure.

Effect of a change in temperature

When the temperature is increased, the *endothermic* reaction is favoured, as this removes the added heat and so opposes the increase in temperature. When the temperature is decreased, the *exothermic* reaction is favoured, as this releases heat energy and so opposes the decrease in temperature.

$$2SO_2(g) + O_2(g) \rightleftharpoons 2SO_3(g) \qquad \Delta H = -189 \text{ kJ mol}^{-1}$$

In this reaction, increasing the temperature pushes the equilibrium to the left.

$$2HI(g) \rightleftharpoons H_2(g) + I_2(g) \qquad \Delta H = +10.4 \text{ kJ mol}^{-1}$$

In this reaction, increasing the temperature pushes the equilibrium to the right.

The ΔH values given are for the *forward* reaction.

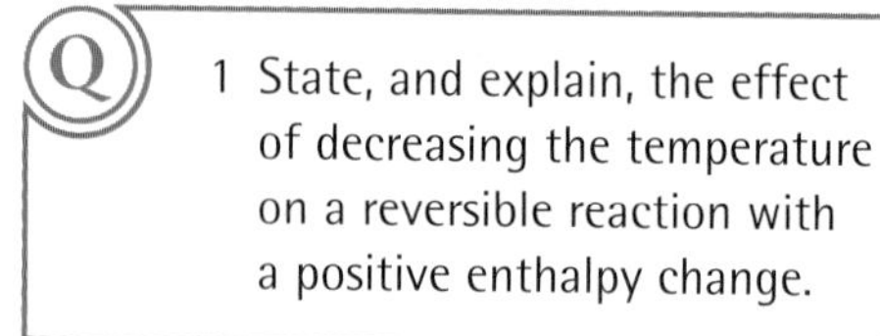
Q

1 State, and explain, the effect of decreasing the temperature on a reversible reaction with a positive enthalpy change.

Effect of a change in concentration

If the concentration of one of the reactants is increased, then the equilibrium will shift to oppose the change. Consider the following reversible reaction:

$$\underset{\text{ethanol}}{CH_3CH_2OH(aq)} + \underset{\text{ethanoic acid}}{CH_3COOH(aq)} \rightleftharpoons \underset{\text{ethyl ethanoate}}{CH_3COOCH_2CH_3(aq)} + H_2O(l)$$

At equilibrium, if more ethanoic acid is added then the equilibrium will shift to the right as this results in the removal of the added ethanoic acid. Similarly, at equilibrium, if some of the ethanoic acid is removed then the equilibrium will shift to the left to replace the removed ethanoic acid.

Effect of a change in pressure

Pressure only affects the position of chemical equilibria that involve gases. If the pressure is increased at equilibrium, then the equilibrium moves to the side of the reaction that has fewer gaseous moles, as this reduces the pressure of the gaseous system. Conversely, if the pressure is decreased then the reaction will move to the side that has more gaseous moles as this increases the pressure of the system.

For example:

$$2SO_2(g) + O_2(g) \rightleftharpoons 2SO_3(g)$$

3 moles → 2 moles

In the forward reaction 3 moles of reactants form 2 moles of product. If the pressure is increased, then the forward reaction is favoured as this reduces the number of moles of gas. If the pressure is decreased then the reverse reaction is favoured, as this increases the number of moles of gas.

$$PCl_5(g) \rightleftharpoons PCl_3(g) + Cl_2(g)$$

1 mole → 2 moles

In this example, the change in pressure has the same effect, but the equilibrium shifts in the opposite direction.

When the pressure is increased, the equilibrium will shift to the left as this involves a reduction in the moles of gas. When the pressure is decreased, then the equilibrium shifts to the right as this involves an increase in the number of moles of gas.

In a reaction that does not involve a change in the moles of gas present, then the equilibrium position does not change when the pressure changes:

$$2HI(g) \rightleftharpoons H_2(g) + I_2(g)$$

2 moles → 2 moles

In this equilibrium there are two moles of gas on either side of the equation and altering pressure has no effect on the position of the equilibrium.

Q

2 State the effect of decreasing the pressure on the position of the following equilibrium reactions.

a) $H_2(g) + Br_2(g) \rightleftharpoons 2HBr(g)$

b) $2NH_3(g) \rightleftharpoons N_2(g) + 3H_2(g)$

c) $CH_4(g) + H_2O(g) \rightleftharpoons CO(g) + 3H_2(g)$

Effect of a catalyst

A catalyst increases the rate of a chemical reaction. In the case of a dynamic equilibrium, a catalyst increases the rate of both the forward and reverse reactions equally. Since both reaction rates are increased equally, a catalyst has no effect on the *position* of the equilibrium. However, it does increase the rate at which equilibrium is reached.

Q

3 State, and explain, the effect on the following chemical equilibrium of increasing separately the temperature and pressure:

$N_2O_4(g) \rightleftharpoons 2NO_2(g) \quad \Delta H = +56\,kJ\,mol^{-1}$

Key Ideas 114 – 115

- **Factors that affect the position of a chemical equilibrium include temperature, pressure and concentration of the reactants and products.**
- **When the conditions in a chemical equilibrium are changed, then the equilibrium will shift to oppose the changes.**

3 It's a compromise ...

Industry has to produce its products as cheaply as possible if they are to be competitive and the firm is to remain in business. Industrial chemists have to use their knowledge and understanding of energetics, kinetics and equilibria to find the best conditions for a reaction.

Conflicting factors

Many chemical industries produce chemicals in equilibrium reactions and it is important that they produce the best possible **yield** bearing in mind the costs and time involved. (The *yield* is the percentage conversion of reactants into products.)

It would make little sense to produce a high yield in an equilibrium reaction if the rate of reaction was very slow. Nor would it be sensible to have a very fast reaction rate if the yield was very poor.

In equilibrium reactions, industries have to achieve a balance between the *yield* in the reaction, the *rate* of the reaction, and the *costs* involved in the reaction.

Haber process

The **Haber process** is a very important process. The available supply of naturally occurring nitrogen-containing compounds is far less than the demand. The Haber process provides a method of 'fixing' nitrogen in the atmosphere to form such compounds. Most of the ammonia formed in the Haber process is used to make fertilisers such as ammonium nitrate and ammonium sulphate. The rest of the ammonia is used to manufacture nitric acid, nylon, synthetic dyes, explosives, and a range of plastics.

The Haber process provides a good case study to illustrate how industrial chemists use their knowledge of the factors that affect chemical equilibria to find the best conditions needed to produce a good yield of products at a reasonable rate.

In the Haber process nitrogen gas (from air) and hydrogen gas (obtained from natural gas) are passed over a heated catalyst of iron.

$$N_2(g) + 3H_2(g) \rightleftharpoons 2NH_3(g) \quad \Delta H = -92.4\ \text{kJ mol}^{-1}$$

Any unreacted gases are recycled. The reaction is a *homogeneous gas reaction*.

Choosing the conditions

The forward reaction is *exothermic*. Applying Le Chatelier's Principle, the forward reaction releases heat energy and is favoured by lower temperatures. At low temperatures the equilibrium will shift to the right and increase the yield at equilibrium.

The reaction involves a *reduction in the number of moles*. Applying Le Chatelier's Principle, the forward reaction is favoured by high pressures.

At high pressures the equilibrium will shift to the right and increase the yield at equilibrium. Higher pressures will also increase the rate of the reaction.

1 State Le Chatelier's Principle.

Figure 1 shows that the yield increases as the pressure *increases* and as the temperature *decreases*. The best yields would therefore be obtained at very low temperatures and at very high pressures.

However, at low temperatures the reaction rate is slow. Although the yields are much better, the rate of production is too slow.

The use of high pressures involves extra costs in energy and in construction of equipment that will withstand the high pressures. Therefore the use of very high pressures is expensive to maintain and potentially more dangerous.

Industrial chemists therefore use **compromise conditions** that produce an *acceptable* yield at an *acceptable* rate. Most ammonia plants operate at 400–450 °C and 200 atmospheres (20 MPa) pressure. In the presence of a catalyst, these conditions give the most economical production of ammonia.

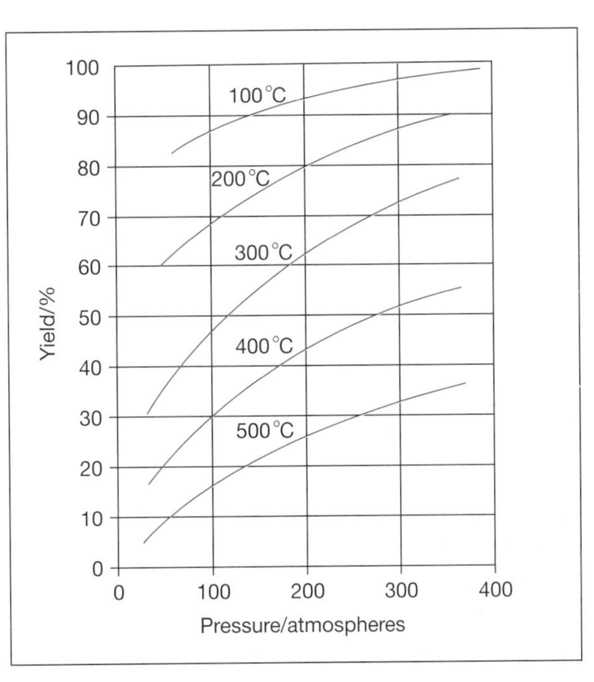

Fig. 1 *Graph of effect of pressure and temperature on yield*

Fig. 2 *A fertiliser production plant.*

Q

2 Explain why the conditions used in the Haber process are called *compromise conditions*.

3 In the manufacture of sulphur trioxide, a pressure of 200 kPa (2 atmospheres), a temperature of 450 °C and the use of a catalyst together produce a 98% yield.

$$2SO_2(g) + O_2(g) \rightleftharpoons 2SO_3(g) \quad \Delta H \text{ -ve}$$

a) Explain why a higher pressure is not used.

b) Explain why a lower temperature is not used.

Key Idea 116 – 117

- **The industrial manufacture of chemicals often involves reversible reactions. Industrial chemists select conditions that will produce a good yield at a reasonable rate and at a reasonable cost. This involves using knowledge of kinetics, energetics and equilibria in deciding the best conditions to be used.**

4

Unit 7 Questions

1. Many chemical reactions are reversible. Explain the meaning of the term *reversible* when applied to a chemical reaction. (2)

2. Explain the meaning of the terms *dynamic* and *equilibrium* when applied to a chemical equilibrium. (2)

3. Chemical equilibria can be homogeneous or heterogeneous. Explain the meaning of the terms *homogeneous* and *heterogeneous*. (2)

4. **a)** State Le Chatelier's Principle. (2)
   b) Use Le Chatelier's Principle to predict the effect of:
   - increasing the pressure
   - increasing the temperature

   on the position of the equilibrium in each of the following reactions:
   (i) $X(g) + Y(g) \rightarrow Z(g)$ ΔH +ve
   (ii) $A(g) + B(g) \rightarrow C(g) + D(g)$ ΔH –ve
   (iii) $S(g) \rightarrow R(g) + T(g)$ ΔH +ve (6)

5. Gaseous phosphorus(V) chloride dissociates reversibly, as shown in the equation:

   $$PCl_5(g) \rightleftharpoons PCl_3(g) + Cl_2(g) \quad \Delta H \text{ +ve}$$

   State and explain the effect on the position of the equilibrium when:
   a) the pressure is increased (2)
   b) the temperature is increased. (2)

6. Carbon monoxide reacts with hydrogen in the presence of a nickel catalyst, as shown in the equation:

   $$CO(g) + 3H_2(g) \rightleftharpoons CH_4(g) + H_2O(g) \quad \Delta H \text{ –ve}$$

   a) Explain why this reaction is described as a *homogeneous* reaction. (1)
   b) State and explain the effect of increasing the pressure on:
   (i) the rate at which equilibrium is reached
   (ii) the position of the equilibrium. (4)
   c) State and explain the effect of increasing the temperature on:
   (i) the rate at which equilibrium is reached
   (ii) the position of the equilibrium. (4)
   d) State and explain the effect of a catalyst on:
   (i) the rate at which equilibrium is reached
   (ii) the position of the equilibrium. (4)
   e) State and explain the effect on the equilibrium position of adding more hydrogen gas when an equilibrium has been reached. (2)

7) The conditions used in industrial processes involving equilibrium reactions are often compromise conditions. Use your knowledge of the conditions used in the Haber Process to explain the meaning of the term *compromise conditions*. (7)

8) The following data were obtained for the percentage yields in a homogeneous gas equilibrium reaction at different temperatures and pressures.

	200 kPa	600 kPa	1000 kPa
400 °C	25%	40%	60%
500 °C	20%	35%	55%
600 °C	15%	30%	50%
700 °C	10%	25%	45%
800 °C	5%	20%	40%

Use the data in the table to help in answering the following questions.

a) (i) State whether the forward reaction involves an increase or a decrease in the number of moles of gas.
(ii) Give the reasons for your answer. (3)

b) (i) State whether the forward reaction is endothermic or exothermic.
(ii) Give reasons for your answer. (3)

9) The equations for four reactions are shown below:

Reaction 1	$A(g) \rightleftharpoons B(g) + C(g)$	ΔH –ve
Reaction 2	$K(g) + L(g) \rightleftharpoons M(g) + N(g)$	ΔH –ve
Reaction 3	$P(g) \rightleftharpoons R(g) + T(g)$	ΔH +ve
Reaction 4	$X(g) + Y(g) \rightleftharpoons Z(g)$	ΔH –ve

Select from these four reactions:

a) A reaction in which the yield increases at lower pressures and at lower temperatures. (1)

b) A reaction in which the yield decreases at higher pressures and at lower temperatures. (1)

c) A reaction in which the yield decreases at lower pressures and at higher temperatures. (1)

d) A reaction in which a change in pressure has no effect on the yield. (1)

1 Electrons on the move

You will have met some ideas at GCSE about oxidation and reduction:

- **oxidation** is used to describe a reaction in which oxygen is added to an element or compound
- **reduction** is the removal of oxygen, the opposite of oxidation.

Redox in terms of electrons

You will remember the burning of magnesium in oxygen. Magnesium burns with a bright white flash, leaving behind a white powder, magnesium oxide.

$$2Mg(s) + O_2(g) \rightarrow 2MgO(s)$$

This can also be written as:

$$Mg(s) + \frac{1}{2}O_2(g) \rightarrow MgO(s)$$

According to the GCSE definition, the magnesium is oxidised to magnesium oxide – but where is the reduction?

Fig. 1 Burning magnesium in oxygen

Magnesium oxide is an ionic compound. Let us look at its formation using half-equations:

$$Mg \rightarrow Mg^{2+} + 2e^- \text{ and } \tfrac{1}{2}O_2 + 2e^- \rightarrow O^{2-}$$

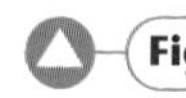

Fig. 2 The magnesium atom loses two electrons to oxygen

Each magnesium atom loses two electrons to form a magnesium ion. Notice that the equation needs two electrons to balance the charges (see page 69).

We have already said that the magnesium atom has been oxidised, so oxidation must involve the *loss* of electrons. These electrons are accepted by the atoms in the oxygen molecule to form oxide ions. But reduction is the opposite of oxidation, so we can now define reduction as *gaining* electrons.

- **oxidation** is the **loss of electrons**.
- **reduction** is the **gain of electrons**.

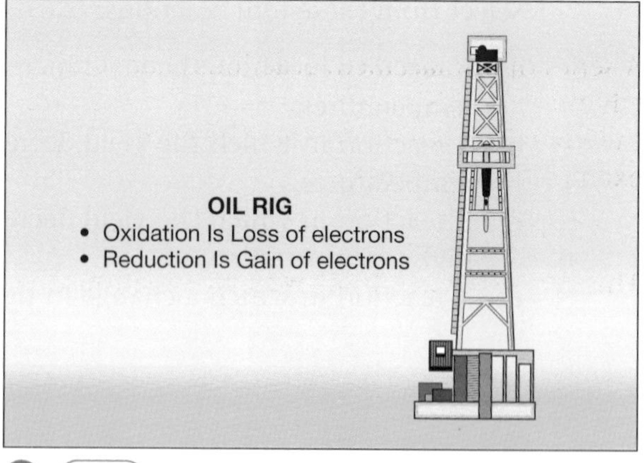

Fig. 3 A mnemonic for oxidation and reduction

Q

1 Decide whether the species in colour is being oxidised or reduced:

a) $Br_2 + 2e^- \rightarrow 2Br^-$
b) $Fe^{3+} + e^- \rightarrow Fe^{2+}$
c) $2F^- \rightarrow F_2 + 2e^-$
d) $Na \rightarrow Na^+ + e^-$

At AS level, all redox reactions are defined in terms of electrons. Even reactions that do not include oxygen are still recognised as redox reactions.

Displacement reactions

Remember displacement reactions? At GCSE you will have put an iron nail into copper(II) sulphate solution (Figure 4).

Since iron is more reactive than copper it displaces copper and forms iron(II) sulphate. The molecular equation for this reaction is:

$$Fe(s) + CuSO_4(aq) \rightarrow FeSO_4(aq) + Cu(s)$$

The ionic equation for this reaction is:

$$Fe(s) + Cu^{2+}(aq) \rightarrow Fe^{2+}(aq) + Cu(s)$$

As two half-equations, this becomes:

$Fe(s) \rightarrow Fe^{2+}(aq) + 2e^-$ loss of electrons: oxidation

$Cu^{2+}(aq) + 2e^- \rightarrow Cu(s)$ gain of electrons: reduction

The iron atoms have been oxidised and the copper ions reduced.

Fig. 4 *Displacement of copper by iron*

Q

2 a) Write an equation to show the displacement reaction between copper metal and silver nitrate solution.

b) Using half-equations, show that this reaction involves the oxidation of copper.

Displacement reactions of the halogens

A series of displacement reactions that you will meet in this module involves the halogens in Group VII of the Periodic Table. An example is shown below:

$$2Br^-(aq) + Cl_2(aq) \rightarrow 2Cl^-(aq) + Br_2(aq)$$

This can be broken into two half-equations:

$2Br^-(aq) \rightarrow Br_2(aq) + 2e^-$ loss of electrons: oxidation

$Cl_2(aq) + 2e^- \rightarrow 2Cl^-(aq)$ gain of electrons: reduction

This is a redox reaction in which bromide ions are oxidised to bromine atoms and chlorine atoms are reduced to chloride ions.

Key Ideas 120 – 121

- **Half-equations can be used to recognise oxidation and reduction.**
- **Oxidation is the loss of electrons.**
- **Reduction is the gain of electrons.**
- **Displacement reactions involve redox.**

2 What a state ...

There is another method of interpreting redox reactions instead of using half-equations. This method involves using *oxidation states* (or *oxidation numbers*). Oxidation states are assigned to atoms in molecules or ions to show how many electrons the atom has used in bonding. This method has the advantage that it can also be applied to covalent substances where complete transfer of electrons does not occur.

Oxidation states

There is a set of rules which you need to learn in order to work out **oxidation states**. These are given below, using the halogen displacement reaction to illustrate the various points.

$$2KBr(aq) + Cl_2(aq) \rightarrow 2KCl(aq) + Br_2(aq)$$
$$2Br^-(aq) + Cl_2(aq) \rightarrow 2Cl^-(aq) + Br_2(aq)$$

Rule 1 For simple ions, the oxidation state is the charge on the ion.

For chloride and bromide ions which each carry one negative charge, the oxidation state is –1. Positively charged ions are given a positive oxidation state.

Rule 2 For atoms in elements, the oxidation state is zero.

Chlorine and bromine are elements, so the oxidation state of each is zero.

Rule 3 For compounds, the sum of the oxidation states is zero.

In KBr and KCl the halide ions have an oxidation state of –1, so for the compound to have an overall zero charge, the oxidation state of potassium must be +1.

You can see from Rule 3 that we can use the molecular equation rather than the ionic equation to work out oxidation states. This is useful when considering covalent compounds, which are not made up of ions.

Rule 4 Covalent molecules are assumed to be ionic.

For example, water is covalent, but if it were ionic, it would contain H^+ and O^{2-} ions. The oxidation state of the hydrogen is +1. There are 2 atoms of hydrogen, giving a total of +2. This value is cancelled out by the oxidation state of oxygen, which is –2, giving the overall oxidation state of water as 0.

Chlorine and bromine, when present in compounds, can show a variety of oxidation states from –1 (as seen in the above examples) to +5. They are said to exhibit a **variable oxidation state**. We will look at some more examples of their compounds in the section on the halogens (see page 138). There are many elements which show a variable oxidation state in their compounds, typically the transition metals. These are dealt with at A2 level: at AS level you will study only the chemistry of the s- and p-block elements of the Periodic Table.

Potassium in componds only ever has an oxidation state of +1. Other elements also have a **fixed oxidation state**. These are shown in Table 1: you need to learn them.

Ion	Oxidation state
H^+, Li^+, Na^+, K^+	+1
Mg^{2+}, Ca^{2+}, Ba^{2+}	+2
Al^{3+}	+3
O^{2-}	–2
F^-	–1

Table 1 *Fixed oxidation states*

When hydrogen, oxygen and fluorine are present in covalent compounds, they are considered as ions and have the same oxidation states as shown in the table.

The exceptions to this table are these: hydrogen in hydrides has an oxidation state of –1, and oxygen in peroxides has an oxidation state of –1.

Working out oxidation states

Fixed oxidation states are used to work out the oxidation states of other elements in compounds. Remember: the sum of the oxidation states of the elements in a compound *must* add up to zero.

Consider sulphur dioxide. This is a covalent compound, but we can still work out the oxidation state of sulphur provided that the formula is known.

The formula is SO_2. Each oxygen has a fixed oxidation state of –2. This gives a total of –4. To cancel out this value, the sulphur must have an oxidation state of +4.

Some more examples are shown in Table 2.

Compound	Element	Fixed oxidation state	Element	Variable oxidation state
KBr	K	+1	Br	–1
CaS	Ca	+2	S	–2
NH_3	H	+1 (× 3) = +3	N	–3
SO_3	O	–2 (× 3) = –6	S	+6
IF_7	F	–1 (× 7) = –7	I	+7

Table 2 *Calculating variable oxidation states*

Element	Elecronegativity value
K	0.8
Ca	1.0
H	2.1
S	2.5
I	2.5
Br	2.8
N	3.0
O	3.5
F	4.0

Table 3 *Electronegativity values for the elements in Table 2*

From Tables 2 and 3 you can see that:

1 The element with the positive oxidation state is *usually* written first in the formula. There are exceptions, including NH_3.

2 The more electronegative element has a negative oxidation state.

3 One oxidation state of sulphur is +6, and one of iodine is +7. These are the **maximum oxidation states** of these elements. The maximum oxidation states equal the Group number in the Periodic Table.

Q

1 Work out the oxidation states of the elements in the following:

a) NaCl b) Li_2O c) H_2S d) Ca e) O_2 f) PF_5

Key Ideas 122 – 123

- **For simple ions, the oxidation state is the charge on the ion.**
- **For elements, the oxidation state is zero.**
- **For compounds, the sum of the oxidation states of the elements present is zero.**

3 What a state ... (2)

So far you have learnt how to work out oxidation states for simple ions and molecules. Now we need to look at compounds that contain more than two elements, and polyatomic ions, such as sulphate (SO_4^{2-}) and nitrate (NO_3^-).

More on oxidation states

Compounds such as sulphuric acid ($H_2\mathbf{S}O_4$), calcium carbonate ($Ca\mathbf{C}O_3$) and sodium nitrate ($Na\mathbf{N}O_3$) contain three elements.
Looking at each one in turn we can work out the oxidation state of the element in bold. Remember that the oxidation states of the other elements are fixed. Table 1 is a reminder.

Ion	Oxidation state
H^+, Li^+, Na^+, K^+	+1
Mg^{2+}, Ca^{2+}, Ba^{2+}	+2
Al^{3+}	+3
O^{2-}	–2
F^-	–1

Table 1 *Fixed oxidation states*

H_2SO_4

Each hydrogen has an oxidation state of +1, giving a total of +2. Each oxygen has an oxidation state of –2. Since there are four oxygen atoms, the total is –8. This leaves –6. Since the molecule has no overall charge, the oxidation state of sulphur must be +6.

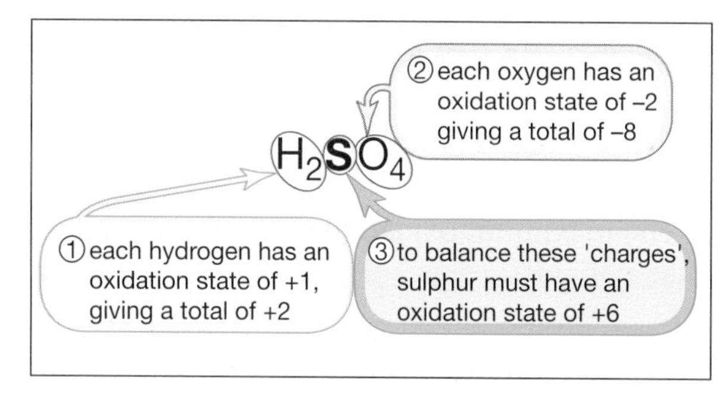

Fig. 1 *Working out the oxidation state of sulphur in sulphuric acid*

$CaCO_3$

The calcium has an oxidation state of +2. Each oxygen has an oxidation state of –2, giving a total of –6. This leaves –4. Since the compound has no overall charge, the oxidation state of carbon must be +4.

$NaNO_3$

The sodium has an oxidation state of +1. There are three oxygen atoms, giving a total of –6. This leaves –5, so the oxidation state of nitrogen must be +5.

Q

1 Work out the oxidation states of the elements shown in colour in the following formulae:

a) $MgSO_3$ b) $KMnO_4$ c) $NaClO_3$ d) $NaNO_2$

Working out oxidation states in polyatomic ions

What do you think is the oxidation state of sulphur in the sulphate ion (SO_4^{2-})?

We have worked out that the oxidation state of sulphur in sulphuric acid (H_2SO_4) is +6.

Since sulphuric acid might also be called hydrogen sulphate, the oxidation state of the sulphur in the sulphate ion must be +6.

Another rule is needed to work out the variable oxidation states in molecular ions:

> The sum of the oxidation states of the elements present must equal the charge on the ion.

In the sulphate ion each oxygen has an oxidation state of –2, giving a total of –8. To retain an overall charge of –2, the sulphur must have an oxidation state of +6 to cancel out the remaining negative charges.

Most of these polyatomic ions will contain oxygen. Some examples of polyatomic ions and how to work out oxidation states are shown in Table 2.

Notice that nitrogen and phosphorus in oxidation state +5 are in their maximum oxidation state. They are both in Group 5 of the Periodic Table.

Ion	No. of oxygen atoms	Total oxidation no. due to oxygen	Charge on ion	Oxidation state of central atom
ClO_3^-	3	–6	–1	+5
ClO^-	1	–2	–1	+1
NO_3^-	3	–6	–1	+5
NO_2^-	2	–4	–1	+3
PO_4^{3-}	4	–8	–3	+5

Table 2 *Oxidation states in polyatomic ions*

The oxidation state of the central ion indicates how many electrons are used to bond it to oxygen.

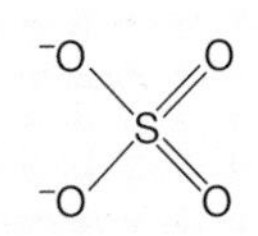

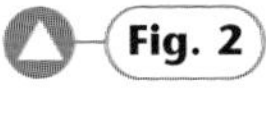
Fig. 2 *The structure of the sulphate ion: the sulphur atom makes six covalent bonds with the oxygen atoms using its six outer electrons*

Q

2 Sulphur forms the following ions: SO_3^{2-}, SO_4^{2-}, HSO_3^-, S^{2-}.

a) Work out the oxidation states of sulphur in each of these ions.

b) What is the maximum oxidation state of sulphur?

Summary of rules for working out oxidation states

1 **For simple ions, the oxidation state is the charge on the ion.**

- If an element has *lost* electrons it is assigned a *positive* oxidation state.
- If an element has *gained* electrons it is assigned a *negative* oxidation state.

  e.g. Cl^- has an oxidation state of –1, Ca^{2+} has an oxidation state of +2.

2 **For atoms in elements, the oxidation state is zero.**

e.g. Ca, Na, S.

This also applies to molecular elements:

e.g. Cl_2, Br_2, O_2.

3 **For compounds, the sum of the oxidation states is zero.**

e.g. In magnesium oxide the oxidation state of magnesium is +2 and that of oxygen is –2.

4 **For polyatomic ions, the sum of the oxidation states must equal the charge on that ion.**

e.g. In the ClO^- ion the oxygen has a fixed oxidation state of –2, whilst the chlorine has an oxidation state of +1, giving the ion an overall charge of –1.

Key Idea 124 – 125

- **For polyatomic ions, the sum of the oxidation states of the elements present must equal the charge on the ion.**

4 What's in a name?

If an element exhibits more than one oxidation state then the oxidation state also becomes part of its name, so $FeCl_2$ is called iron(II) chloride and Fe^{2+} is called the iron(II) ion. The oxidation state is given as a Roman numeral.

Names that indicate variable oxidation states

What is copper oxide?

At GCSE you may have called the compound with the formula CuO 'copper oxide', but as well as CuO there is also Cu_2O. Both compounds cannot be given the same name. How can we distinguish between them? The answer is to include the oxidation state of copper in the name, so CuO becomes copper(II) oxide and Cu_2O is copper(I) oxide. For compounds containing elements with a variable oxidation state, it is essential at AS level to include the oxidation state in the name.

Finding the formula of phosphorus(V) oxide

Oxidation states idea also help in writing the formula of unfamiliar compounds, such as phosphorus(V) oxide. Oxygen always has an oxidation state of –2 in oxides, and the name tells us that phosphorus has an oxidation state of 5. It must be an oxidation state of +5, to balance the negative value of the oxygen. Since the sum of the oxidation states of a molecule must be zero, then the formula for phosphorus(V) oxide must be P_2O_5.

i.e. $(2 \times +5) + (5 \times -2) = 0$

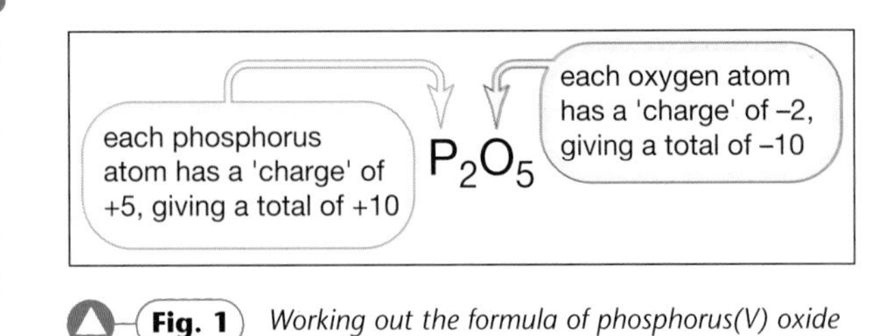

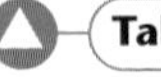

Fig. 1 *Working out the formula of phosphorus(V) oxide*

A more difficult example

Some compounds may contain more than one element with a variable oxidation state (Figure 2).

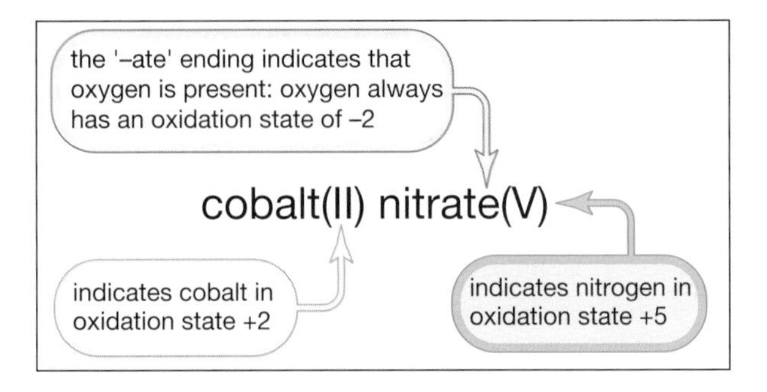

Fig. 2 *Working out the oxidation states of cobalt nitrate*

For elements which never change their oxidation state it is not necessary to include the Roman numeral in the name. For example Na_2O can simply be called sodium oxide rather than sodium(I) oxide(II).

Table 1 gives some more examples.

Name	Formula	Oxidation state (of the element in bold)
Iron(II) chloride	$\mathbf{Fe}Cl_2$	+2
Sodium chlorate(V)	$Na\mathbf{Cl}O_3$	+5
Sodium dichromate(VI)	$Na_2\mathbf{Cr}_2O_7$	+6
Potassium manganate(VII)	$K\mathbf{Mn}O_4$	+7

Table 1 *Variable oxidation states*

Variable oxidation states in cations

The *position* of the Roman numeral within the name depends on whether the element showing a variable oxidation state is part of the cation or anion. Thus in iron(II) chloride it is the iron (Fe^{2+}) that shows the variable oxidation state, so the Roman numeral comes directly after iron in the name.

Table 2 shows some elements which show variable oxidation in cations. These are usually transition metal ions. You need to learn only these for A2 Chemistry.

Name	Formula
Copper(II)	Cu^{2+}
Iron(II)	Fe^{2+}
Iron(III)	Fe^{3+}
Cobalt(II)	Co^{2+}
Chromium(III)	Cr^{3+}

Table 2 *Variable oxidation states: cations*

Variable oxidation states in anions

For sodium dichromate(VI) it is the chromium in the anion ($Cr_2O_7^{2-}$) that exhibits the variable oxidation state, so the Roman numeral, representing the oxidation state for chromium, comes at the end of the name.

Formula	Systematic name	Trivial name
NO_3^-	nitrate(V)	nitrate
NO_2^-	nitrate(III)	nitrite
SO_4^{2-}	sulphate(VI)	sulphate
SO_3^{2-}	sulphate(IV)	sulphite
Cl^-	chloride(I)	chloride
ClO^-	chlorate(I)	chlorite
ClO_3^-	chlorate(V)	chlorate

Table 3 *Variable oxidation states: anions*

Table 3 shows some elements which show variable oxidation in anions. Some of these are still known by their trivial names – for instance, cobalt(II) nitrate(V) is generally known as cobalt(II) nitrate.

Q

1 Write formulae for these compounds:

a) iron(II) sulphate
b) iron(III) chloride
c) copper(II) carbonate
d) phosphorus(III) oxide

2 Copy out and complete the table below.

Name	Formula	Variable oxidation state
Lead(II) oxide		
	PbO_2	+4
Sodium chlorate(I)	NaClO	
Iron(III) hydroxide		+3

3 Use the information in Tables 2 and 3 to write formulae for:

a) iron(II) nitrate
b) chromium(III) chloride
c) copper(II) sulphite

4 Name the following compounds:

a) $Fe_2(SO_4)_3$
b) $CoCl_2$
c) $NaNO_2$

Key Ideas 126 – 127

- **Some elements show variable oxidation states.**
- **The variable oxidation state of such an element is part of the name of the compound that contains it.**
- **When the names of compounds containing variable oxidation states are given, this makes it easier to work out the formulae.**

5 The state of oxidation

Having learnt how to work out oxidation states, you are now in a position to apply this learning to chemical equations in order to establish which species are oxidised and which are reduced.

Linking oxidation states with redox reactions

Looking back at the halogen displacement reactions, we can summarise what has been learnt by writing the variable oxidation states of the elements under each formula, in either the molecular or ionic equation:

$$\underset{-1}{2KBr(aq)} + \underset{0}{Cl_2(aq)} \rightarrow \underset{-1}{2KCl(aq)} + \underset{0}{Br_2(aq)}$$

$$\underset{-1}{2Br^-(aq)} + \underset{0}{Cl_2(aq)} \rightarrow \underset{-1}{2Cl^-(aq)} + \underset{0}{Br_2(aq)}$$

How does this help us decide which species has been oxidised and which reduced?

- An *increase* in oxidation state is *oxidation*.
- A *decrease* in oxidation state is *reduction*.

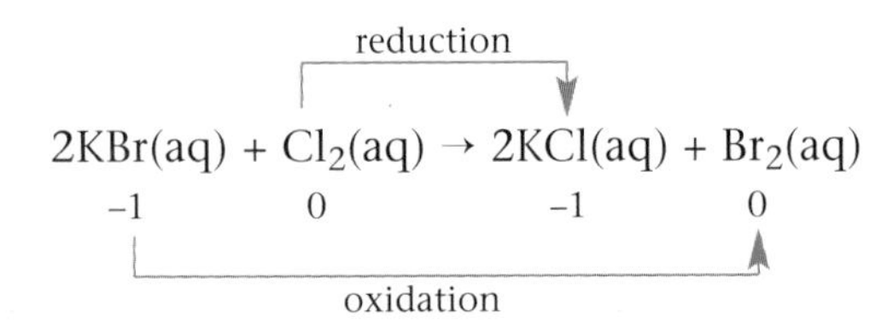

The bromide ions have been oxidised to bromine atoms, and the chlorine atoms reduced to chloride ions.

This information exactly matches the information from half-equations (see page 121).

Let us look at another example from halogen chemistry.

$$2HBr(g) + H_2SO_4(aq) \rightarrow 2H_2O(l) + SO_2(g) + Br_2(l)$$

1 Hydrogen and oxygen have fixed oxidation states (H = +1 and O = –2), so you should be able to use these to decide that bromine in HBr has an oxidation state of –1, and sulphur in SO_2 has an oxidation state of +4.

2 Br_2 is an element, so its oxidation state is 0.

3 We have already worked out the oxidation state of sulphur in H_2SO_4.

Each hydrogen has an oxidation state of +1, giving a total of +2; and each oxygen has an oxidation state of –2, giving a total of –8. This leaves –6. Since the molecule has no overall charge, the oxidation state of sulphur must be +6.

Summarising this information, and ignoring hydrogen and oxygen which have not changed oxidation states, we get:

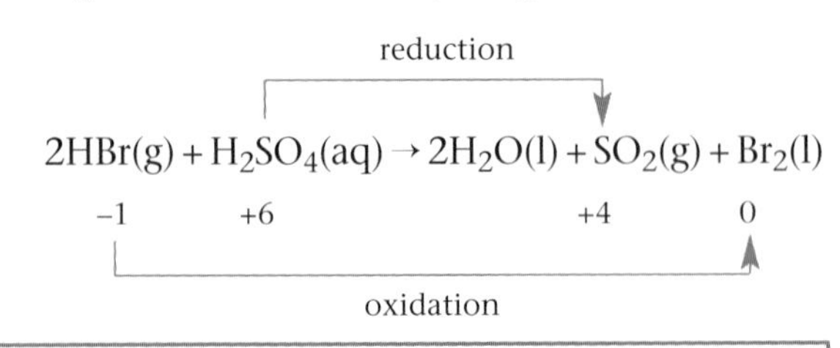

Q

1 For each of the equations given below, work out the oxidation states for each species present, and use this information to decide which species has been oxidised and which species reduced.

a) $2I^- + Cl_2 \rightarrow I_2 + 2Cl^-$
b) $8HI + H_2SO_4 \rightarrow H_2S + 4H_2O + I_2$
c) $SO_2 + 2H_2O + 2Cu^{2+} + 2Cl^- \rightarrow H_2SO_4 + 2H^+ + 2CuCl$
d) $Cr_2O_7^{2-} + 8H^+ + 3SO_3^{2-} \rightarrow 2Cr^{3+} + 4H_2O + 3SO_4^{2-}$
e) $5Fe^{2+} + MnO_4^- + 8H^+ \rightarrow 5Fe^{3+} + Mn^{2+} + 4H_2O$

Oxidising and reducing agents

In redox reactions an increase in oxidation state means that the species has been oxidised. Linking this with half-equations, the species which has been oxidised has *lost* electrons: this is called a **reducing agent**. The species which has been reduced has *gained* electrons: this is called an **oxidising agent**. Reducing agents are themselves oxidised, whilst oxidising agents are themselves reduced.

- An *oxidising* agent is an electron *acceptor*.
- A *reducing* agent is an electron *donor*.

To identify oxidising and reducing agents, you can look either at the overall equation or at the half-equations, as in the examples on the right.

Vitamin C (ascorbic acid) is a reducing agent. Recent evidence suggests that a high daily dose of vitamin C in the diet may have beneficial effects, including an increased resistance to infection.

Example 1

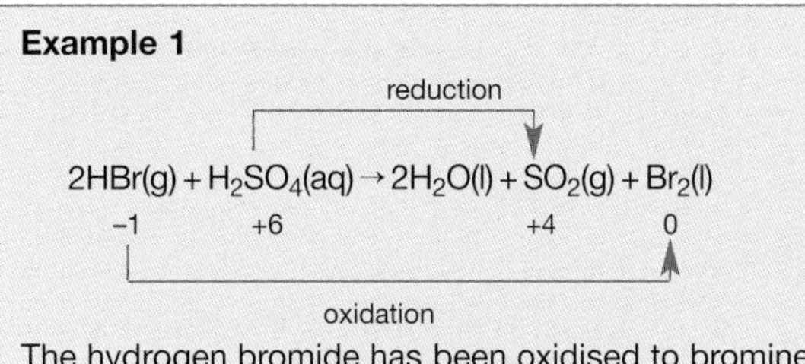

The hydrogen bromide has been oxidised to bromine, so the hydrogen bromide is a reducing agent in this reaction. The sulphuric acid has been reduced to sulphur dioxide, so in this reaction sulphuric acid is behaving as an oxidising agent.

Example 2

$$2Br^-(aq) + Cl_2(aq) \rightarrow 2Cl^-(aq) + Br_2(aq)$$

The half-equations for this reaction are:

$2Br^-(aq) \rightarrow Br_2(aq) + 2e^-$	loss of electrons: oxidation
$Cl_2(aq) + 2e^- \rightarrow 2Cl^-(aq)$	gain of electrons: reduction

The bromide ions have lost electrons so they are behaving as reducing agents in this reaction. The chlorine atoms having gained electrons are defined as oxidising agents; they are themselves reduced.

Fig. 1

Some vitamin C tablets are sold under the tradename Redoxon, indicating that the action of this vitamin involves redox

Q

2 Decide whether each species in colour in the following equations is acting as an oxidising agent or a reducing agent:

a) $Fe_2O_3 + 3CO \rightarrow 2Fe + 3CO_2$
b) $2Mg + O_2 \rightarrow 2MgO$
c) $Zn + Fe^{2+} \rightarrow Zn^{2+} + Fe$
d) $SO_2 + 2H_2O + 2Cu^{2+} + 2Cl^- \rightarrow H_2SO_4 + 2H^+ + 2CuCl$
e) $Cr_2O_7^{2-} + 8H^+ + 3SO_3^{2-} \rightarrow 2Cr^{3+} + 4H_2O + 3SO_4^{2-}$

Key Ideas 128 – 129

- **Oxidation states can be used to decide which species in a reaction has been oxidised and which species has been reduced, because an increase in oxidation state is oxidation and a decrease in oxidation state is reduction.**
- **An oxidising agent is an electron acceptor.**
- **A reducing agent is an electron donor.**

6 In the balance

We started the section on redox by looking at half-equations. Provided that you are given at least one reactant and one product, you should now be in a position to construct half-equations yourself.

Constructing simple half-equations

A half-equation shows whether a species has been oxidised or reduced. Remember that:

- oxidation is the loss of electrons
- reduction is the gain of electrons.

Here are some examples of simple half-equations which you have already met in this module:

Oxidation	Reduction
$2Br^-(aq) \rightarrow Br_2(aq) + 2e^-$	$Cl_2(aq) + 2e^- \rightarrow 2Cl^-(aq)$
$Na(s) \rightarrow Na^+(aq) + e^-$	$Fe^{3+}(aq) + e^- \rightarrow Fe^{2+}(aq)$
$Fe(s) \rightarrow Fe^{2+}(aq) + 2e^-$	$Mg^{2+}(aq) + 2e^- \rightarrow Mg(s)$

These equations are constructed by using the formulae of the ions (see Appendix B), and adding electrons for reduction to the left-hand side (LHS) of the equation or adding electrons to the right-hand side (RHS) of the equation for oxidation. The number of electrons added must cancel out the overall charge on the ions so that the equation is balanced (see page 69).

Q

1 Construct half-equations to show the oxidation of:
   a) a calcium atom to a calcium ion
   b) an iron(II) ion to an iron(III) ion.

2 Construct half-equations to show the reduction of:
   a) a copper(II) ion to a copper atom
   b) an aluminium ion to an aluminium atom.

Constructing more complex half-equations

You will need to follow these rules:

1. Write down the formulae for the reactants and products, and balance the atoms undergoing redox.
2. Balance any oxygen atoms present by adding water.
3. Balance any hydrogen atoms present by adding H^+ ions.
4. Balance the charges by adding electrons.
5. Check that the number of atoms are balanced.
6. Check that the number of charges are balanced. (The total number of electrons transferred will be equal to the total change in oxidation states.)
7. Add state symbols.

The examples shown below explain how to apply these rules.

Fig. 1 *Sodium chlorate(V) is present in weedkillers*

1 Write a half-equation for the conversion of nitric acid into nitrogen dioxide gas.

1. Start with the formulae of nitric acid and nitrogen dioxide:

   $HNO_3 \rightarrow NO_2$

   The numbers of nitrogen atoms on the two sides of the equation are already balanced.

2. Since there are three atoms of oxygen on the left-hand side (LHS) and only two atoms of oxygen on the right-hand side (RHS), we need to add a molecule of water on the RHS to balance the oxygen atoms:

   $HNO_3 \rightarrow NO_2 + \mathbf{H_2O}$

3. Now the hydrogen atoms need balancing. We need to add one H^+ ion to the LHS to balance these:

   $HNO_3 + \mathbf{H^+} \rightarrow NO_2 + H_2O$

4. We have introduced a positive charge, so this must be balanced by adding an electron which carries a negative charge:

   $HNO_3 + H^+ + \mathbf{e^-} \rightarrow NO_2 + H_2O$

5. Now the numbers of atoms are balanced.
6. Now the numbers of charges are balanced. *Check*: The nitrogen has been reduced (electron gain) from an oxidation state of +5 to an oxidation state of +4. This represents an overall change in oxidation states of 1, hence one electron is added.
7. Add the state symbols. The half-equation is:

   $HNO_3(aq) + H^+(aq) + e^- \rightarrow NO_2(g) + H_2O(l)$

Nitrogen dioxide is a brown gas released into the atmosphere from car exhaust fumes. It causes acid rain (see page 190).

Q

3 Sulphur dioxide is another gas which pollutes the atmosphere by causing acid rain. It dissolves in rainwater, producing a mixture of sulphurous (H_2SO_3) and sulphuric (H_2SO_4) acids. Construct a half-equation for the conversion of sulphur dioxide into sulphuric acid.

2 Write a half-equation for the conversion of chlorine gas into chlorate(V) ions.

1. Write the formulae for chlorine gas and the chlorate(V) ion:

   $Cl_2 \rightarrow ClO_3^-$

   As there are two atoms of chlorine on the LHS and only one atom of chlorine on the RHS, we need to balance these:

   $Cl_2 \rightarrow \mathbf{2ClO_3^-}$

2. There are now six atoms of oxygen on the RHS and none on the LHS, so six molecules of water must be added to the LHS:

   $Cl_2 + \mathbf{6H_2O} \rightarrow 2ClO_3^-$

3. Now we balance the hydrogen atoms. We have introduced twelve of these, so the RHS needs twelve hydrogen ions:

   $Cl_2 + 6H_2O \rightarrow 2ClO_3^- + \mathbf{12H^+}$

4. Next we balance the charges. There is no total charge on the LHS, but there are two negative and twelve positive charges on the RHS. This gives a total of ten positives on the RHS, so ten electrons must be added:

   $Cl_2 + 6H_2O \rightarrow 2ClO_3^- + \mathbf{10e^-} + 12H^+$

5. The numbers of atoms are balanced.
6. The number of charges are balanced. Check: The chlorine has been oxidised (electron loss) from an oxidation state of 0 to +5. Since two chlorine atoms are involved, this represents an overall change in oxidation state of 10, hence ten electrons are lost.
7. Add state symbols. The half-equation is:

   $Cl_2(g) + 6H_2O(l) \rightarrow 2ClO_3^-(aq) + 10e^- + 12H^+(aq)$

Chlorate(V) ions are present in weedkillers (Figure 1), usually as sodium chlorate(V).

Q

4 Chlorate(V) ions are powerful oxidising agents. Construct a half-equation to show the reduction of chlorate(V) ions to chloride ions.

Key Idea 130 – 131

- **Half-equations can be constructed using a set of rules which identify the oxidation and reduction processes in redox reactions, provided that the reactants and products are specified.**

7 Two halves make a whole

The last stage in this work on redox equations is to learn how to construct an ionic equation for a redox reaction by combining two half-equations.

Constructing equations for redox reactions from half-equations

We have seen in this module that one way to decide whether a species is oxidised or reduced is to look at the half-equations. These are however not reactions that can occur in isolation; instead, two half-equations can be combined to represent an overall redox reaction.

When combining half-equations there are five rules:

1. One half-equation must show oxidation and the other half-equation must show reduction – that is, one species must be losing electrons and another species gaining them.
2. The number of electrons being lost and gained must be the same. This may require some multiplication.
3. The two half-equations are then added together to produce the overall equation, which should not contain any electrons.
4. The final equation must balance for atoms.
5. The final equation must balance for charges on the ions.

The examples below illustrate how to combine two half-equations. These equations are both taken from halogen chemistry, which is an important part of this module (see page 136).

1 Displacement reaction between bromine and iodide ions

1. Write down the two half-equations. Check that one shows oxidation and the other shows reduction:

   $Br_2 + 2e^- \rightarrow 2Br^-$ reduction

   $2I^- \rightarrow I_2 + 2e^-$ oxidation

2. Balance the electrons on the two sides of the equation. The numbers of electrons are the same in both half-equations, so no multiplication is required here.
3. Add together the two half-equations:

   $$Br_2 + 2e^- \rightarrow 2Br^-$$
   $$2I^- \rightarrow I_2 + 2e^-$$
   $$Br_2 + 2e^- + 2I^- \rightarrow 2Br^- + I_2 + 2e^-$$

   Cancel out the electrons:

   $$Br_2 + 2I^- \rightarrow 2Br^- + I_2$$

4. Check that the final equation balances for numbers of atoms.
5. Check that the final equation balances for numbers of charges.
6. Add state symbols if possible. The final answer is:

   $Br_2(l) + 2I^-(aq) \rightarrow 2Br^-(aq) + I_2(s)$

   This equation shows that bromine oxidises iodide ions to iodine. You can check this by using oxidation states:

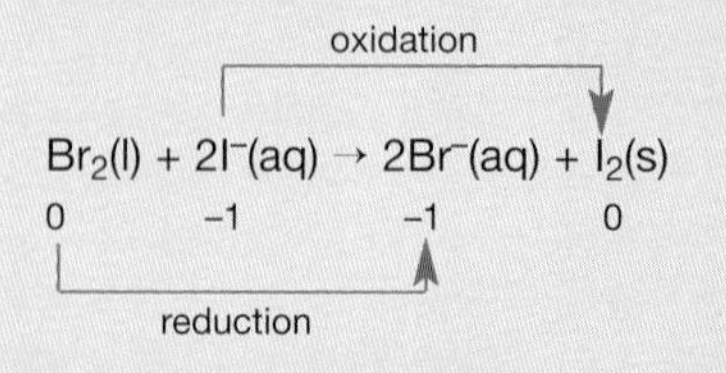

Bromine is a more powerful oxidising agent than iodine (see page 138).

2 Oxidation of hydrogen iodide to iodine using concentrated sulphuric acid

1 Write down the two half-equations:

$2HI \rightarrow I_2 + 2H^+ + 2e^-$ oxidation
$H_2SO_4 + 6H^+ + 6e^- \rightarrow S + 4H_2O$ reduction

2 Balance the electrons. Since the reduction of sulphuric acid requires six electrons and the oxidation process only supplies two, then the half-equation for oxidation must be multiplied throughout by three:

$$\mathbf{6}HI \rightarrow \mathbf{3}I_2 + \mathbf{6}H^+ + \mathbf{6}e^-$$

3 Add together the two half-equations:

$$6HI \rightarrow 3I_2 + 6H^+ + 6e^-$$
$$H_2SO_4 + 6H^+ + 6e^- \rightarrow S + 4H_2O$$

$$H_2SO_4 + 6H^+ + 6e^- + 6HI \rightarrow S + 4H_2O + 3I_2 + 6H^+ + 6e^-$$

Cancel out the electrons:

$$H_2SO_4 + 6H^+ + 6HI \rightarrow S + 4H_2O + 3I_2 + 6H^+$$

Since there are six hydrogen ions on each side of the equation, these can also be cancelled:

$$H_2SO_4 + 6HI \rightarrow S + 4H_2O + 3I_2$$

4 Check that the atoms balance. There are eight hydrogen atoms, one sulphur, four oxygen atoms and six iodine atoms on each side of the equation.

5 Check that the charges on the ions balance. In this case there are no ions present.

6 Add state symbols to give the final equation:

$$H_2SO_4(aq) + 6HI(g) \rightarrow S(s) + 4H_2O(l) + 3I_2(s)$$

This is an important redox reaction showing the ability of hydrogen iodide to behave as a reducing agent. The concentrated sulphuric acid may be reduced further to form hydrogen sulphide (see page 141).

Q

1 Check that this equation

$$H_2SO_4(aq) + 6HI(g) \rightarrow S(s) + 4H_2O(l) + 3I_2(s)$$

is a redox reaction by working out the changes in oxidation states of sulphur and iodine.

2 Construct redox equations for the following reactions using the half-equations provided:

a) The oxidation of chloride ions using manganese(IV) oxide (MnO_2):

$2Cl^- \rightarrow Cl_2 + 2e^-$

$MnO_2 + 4H^+ + 2e^- \rightarrow Mn^{2+} + 2H_2O$

b) The reduction of iron(III) using sulphur dioxide:

$SO_2 + 2H_2O \rightarrow SO_4^{2-} + 4H^+ + 2e^-$

$Fe^{3+} + e^- \rightarrow Fe^{2+}$

c) The reaction of iodate(V) with iodide ions to produce iodine:

$2IO_3^- + 12H^+ \rightarrow 10e^- \rightarrow I_2 + 6H_2O$

$2I^- \rightarrow I_2 + 2e^-$

For each equation, decide which is the oxidising agent and which is the reducing agent.

Key Idea 132 – 133

- **Two half-equations can be combined to give an overall redox equation.**

Unit 8 Questions

8

1. **a)** Define *oxidation* in terms of electrons. (1)
   b) Write the half-equation for the oxidation of bromide ions to bromine. (1)
   c) Write the half-equation for the reduction of sulphuric acid to sulphur dioxide. (2)
   d) Use your answers to parts **b)** and **c)** to construct the overall equation that represents the redox reaction between bromide ions and sulphuric acid. (2)

2. Two compounds of potassium decompose on heating to produce oxygen:

   $2KBrO_3(s) \rightarrow 2KBr(s) + 3O_2(g)$

   $2KNO_3(s) \rightarrow 2KNO_2(s) + O_2(g)$

   a) What is the oxidation state of bromine in: (i) $KBrO_3$; (ii) KBr? (2)
   b) What is the oxidation state of nitrogen in: (i) KNO_3; (ii) KNO_2? (2)
   c) In both cases the potassium compound is reduced when it decomposes. Explain *reduction* in terms of electron transfer. (1)
   d) $KBrO_3$ is the oxidising agent in matches and some fireworks. Explain what is meant by an *oxidising agent* in terms of electron transfer. (1)
   e) Give the systematic names for: (i) $KBrO_3$; ii) KNO_2. (2)

3. When sulphur dioxide is bubbled through a solution of copper(II) chloride the following reaction occurs:

   $SO_2(g) + 2H_2O(l) + 2Cu^{2+}(aq) + 2Cl^-(aq) \rightarrow SO_4^{2-}(aq) + 4H^+(aq) + 2CuCl(s)$

   a) Identify the reducing agent in this reaction (1)
   b) Identify the oxidising agent in this reaction. (1)
   c) Write a half-equation for the oxidation reaction. (2)

4. In volumetric analysis a solution containing cerium(IV) ions (Ce^{4+}) is titrated against a solution of sodium ethanedioate ($Na_2C_2O_4$). The half-equations for the reaction are:

   $Ce^{4+}(aq) + e^- \rightarrow Ce^{3+}(aq)$

   $C_2O_4^{2-}(aq) \rightarrow 2CO_2(g) + 2e^-$

   a) Construct an overall ionic equation for this redox reaction. (2)
   b) Calculate the volume of 0.100 M cerium(IV) solution needed to react exactly with a solution containing 0.0025 mol of ethanedioate ions. (3)

5. When calcium metal is dropped onto cold water, the following reaction occurs:

$$Ca(s) + 2H_2O(l) \rightarrow Ca(OH)_2(aq) + H_2(g)$$

a) What would you observe during this reaction? (2)
b) Write a half-equation for the changes to the calcium atoms in this reaction. (1)
c) Hence explain the role played by water in this reaction. (1)

6. Sodium iodate(V) is one of the main sources of iodine. It occurs in the sodium nitrate deposits in Chile. Once separated from the sodium nitrate, the sodium iodate(V) is treated with sodium hydrogen sulphite ($NaHSO_3$). The overall equation for the reaction is:

$$2IO_3^-(aq) + 5HSO_3^-(aq) \rightarrow 3HSO_4^-(aq) + 2SO_4^{2-}(aq) + I_2(aq) + H_2O(l)$$

a) What is the oxidation state of sulphur in the ions: (i) HSO_3^-; (ii) HSO_4^-; (iii) SO_4^{2-}? (3)
b) Construct a half-equation for the conversion of iodate(V) ions into iodine. (2)
c) What is the role of the hydrogen sulphite ions in this reaction?
d) Crude iodine may be contaminated with iodine monochloride (ICl). It is purified using potassium iodide.
   (i) What is the oxidation state of iodine in ICl?
   (ii) Give the systematic name for this compound.
   (iii) Potassium chloride and iodine are the products of this reaction. Construct an ionic equation to represent the purification process. (3)

7. a) Construct a half-equation to show the reduction of chlorine to chloride ions. (1)
b) Construct a half-equation to show the oxidation of sulphate(IV) ions to sulphate(VI) ions. (2)
c) Hence construct an overall equation to show the redox reaction between chlorine and sulphate(IV). (1)
d) In the above equation the sulphate(IV) ions behave as a reducing agent. The equation for the reaction between hydrogen sulphide and sulphate(IV) ions is shown below.

$$SO_3^{2-}(aq) + 2H^+(aq) + 2H_2S(g) \rightarrow 3H_2O(l) + 3S(s)$$

Hydrogen sulphide is a reducing agent also, but it is stronger than sulphate(IV). Explain how this equation illustrates this statement. (2)

8. When solid sodium iodide is warmed with concentrated sulphuric acid, a variety of products are formed. The formulae of some of these are shown below:

$NaHSO_4$ Na_2SO_4 HI H_2O SO_2 I_2 S H_2S

a) Which of these products is/are formed by reduction?
b) Which of these products is/are formed by oxidation?
c) Which of these products is/are formed by neither oxidation nor reduction?
d) Construct a half-equation for the conversion of sulphuric acid into hydrogen sulphide (H_2S). (10)

1 Trends in the halogens

Group VII of the Periodic Table forms a family of elements known as the **halogens**. Due to the similarity in the electronic structure of their atoms, the halogens show similar chemical properties as well as definite trends in properties down the Group as the atomic numbers of the elements increase. This section should look familiar because you have met all these ideas before, at GCSE or in Module 1.

Trends in physical properties

From GCSE you will know that the halogens:

- are non-metals with toxic, coloured vapours
- consist of molecules made up of pairs of atoms (diatomic molecules)
- form ionic salts such as Na^+Cl^-, with most metals, and covalent compounds, such as SF_6, with other non-metals.

The halogens are toxic **Fig. 1**

Table 1 *The halogens*

Element	Appearance at room temperature and pressure	Atomic number	Outer electronic structure of the atoms	Atomic radius/nm	Boiling point/°C	Electronegativity value
F_2	pale yellow gas	9	$2s^2, 2p^5$	0.071	–188	4.0
Cl_2	pale green gas	17	$3s^2, 3p^5$	0.099	–35	3.0
Br_2	dark red liquid	35	$4s^2, 4p^5$	0.114	59	2.8
I_2	dark purple solid	53	$5s^2, 5p^5$	0.133	184	2.5

The last member of the halogens, astatine, is not included in Table 1 because it is an artificial radioactive element, with a half-life of only 7.5 hours. A discussion of its chemistry is not relevant here.

Increasing atomic radius

Down Group VII the atomic radius of the atoms increases. As the atomic number increases (see Table 1), so the number of electrons in the atom increases. This means that more principal energy levels must be occupied by electrons. You can see from the electronic structures which principal energy levels are being filled. The outer energy levels are increasingly further from the nucleus, so as more energy levels are occupied the atom increases in size.

Decreasing electronegativity

Down Group VII the electronegativity of the atoms decreases. You first met the term 'electronegativity' in the section on bonding (page 28): it is a measure of the relative ability of an atom to attract a pair of electrons in a covalent bond.

The size of the atom influences its electronegativity. The bigger the atom, the more the nucleus is shielded by the extra inner electrons. This extra shielding means that the effect of the nuclear charge is reduced, so the ability of the nucleus to attract electrons is also reduced.

Increasing boiling point

Down Group VII the boiling point of the elements increases. The state of the halogens at room temperature and pressure depends on their boiling points (Figure 2). The halogen molecules are non-polar, so the attraction between these molecules is due to weak van der Waals forces. Down the Group the boiling points increase because as the atomic radius increases, so does the size of the molecules. As the molecules get bigger, the strength of the van der Waals forces between the molecules increases.

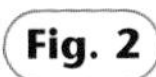

Fig. 2

At room temperature and pressure chlorine is a pale green gas, bromine is a dark red liquid, and iodine is a purple solid

Q

1 Explain why fluorine is a gas and iodine is a solid at room temperature and pressure.

2 Why is fluorine more likely to form ionic compounds than iodine?

Trends in chemical properties

The halogens can all act as oxidising agents. (An oxidising agent is an electron acceptor – see page 129.) When a halogen acts as an oxidising agent, each halogen atom gains an electron to form the corresponding halide ion: the halogen atom is itself reduced. For example:

$$I_2 + 2e^- \rightarrow 2I^-$$

Down Group VII the ability of the halogen to act as an oxidising agent decreases. This is because as the atomic radius increases and the effect of the nuclear charge is reduced (because of the shielding by the inner electrons), the halogen atom gains an electron less readily.

The bigger the atom, the weaker the oxidising agent **Fig. 3**

Q

3 Why is fluorine the most electronegative element?

4 Why is bromine a more powerful oxidising agent than iodine?

Key Ideas 136 – 137

- **Group VII of the Periodic Table includes the elements chlorine, bromine and iodine.**
- **The family name for this Group is the halogens.**
- **Down the Group there is a trend in physical properties – the boiling point increases and the electronegativity decreases.**
- **Down the Group the ability of the halogens to act as oxidising agents decreases.**

2 The halogens are oxidising

The trend in the ability of the halogens to act as oxidising agents can be demonstrated by a series of simple test-tube reactions in which aqueous solutions of the halogens are added separately to aqueous solutions of halide salts. These are called **displacement reactions**.

Displacement reactions

Aqueous solutions of the halogens are added dropwise to colourless solutions containing halide ions; solutions of potassium chloride, potassium bromide and potassium iodide are usually used. Aqueous solutions of the halogens are used because the undiluted halogens would be too toxic. The results are shown in Table 1.

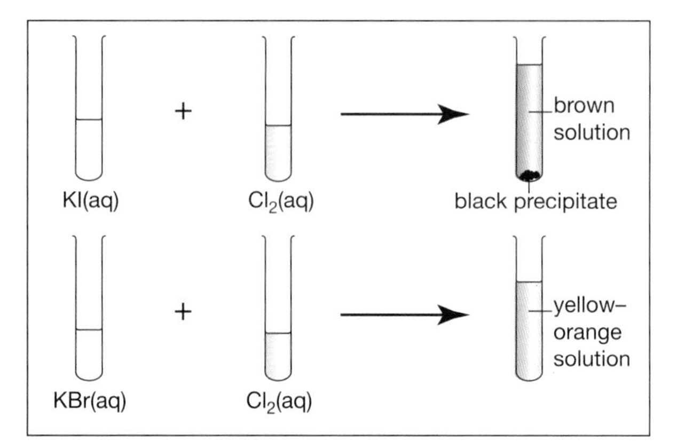

The displacement reactions of chlorine **Fig. 1**

	Cl^-(aq)	Br^-(aq)	I^-(aq)
Chlorine(aq)	no change	yellow–orange solution	black precipitate + brown solution
Bromine(aq)	no change	no change	black precipitate + brown solution
Iodine(aq)	no change	no change	no change

Table 1 *Observations made when aqueous solutions of the halogens are added to aqueous solutions containing halide ions*

The potassium ions are not included as they do not undergo redox reactions (see page 122).

From these results we can draw three conclusions.

Chlorine displaces bromide and iodide ions

$$Cl_2(aq) + 2Br^-(aq) \rightarrow 2Cl^-(aq) + Br_2(aq)$$

The bromine is observed as a yellow–orange solution.

$$Cl_2(aq) + 2I^-(aq) \rightarrow 2Cl^-(aq) + I_2(s)$$

The iodine is observed as a black precipitate. Some of this precipitate dissolves in aqueous solution, causing a brown coloration. This brown solution will be familiar to you as 'iodine solution'. It is used to test for the presence of starch: in the presence of starch a distinctive blue–black colour results (see page 146).

Bromine displaces iodide ions but not chloride ions

$$Br_2(aq) + 2I^-(aq) \rightarrow 2Br^-(aq) + I_2(s)$$

$$Br_2(aq) + 2Cl^-(aq) \rightarrow \text{no reaction}$$

Iodine does not displace either chloride or bromide ions

$$I_2(aq) + 2Cl^-(aq) \rightarrow \text{no reaction}$$

$$I_2(aq) + 2Br^-(aq) \rightarrow \text{no reaction}$$

During the displacement a redox reaction occurs. We have already considered these reactions before (see page 121): here is a reminder.

Looking at the displacement of iodide ions by bromine, the ionic equation for this reaction is:

$$Br_2(aq) + 2I^-(aq) \rightarrow 2Br^-(aq) + I_2(s)$$

If the equation is broken down into two half-equations we get:

$Br_2(aq) + 2e^- \rightarrow 2Br^-(aq)$ gain of electrons: reduction

$2I^-(aq) \rightarrow 2e^-(aq) + I_2(s)$ loss of electrons: oxidation

Since an oxidising agent is an electron acceptor it gains electrons. So in the above example, bromine is the oxidising agent: it oxidises iodide ions to iodine. Table 2 summarises the results in terms of oxidation.

	$Cl^-(aq)$	$Br^-(aq)$	$I^-(aq)$
Chlorine(aq)	–	chlorine oxidises bromine ions	chlorine oxidises iodine ions
Bromine(aq)	bromine does not oxidise chloride ions	–	bromine oxidises iodide ions
Iodine(aq)	iodine does not oxidise chloride ions	iodine does not oxidise bromide ions	–

Table 2 *Oxidation in displacement reactions*

The results of the displacement reactions therefore confirm the trend in oxidising power down Group VII:

A *more* powerfully oxidising halogen will displace a *less* powerfully oxidising halogen from a solution of its salt.

Q

1 'Chlorine is a better oxidising agent than iodine.' Explain why this statement is true, and describe a reaction to illustrate your answer.

2 When chlorine gas is bubbled through solution X, the solution changes from colourless to orange. Some brown fumes are also observed. Suggest a name for solution X, and write an equation for the reaction.

Extracting bromine from sea water

Most of the world's supply of bromine is obtained from sea water. The displacement reaction between bromide ions and chlorine forms the basis of this extraction:

$$Cl_2(aq) + 2Br^-(aq) \rightarrow 2Cl^-(aq) + Br_2(aq)$$

The bromide ions are oxidised by chlorine to bromine.

Q

3 Show, using a half-equation, that bromide ions are oxidised to bromine in this extraction process.

Key Ideas 138 – 139

- **Down Group VII the ability of the halogens to act as oxidising agents decreases.**
- **The trend in oxidising ability can be demonstrated using displacement reactions.**

3 The halides are reducing

When the halogens are reduced they form halide ions. This transfer of electrons makes a considerable difference to their properties. The halogens exist as molecules which produce coloured, toxic vapours, whereas the halides exist naturally as colourless ions in rock salt and sea salt. The word 'halogen' means 'salt producer'.

Halide ions as reducing agents

The halide ions act as *reducing agents* because they lose electrons.

$$2Br^- \rightarrow Br_2 + 2e^-$$

The ability of a halide ion to act as a reducing agent *increases* down Group VII. As the size of halide ion increases, so the outer electrons are further from the influence of the nuclear charge, and the ions are able to lose an electron more readily.

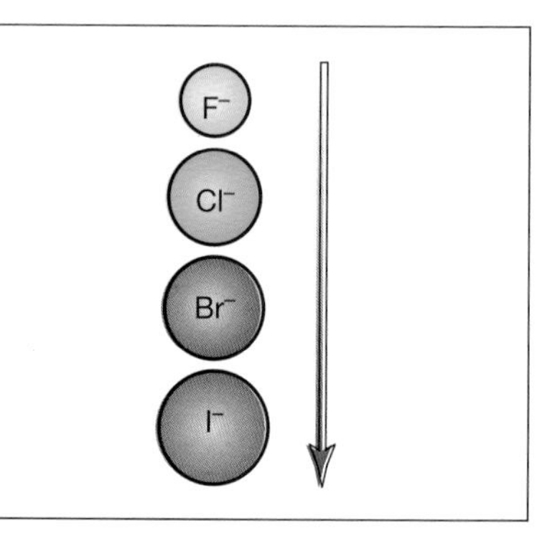

The bigger the halide ion, the stronger the reducing agent **Fig. 1**

Reactions between halide ions and concentrated sulphuric acid

The trend in reducing ability is shown in the reaction of solid halides with concentrated sulphuric acid. These reactions are usually studied using sodium or potassium halides. *Take care!* These reactions must be carried out in a fume cupboard.

The reaction occurs in two stages, displacement and redox.

Stage 1: Displacement

When concentrated sulphuric acid is added to a solid halide at room temperature, it will displace the hydrogen halide. For example:

$$NaCl(s) + H_2SO_4(aq) \rightarrow NaHSO_4(aq) + HCl(g)$$

The hydrogen halide fumes in moist air.

This reaction will occur with chlorides, bromides and iodides. With chlorides this is the *only* reaction that occurs; chloride ions do not reduce sulphuric acid.

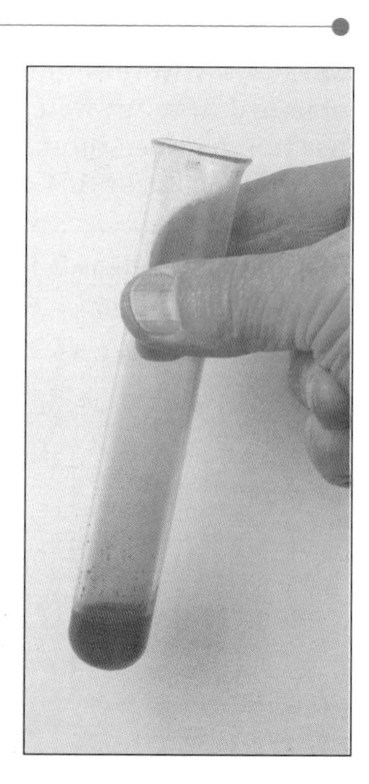

Brown fumes of bromine are produced when concentrated sulphuric acid is added to sodium bromide **Fig. 2**

Stage 2: Redox

With bromides and iodides a redox reaction then occurs between the concentrated sulphuric acid and the halide ion.

For *bromides* with concentrated sulphuric acid, the equation for the reaction is:

$$2HBr(aq) + H_2SO_4(aq) \rightarrow Br_2(l) + SO_2(g) + 2H_2O(l)$$

During this reaction, the sulphuric acid acts as an oxidising agent and the halide ion acts as a reducing agent.

> **Q**
> 1 Define *reducing agent* in terms of electrons.
> 2 Work out the change in oxidation state of the sulphur in the reaction above.

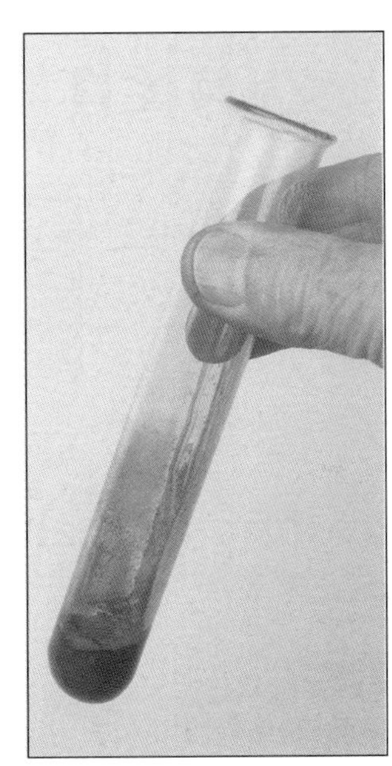

Fig. 3

Fumes of iodine are produced when sulphuric acid is added to sodium iodide

When *iodides* react with concentrated sulphuric acid, the iodide ions are oxidised to iodine and the sulphur in sulphuric acid is reduced, firstly to sulphur dioxide, then to sulphur, and finally to hydrogen sulphide. The overall equation for the reaction is:

$$8HI(aq) + H_2SO_4(aq) \rightarrow 4I_2(s) + H_2S(g) + 4H_2O(l)$$

Table 1 summarises all the observations made when the sodium halides react with concentrated sulphuric acid.

Table 1 *Reactions between sodium halides and sulphuric acid*

Halide	Reaction products	Observations
Sodium chloride	HCl	misty fumes
Sodium bromide	HBr	misty fumes
	Br_2	brown fumes
	SO_2	colourless gas with choking odour
Sodium iodide	HI	misty fumes
	I_2	purple fumes + black precipitate
	H_2S	colourless gas with a smell of bad eggs
	SO_2	colourless gas with choking odour
	S	yellow precipitate

> **Q**
> 3 Write half-equations to show the reduction of sulphuric acid to:
> a) sulphur dioxide b) sulphur c) hydrogen sulphide.
> 4 Use these half-equations to explain why iodide ions are stronger reducing agents than bromide ions.

Key Ideas 140 – 141

- **The halide ions are reducing agents.**
- **The reducing ability of the halides increases down Group VII.**
- **The trend in reducing ability can be demonstrated by reacting solid halides with concentrated sulphuric acid, and studying the oxidation states of the reduction products.**

4 Which halide?

From GCSE you will know that the silver halides are reduced to silver by the action of light. You will also know that these reactions are important in photography. Here we investigate the use of the silver salts to distinguish between the halide ions.

Testing for halide ions

The halide ions exist in most natural water supplies. They also occur as minerals in rocks – for example, fluoride ions are present in fluorite and chloride ions occur in rock salt, which is mainly sodium chloride.

How can we prove the presence of halide ions? The test is in two parts.

Fig. 1 *Fluorite is calcium fluoride*

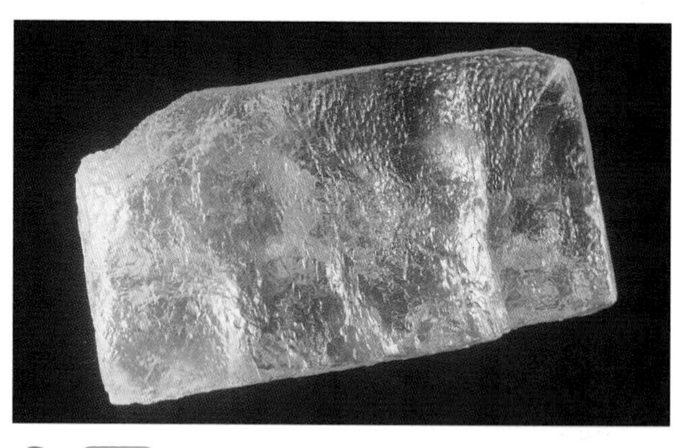

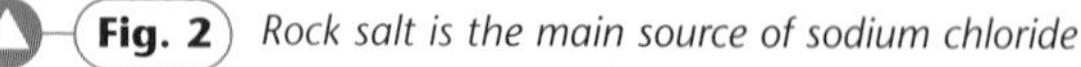

Fig. 2 *Rock salt is the main source of sodium chloride*

Precipitating the silver halide

Chloride ions are detected in solution by adding aqueous silver nitrate. The silver ions react with the chloride ions, producing a white precipitate of silver chloride:

$$Ag^+(aq) + Cl^-(aq) \rightarrow AgCl(s)$$

Similarly, bromide ions react with silver ions to produce a cream precipitate of silver bromide, and iodide ions produce a yellow precipitate of silver iodide:

$$Ag^+(aq) + Br^-(aq) \rightarrow AgBr(s)$$
$$Ag^+(aq) + I^-(aq) \rightarrow AgI(s)$$

Because other silver salts are also insoluble, dilute nitric acid is added prior to the silver nitrate solution to prevent the formation of unwanted precipitates.

Silver fluoride is soluble, so no precipitate is formed if fluoride ions are added to silver nitrate solution.

Solubility of the silver halide in ammonia

It is often difficult to distinguish the cream precipitate of silver bromide from the white precipitate of silver chloride or the yellow precipitate of silver iodide just by appearance.

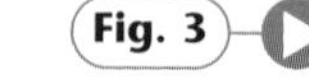

Fig. 3 *Aqueous silver nitrate reacts with the halide ions to form precipitates*

But the silver halides have different solubilities in ammonia solution, and this helps to further distinguish between them:

- silver chloride readily dissolves in dilute ammonia solution, forming a colourless solution
- silver bromide dissolves in concentrated ammonia solution, but is virtually insoluble in dilute ammonia
- silver iodide is insoluble in concentrated ammonia solution.

Q

1 A solution containing the anion X^- is reacted with silver nitrate solution. An off white precipitate forms, which is insoluble in dilute ammonia solution but dissolves readily in concentrated ammonia solution. Identify X^-.

2 What would you observe when a solution of sodium chloride is mixed with a solution of silver nitrate, followed by dilute aqueous ammonia?

	F^-(aq)	Cl^-(aq)	Br^-aq)	I^-(aq)
Addition of Ag^+(aq)	no reaction	white precipitate	cream precipitate	yellow precipitate
Addition of dilute NH_3(aq)	no reaction	soluble	insoluble	insoluble
Addition of concentrated NH_3(aq)	no reaction	soluble	soluble	insoluble

Table 1 *Testing for halide ions – a summary*

Black-and-white photography

Silver salts decompose on exposure to light. This is why the silver nitrate solutions you use in the laboratory are stored in brown bottles. You may have noticed that when you react silver nitrate solution with a solution of chloride ions, the white precipitate which forms rapidly darkens to a grey–violet colour. This is due to the formation of silver:

$$2AgCl(s) \rightarrow 2Ag(s) + Cl_2(g)$$

This is called a **photochemical decomposition reaction**. It is the presence of light that brings about the change. The change which occurs when silver bromide is exposed to light is similar:

$$2AgBr(s) \rightarrow 2Ag(s) + Br_2(g)$$

This reaction is used in black-and-white photography. The silver ions are reduced to silver metal, which remains as an opaque image on the photographic film.

Fig. 4 *Silver bromide is used in the production of photographic film*

Q

3 By stating reagents and observations, describe a chemical test which you could use to distinguish between aqueous solutions of HCl and HBr.

Key Idea 142 – 143

- **The halide ions can be distinguished from each other by using a solution of silver nitrate followed by ammonia.**

5 The killer halogen

At GCSE you will have studied how chlorine is manufactured by the electrolysis of brine. About 36 megatonnes of chlorine are produced annually and converted into plastics, solvents, insecticides and bleaches. Here we look at the use of chlorine in bleach, and study some of its reactions in more detail. Many of these are redox reactions.

Chlorine water

When chlorine dissolves in water it forms a mixture of hydrochloric acid and chloric(I) acid. There are two important aspects of this reaction to consider, equilibrium and redox.

An equilibrium is established

$$Cl_2(g) + H_2O(l) \rightleftharpoons HCl(aq) + HClO(aq)$$

or $$Cl_2(g) + H_2O(l) \rightleftharpoons 2H^+(aq) + Cl^-(aq) + ClO^-(aq)$$

This equilibrium mixture is often called **chlorine water**. It is pale green in colour, showing the presence of chlorine. Addition of alkali to this equilibrium will remove the hydrogen ions and make more water molecules:

$$H^+(aq) + OH^-(aq) \rightarrow H_2O(l)$$

The removal of hydrogen ions and the addition of water will both, by Le Chatelier's Principle, shift the equilibrium to the right, and the green colour will fade (see page 114). Alternatively the addition of acid will shift the equilibrium to the left, and the green colour will intensify.

The reaction is a redox reaction

The oxidation states of the chlorine are shown under the equation:

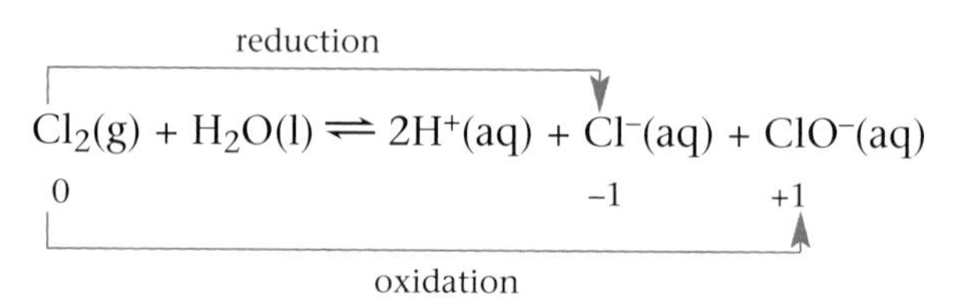

In this reaction chlorine has been both oxidised and reduced simultaneously. This is an example of a special type of redox reaction called **disproportionation**.

> **Q**
> 1 Why might it be dangerous to tip salt (sodium chloride) into chlorine water?
> 2 Write half-equations to show the conversion of chlorine into:
> a) chloride ions
> b) chlorate(I) ions.

Testing for chlorine

If chlorine water is tested with a piece of blue litmus paper, the paper first turns red and then white. The paper turns red because the solution is acidic. It turns white because chloric(I) acid is a bleach.

Water treatment

Chlorine and its compounds have been used for many years to sterilise drinking water and water in swimming baths. Since chlorine is toxic it can only be used in very small amounts. Usually chloric(I) acid or sodium chlorate(I) is added to the water in concentrations between 1.5 and $3.0\,mg\,dm^{-3}$. Even in this low concentration it kills bacteria, making the water safe.

The reaction between chlorine and alkali

We have already seen that chlorine in water produces an equilibrium mixture:

$$Cl_2(g) + H_2O(l) \rightleftharpoons 2H^+(aq) + Cl^-(aq) + ClO^-(aq)$$

If chlorine is reacted with an excess of cold, dilute sodium hydroxide solution, the equilibrium shifts so far to the right that the reaction goes to completion. The equation for the reaction can be written as:

$$Cl_2(g) + 2OH^-(aq) \rightarrow H_2O(l) + Cl^-(aq) + ClO^-(aq)$$

or $$Cl_2(g) + 2NaOH(aq) \rightarrow H_2O(l) + NaCl(aq) + NaClO(aq)$$

This reaction is used industrially to make bleach, which contains a mixture of sodium chloride and sodium chlorate(I).

Q

3 Sodium chlorate can have the formula NaClO or $NaClO_3$. Work out the oxidation states of the chlorine in these two compounds, and assign each a systematic name which can be used to distinguish between them.

The uses of chlorine

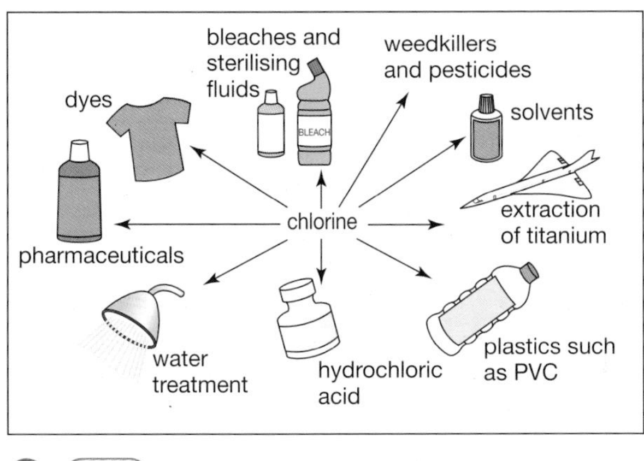

Fig. 1 *The various uses of chlorine*

Fig. 2 *The water in swimming baths must be sterilised to make it safe to swim in*

The various uses of chlorine are summarised in Figure 1.

Recent concerns about the environment have reduced the uses of some organic chloro-compounds, including solvents and chlorofluorocarbons (CFCs). CFCs had many uses, ranging from aerosol propellents to refrigerants, but they escape into the atmosphere and eventually decompose to chlorine atoms. These trigger a chain reaction which is causing the destruction of the ozone layer, thus exposing us to harmful UV light from the sun (see page 183).

Key Ideas 144 – 145

- **Chlorine dissolves in water or alkali to form a solution containing chloride and chlorate(I) ions.**
- **This solution can be used as a bleach or sterilising agent.**

6 Getting value for money

The cheapest bleach is not necessarily good value for money. The amount of chlorine in a known volume of bleach can be estimated by titration. This can be repeated for different bleaches to see which one contains the most chlorine.

What is bleach?

Bleach contains a mixture of sodium chloride and sodium chlorate(I). Sodium chlorate(I) is often still known by its trivial name of sodium hypochlorite.

When bleach is acidified, the following reaction occurs:

$$2H^+(aq) + Cl^-(aq) + ClO^-(aq) \rightleftharpoons Cl_2(g) + H_2O(l)$$

We have already considered this reaction. Addition of acid (containing H^+ ions) will, by Le Chatelier's Principle, shift the equilibrium to the right. This reaction explains why the label on the bottle of bleach contains the warning 'Do not use with other products'. The 'other product' might be a lavatory cleaner which contains acid. Mixing the bleach with the lavatory cleaner would release chlorine gas, and this could be fatal.

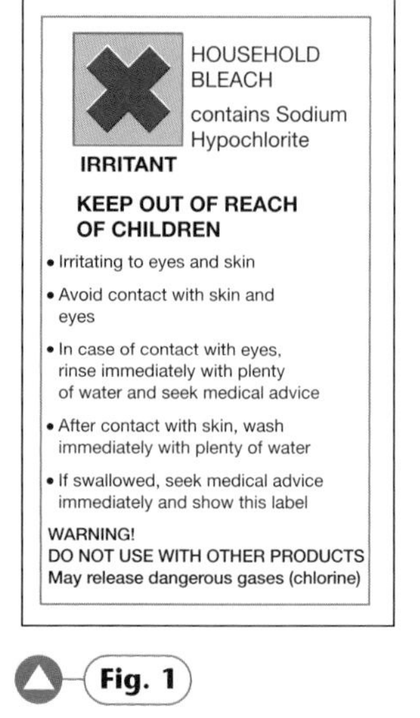

Fig. 1

The active ingredient in bleach is sodium hypochlorite

Estimating the amount of chlorate(I) in bleach

If an excess of potassium iodide is added to bleach after the addition of acid, a displacement reaction will occur:

$$Cl_2(aq) + 2I^-(aq) \rightarrow 2Cl^-(aq) + I_2(s)$$

The iodine produced will dissolve in solution. The amount of iodine produced will depend on the amount of chlorine released from the bleach. It can be seen from the equation that one mole of chlorine produces one mole of iodine.

The amount of iodine produced in solution can be determined by titration against a standard solution of sodium thiosulphate, which is added from a burette. The sodium thiosulphate reacts with iodine according to the following equation:

$$I_2(aq) + 2S_2O_3^{2-}(aq) \rightarrow 2I^-(aq) + S_4O_6^{2-}(aq)$$

The iodine can be detected as a yellow solution. At the end-point of the titration, when all the iodine has reacted, the solution will become colourless. This colour change is not very distinct, so the end-point of the titration is detected by adding starch solution. This gives a blue colour when it reacts with iodine, which disappears at the end-point. The starch is added near to the end-point, otherwise iodine becomes trapped in the starch molecules and the titration is less accurate.

By calculating the amount of thiosulphate used in the titration, we can determine the amount of iodine produced. From this we can determine how much chlorine was present in the original bleach. The report below gives the experimental details and shows you how to calculate the chlorine present in bleach.

Experiment to determine the amount of available chlorine in commercial bleach

Bleach is a mixture of sodium chloride and sodium chlorate(I). In acid solution chlorine is produced according to the equation:

$$2H^+(aq) + Cl^-(aq) + ClO^-(aq) \rightleftharpoons Cl_2(g) + H_2O(l)$$

Method

1. 10.0 cm^3 of bleach is measured out using a pipette with a pipette filler.
2. This is diluted with water by making up to 250 cm^3 in a standard flask.
3. 25 cm^3 of this diluted bleach is transferred into a conical flask using a pipette, and 1.5 g of potassium iodide and 10 cm^3 of dilute sulphuric acid are added.
4. The contents of the flask are titrated against 0.010 M sodium thiosulphate solution from the burette.
5. Near the end-point, 2 cm^3 of starch solution are added.
6. Once the blue coloration disappears, the volume of thiosulphate solution used are noted.
7. The titration is repeated until concordant results are achieved.

Results	Titre 1 (rough)	Titre 2	Titre 3
Final burette reading/cm^3	26.2	25.50	25.65
Initial burette reading/cm^3	0.0	0.00	0.05
Volume of thiosulphate used/cm^3	26.2	25.50	25.60

Average of the concordant results $= \dfrac{25.50 + 25.60}{2} = 25.55$

Calculation

The number of moles of thiosulphate used in the

$$\text{Titration} = \frac{\text{volume} \times \text{molarity}}{1000} = \frac{25.55 \times 0.010}{1000} = 2.55 \times 10^{-4}$$

Using the equation:

$$I_2(aq) + 2S_2O_3^{2-}(aq) \rightarrow 2I^-(aq) + S_4O_6^{2-}(aq)$$

2 mol of thiosulphate reacts with 1 mol of iodine

So the number of moles of iodine produced

$$= \frac{2.55 \times 10^{-4}}{2} = 1.275 \times 10^{-4}$$

But $Cl_2(aq) + 2I^-(aq) \rightarrow 2Cl^-(aq) + I_2(s)$
1 mol (of Cl_2) ... 1 mol (of I_2)

This means that 1.275×10^{-4} is also the number of moles of chlorine produced from 25 cm^3 of the diluted bleach.

In 250 cm^3 of diluted bleach there are 1.275×10^{-3} mol of chlorine.

10 cm^3 of bleach were diluted to 250 cm^3.

So the number of moles of chlorine in the original 10 cm^3 sample of bleach is 1.275×10^{-3}.

Concentration is measured in mol dm^{-3}. So in 1000 cm^3 of bleach there are $1.275 \times 10^{-3} \times 100$ mol of chlorine.

The concentration of chlorine in the bleach is 0.1275 mol dm^{-3}.

Q

1. Why is it dangerous to mix bleach with other cleaners?
2. Write the equations to show the reactions between:
   a) iodide ions and chlorine b) iodine and thiosulphate ions.
3. Describe an experiment to compare the effectiveness of a range of household bleaches.

Key Idea 146 – 147

- **The amount of chlorine in commercial bleaches can be estimated by titration, using potassium iodide and a standard solution of sodium thiosulphate.**

Unit 9 Questions

1. **a)** Describe how you would make bromine in the laboratory, given a supply of aqueous chlorine and a solution of sodium bromide. (1)
   b) What would you observe during this reaction? (1)
   c) Write an ionic equation to represent the reaction, and explain why this is a redox reaction. (3)

2. State and explain the trend in boiling points of the halogens down Group VII of the Periodic Table. (3)

3. **a)** Define the term *electronegativity*. (2)
   b) State and explain the trend in electronegativity down Group VII. (3)

4. When concentrated sulphuric acid is added to a solid sample of potassium bromide and the mixture warmed,, bromine and sulphur dioxide are produced.

   a) What is the oxidation state of sulphur in sulphuric acid? (1)
   b) What is the oxidation state of sulphur in sulphur dioxide? (1)
   c) Write a half-equation for the formation of sulphur dioxide from sulphuric acid. (2)
   d) Write a half-equation for the formation of bromine from bromide ions. (1)
   e) What is the role of the sulphuric acid in this conversion? (1)
   f) Use your answers to parts **c)** and **d)** to construct the overall equation for this redox reaction. (2)

5. Identify a halide ion (Cl^-, Br^-, I^-) which:

   a) forms a white precipitate with silver nitrate solution (1)
   b) forms a cream precipitate with silver nitrate solution (1)
   c) reduces concentrated sulphuric acid to sulphur (1)
   d) does not reduce concentrated sulphuric acid (1)
   e) is displaced from solution as the halogen when treated with aqueous bromine. (1)

6. Explain how ammonia solution can be used to distinguish between the precipitates formed when aqueous solutions of halide ions are reacted with acidified silver nitrate solution. (4)

7. How would you test for the presence of:

   a) chlorine?
   b) chloride ions?

   Name the chemicals you would use, and give the positive results for each test. (6)

(8) **a)** Write equations to show how chlorine reacts with:

(i) water

(ii) cold, dilute sodium hydroxide solution. (2)

b) (i) Name the compound produced in reaction **a)** (ii) which is of commercial importance.

(ii) What is this compound used for commercially? (2)

(9) **a)** When potassium iodide is added to a solution containing copper(II) ions, a redox reaction occurs. One of the products of this reaction is a white solid. Answer the following questions, given that the equation for the reaction is:

$$2Cu^{2+}(aq) + 4I^{-}(aq) \rightarrow 2CuI(s) + I_2(s)$$ *equation 1*

(i) Apart from the appearance of a white solid, what other observations would you make?

(ii) Deduce the name of the white solid.

(iii) What is the role of the iodide ions in this reaction? (5)

b) $50\,cm^3$ of 0.020 M copper(II) sulphate solution is reacted with an excess of potassium iodide, according to equation 1 shown above. The iodine that is liberated is titrated against 0.100 M sodium thiosulphate solution.

(i) Calculate the number of moles of copper(II) in $50\,cm^3$ of 0.020 M copper(II) sulphate solution.

(ii) How many moles of iodine are liberated from $50\,cm^3$ of 0.020 M copper(II) sulphate solution?

(iii) Iodine reacts with sodium thiosulphate according to the equation:

$$I_2(aq) + 2S_2O_3^{2-}(aq) \rightarrow 2I^{-}(aq) + S_4O_6^{2-}(aq)$$ *equation 2*

Calculate the volume of 0.100 M sodium thiosulphate used.

(iv) Name the indicator used in this titration. (5)

(10) An equilibrium is established when chlorine gas is dissolved in water. The resulting solution is pale green. It is commonly called 'chlorine water'.

a) (i) Write an equation for this equilibrium reaction.

(ii) How would you prove the presence of chlorine in the chlorine water? (4)

b) (i) What would you observe when dilute sodium hydroxide solution was added, in excess, to the chlorine water?

(ii) Use Le Chatelier's Principle to explain this observation.

(iii) Write an equation for the reaction. (4)

c) (i) What would you observe when dilute hydrochloric acid was added, in excess, to the chlorine water?

(ii) Use Le Chatelier's Principle to explain this observation.

(iii) Write an equation for the reaction. (4)

1 Choosing the extraction method ...

You will have studied the extraction of some metals at GCSE. There is an ever-increasing demand for metals for the construction, transport and manufacturing industries. Metals are found in the Earth's crust as ores. Ores contain metal compounds, often in the form of the metal oxides. This unit studies the extraction processes involved in the production of some metals.

Occurrence of metals

There is a great variation in the abundance of elements in the Earth's crust. Table 1 provides the percentage composition by mass of some elements in the Earth's crust.

Element	% by mass	Element	% by mass
Oxygen	49.5	Magnesium	1.9
Silicon	25.7	Hydrogen	0.88
Aluminium	7.5	Titanium	0.58
Iron	4.7	Chlorine	0.19
Calcium	3.4	Phosphorus	0.12
Sodium	2.6	Carbon	0.09
Potassium	2.4	Manganese	0.08

Table 1 *Abundance of elements in the Earth's crust*

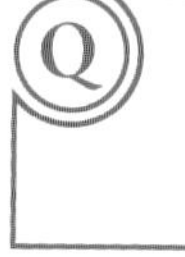

1 Which are the two most abundant metals in the Earth's crust?

The remaining elements have a combined percentage abundance of 0.36%. Some of the rarest metals are gold, silver and copper. Fortunately they are found in relatively concentrated forms and in only a few places, and this reduces the costs of extraction.

Extraction methods for metals

The extraction of a metal from its compound is a *reduction* process. Most metals are extracted by the reduction of their oxides. The method of extraction used depends upon several factors:

- the required purity of the metal
- the cost of the reducing agent used (the **reductant**)
- the energy requirements.

The main consideration is **cost**, but if a metal is required to be of very high **purity** then a more expensive process will be used. One example is the extraction of titanium (Ti). This could be produced cheaply by reduction with carbon (an abundant and low-cost reductant), but this results in the formation of titanium carbide. Tungsten (W) also forms a carbide when the metal is extracted from its oxide by carbon reduction.

Methods used to extract metals include:

- carbon reduction (for example iron)
- electrolysis (for example aluminium)
- displacement by a more reactive metal (for example titanium).

These extraction methods are described in detail in the following pages.

Environmental factors and recycling

Extraction of metals involves the consumption of large amounts of **energy**, and often produces a large range of gaseous by-products that can cause **pollution** problems. Many ores contain sulphides, which in the extraction process form oxides of sulphur. If these sulphur oxides escape into the atmosphere they react with the moisture in the air to form **acid rain**, which causes damage to the environment (see page 190).

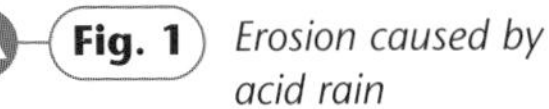

Fig. 1 *Erosion caused by acid rain*

The use of carbon as a reductant results in the formation of carbon monoxide and carbon dioxide. Carbon monoxide is a **toxic** gas; carbon dioxide is a **'greenhouse' gas** and is thought to increase **global warming**.

Some extraction processes use potentially dangerous chemicals that could cause health problems if they were released into the environment. For example, the extraction of titanium uses large quantities of chlorine.

Many metals are now recycled. **Recycling** helps to reduce the exploitation of finite mineral resources. Recycling also helps to reduce energy demands. The two most widely used metals are iron and aluminium, and both metals are now extensively recycled.

- High-quality 'scrap iron' is used in the manufacture of steel. The scrap iron is mixed with the iron from the blast furnace and is converted into steel.
- The extraction of aluminium by electrolysis uses a great deal of energy, and it makes sense to recycle used aluminium objects. Nearly every town has an aluminium-can collection point. The crushed cans are melted down and the aluminium re-used.

Fig. 2 *Aluminium cans may be recycled*

Q

2 Name three gaseous pollutants produced during metal extraction.

Key Ideas 150 – 151

- **Metals are found in ores in the Earth's crust.**
- **The extraction of a metal from its naturally occurring compounds is a reduction process.**
- **The method used to extract a metal depends upon several factors, including the required purity of the metal, the cost of the reductant used, and the energy considerations.**
- **The extraction of metals consumes a lot of energy and can cause pollution problems.**

2 A very useful metal ...

Iron is the most used metal in modern society. It is a strong metal with a wide range of uses both as the metal and as steel. Iron is abundant in the Earth's crust, and its method of extraction is relatively cheap.

Extraction of iron

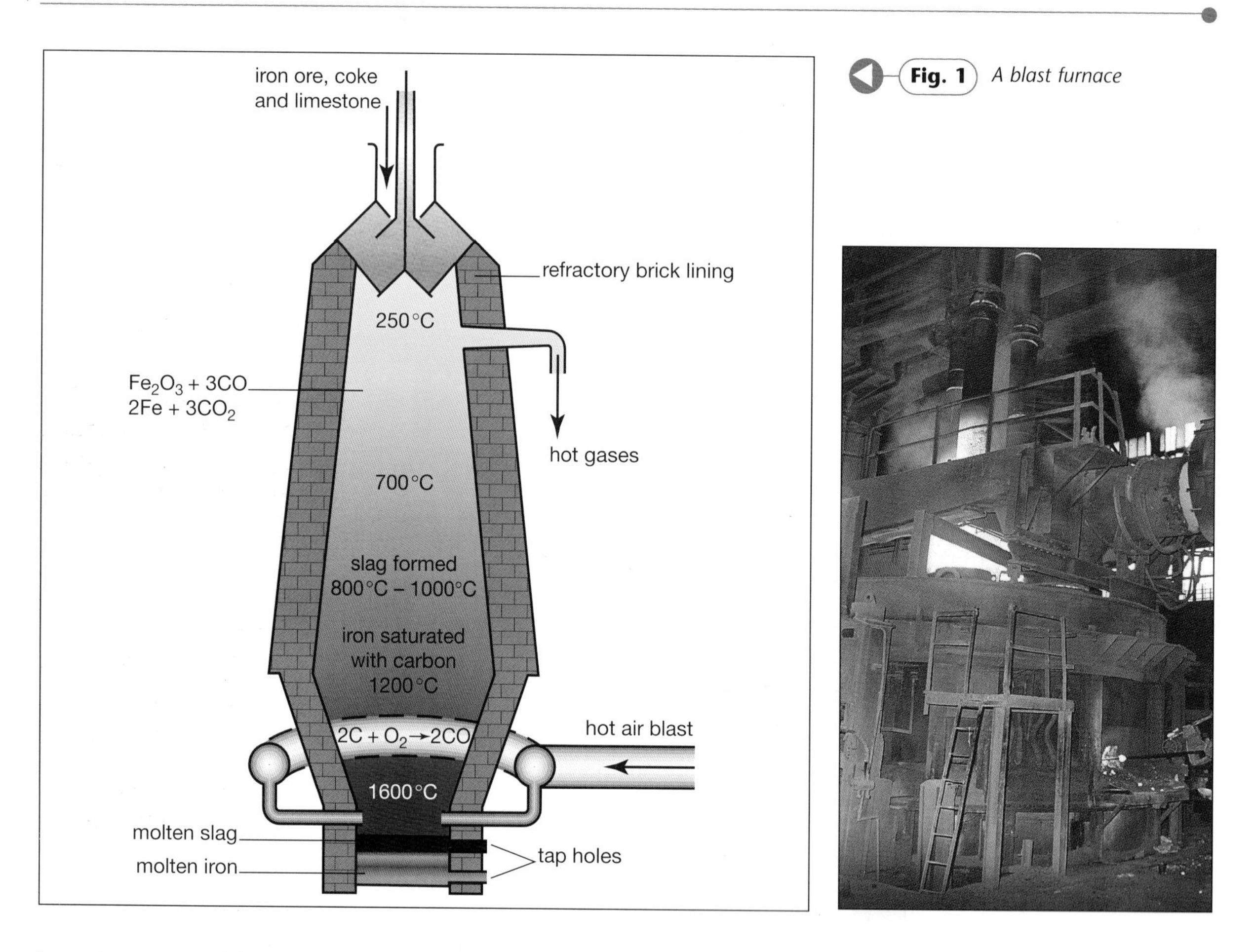

Fig. 1 *A blast furnace*

Iron is extracted by carbon reduction in a **blast furnace**. The raw materials are iron ore (**haematite**), coke, limestone and air. The coke burns in the oxygen in the hot air that is blasted into the furnace. The energy released from the combustion of the coke heats up the furnace, allowing the formation of carbon monoxide in an endothermic reaction between carbon and carbon dioxide:

$$C + O_2 \rightarrow CO_2$$
$$C + CO_2 \rightarrow 2CO$$

At the high temperatures in the blast furnace, the carbon and carbon monoxide reduce the iron(III) oxide in the iron ore to iron:

$$Fe_2O_3 + 3CO \rightarrow 2Fe + 3CO_2$$
$$Fe_2O_3 + 3C \rightarrow 2Fe + 3CO$$

At the high temperatures the molten iron formed flows to the bottom of the furnace, where it is tapped off. The impure iron formed in the blast furnace is called **pig iron**. Some of this pig iron is used to make cast iron materials such as engine blocks.

The limestone is added to the blast furnace to remove acidic oxide impurities, including silicon dioxide (silica). The limestone decomposes in the blast furnace, forming a basic oxide, calcium oxide. The calcium oxide (lime) reacts with acidic oxide impurities including silica, SiO_2:

$$CaCO_3 \rightarrow \underset{\text{basic}}{CaO} + CO_2$$

$$CaO + SiO_2 \rightarrow \underset{\text{'slag'}}{CaSiO_3}$$

The calcium silicate ($CaSiO_3$) slag forms a layer on top of the molten iron and is removed from the furnace. It is used in the construction industry, mainly in road building.

The blast furnace is a **continuous process**: more raw materials are added to the top of the furnace as the products are removed from the bottom of the furnace.

Q

1 Name the raw materials used in the blast furnace.

2 Write equations for the reduction of Fe_3O_4 by carbon and by carbon monoxide.

Steel

Most of the iron from the blast furnace is used to make steel. The iron obtained from the blast furnace contains many impurities including carbon, sulphur and phosphorus: these are removed in a **Basic Oxygen Converter**.

1 First, some magnesium metal is added to molten iron to remove sulphur.
2 Secondly, the molten impure iron is mixed with high-quality scrap iron and pure oxygen gas is forced into the molten mixture. The carbon and any remaining sulphur burn to form gaseous oxides, which escape into the atmosphere.
3 Lime, CaO, is then added to react with any solid acidic oxides, including silica and phosphorus(V) oxide, forming a slag which is removed from the surface.
4 The pure iron can then be converted into a range of steels by the addition of small quantities of carbon or other metals.

The blast furnace can cause pollution problems, because it produces carbon monoxide and carbon dioxide from the coke, and oxides of sulphur from the use of sulphide ores.

Theoretically, any metal can be extracted using carbon reduction, but often the temperatures required to do so are too high. In many cases reduction using carbon also leads to the formation of carbides.

Q

3 Explain, using equations, why limestone is used in the blast furnace.

4 Describe briefly how steel is made from the iron formed in a blast furnace.

5 The reactions in a blast furnace produce a mixture of hot gases. Suggest one use for these hot gases.

Key Ideas 152 – 153

- **Iron is extracted by carbon reduction in a blast furnace.**
- **Limestone is added to remove impurities from the iron ore.**
- **Iron obtained from the blast furnace contains many impurities, including carbon.**
- **Steels are alloys of iron. They are made in a Basic Oxygen Converter.**

3 Other useful metals

Aluminium and titanium are very useful metals and are extracted from their ores by different methods. Aluminium is extracted by the electrolysis of aluminium oxide; titanium is extracted by reaction with a more reactive metal.

Extraction of aluminium

Aluminium is extracted by the electrolysis of purified aluminium oxide (**alumina**). The main ore of aluminium is **bauxite**, which contains aluminium oxide, Al_2O_3. The ore is crushed and sodium hydroxide is used to extract the aluminium oxide from the ore.

The aluminium oxide is dissolved in molten cryolite in an electrolysis cell that has carbon electrodes. The melting point of aluminium oxide is above 2000 °C, and the energy costs are reduced by dissolving the aluminium oxide in molten cryolite at 1000 °C.

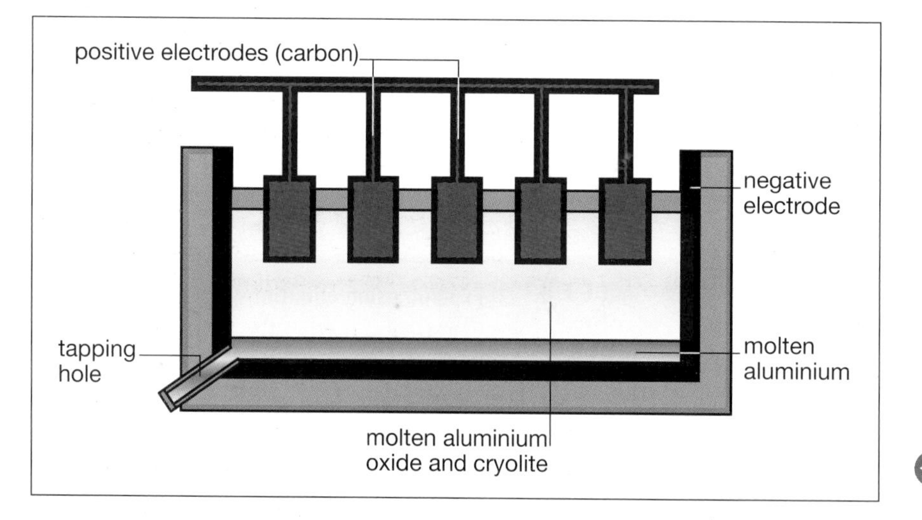

Fig. 1 *Aluminium cell*

The overall equation for the reaction in the electrolysis cell is:

$$2Al_2O_3 \rightarrow 4Al + 3O_2$$

The aluminium forms at the negative electrode:

$$Al^{3+} + 3e^- \rightarrow Al$$

The oxide ions are discharged at the positive electrode:

$$2O^{2-} \rightarrow O_2 + 4e^-$$

The balanced electrode reactions are:

$$4Al^{3+} + 12e^- \rightarrow 4Al$$
$$6O^{2-} \rightarrow 3O_2 + 12e^-$$

The positive electrodes are made of carbon, and at the high temperatures used in the electrolysis cell the carbon electrodes will burn in the oxygen formed at the electrode:

$$2C + O_2 \rightarrow 2CO$$

and $$C + O_2 \rightarrow CO_2$$

The carbon monoxide and carbon dioxide are removed using fume hoods.

Because the positive electrodes burn away they have to be replaced from time to time, and this increases the cost of the extraction of aluminium.

The extraction of aluminium by electrolysis is a **continuous process**. The molten aluminium formed in the electrolysis cell is tapped off, and more aluminium oxide is added to the cell.

The extraction of aluminium is expensive because of the electricity required for the electrolysis and for heating the electrolysis cell. Therefore most aluminium smelters are built where there is a cheap source of electricity (usually hydroelectricity).

Q

1. Why is cryolite used in the extraction of aluminium?
2. Give two reasons why the extraction of aluminium is an expensive process.
3. Write equations for the electrode reactions that occur in the electrolysis of aluminium oxide.

Extraction of titanium

Titanium is extracted by reduction by a more reactive metal, either sodium or magnesium. Titanium is found as titanium oxide in the ores **rutile** (TiO_2) and **ilmenite** ($FeTiO_3$).

Titanium oxide is converted into titanium(IV) chloride using chlorine and carbon at 900 °C:

$$TiO_2 + C + 2Cl_2 \rightarrow TiCl_4 + CO_2$$

Titanium(IV) chloride is covalent and is removed by distillation. The titanium(IV) chloride is then reduced by sodium or magnesium in an atmosphere of argon gas at an initial temperature of 600 °C.

$$TiCl_4 + 4Na \rightarrow Ti + 4NaCl$$

or $$TiCl_4 + 2Mg \rightarrow Ti + 2MgCl_2$$

The argon atmosphere is used to prevent any air from reacting with the titanium. At the high temperatures used in the process, the titanium metal formed would react with the oxygen and nitrogen in the air.

The extraction of titanium by metal reduction by sodium or magnesium is a very expensive **batch process**. As a result, despite its relatively high abundance titanium is a very expensive metal and is used only where its useful properties outweigh the high cost of production.

Titanium can be extracted by carbon reduction but this results in the formation of titanium carbide, which makes the titanium brittle.

Titanium is a very strong and corrosion-resistant metal. If a cheaper method of extraction could be developed then it would replace iron and steel in most uses.

Q

4. Write equations to show how titanium is extracted from titanium oxide.
5. Explain why the extraction of titanium is an expensive process.

Key Ideas 154 – 155

- **Aluminium is extracted by the electrolysis of molten aluminium oxide.**
- **Titanium is extracted from titanium oxide by reduction, using a more reactive metal.**

Unit 10 Questions

4

1. Figure 1 shows a typical aluminium electrolysis cell.

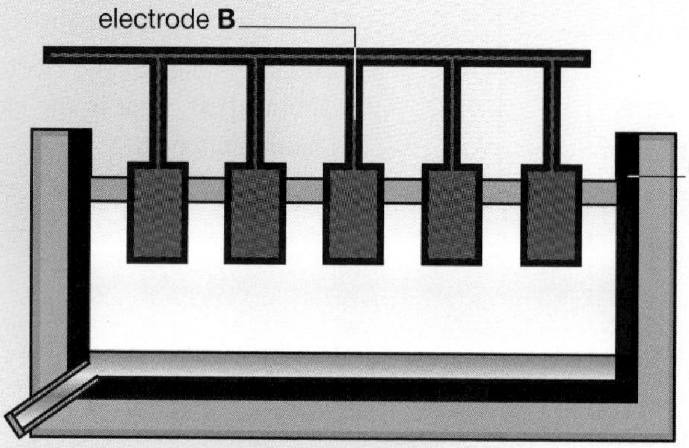

Fig. 1

a) Name the chief ore from which aluminium is extracted. (1)
b) Name the aluminium compound found in this ore. (1)
c) Give the charge on electrode **A**. (1)
d) Write the equations for the reactions occurring at:
(i) the positive electrode
(ii) the negative electrode. (4)
e) Explain why the positive electrodes have to be replaced at regular intervals. (2)
f) Explain why aluminium is an expensive metal to extract. (2)

2. Iron is extracted in a blast furnace.

a) Name the raw materials used in the blast furnace. (4)
b) Write equations for the reactions in which iron is formed in the blast furnace. (2)
c) Write equations to show the reactions that occur when limestone is used to remove the main impurity in the iron ore. (2)
d) Name the chief impurity in the iron formed in the blast furnace. (1)

3. The iron obtained from the blast furnace contains many impurities. Outline the essential chemistry involved in producing steel in a Basic Oxygen Converter. (7)

4. **a)** Name the main ore of titanium. (1)
b) Name the titanium compound obtained from this ore. (1)
c) Write an equation to show the conversion of this titanium compound into titanium(IV) chloride. (2)
d) (i) Write an equation for the reduction of titanium(IV) chloride into titanium metal.
(ii) Give the essential conditions for this reaction. (4)

e) Give *two* properties of titanium that make it a very useful metal. (2)

f) Explain why the extraction of titanium is an expensive process. (3)

g) Titanium can be extracted cheaply by carbon reduction. Explain why this method is not used. (2)

5) Explain the difference between a *batch process* and a *continuous process*. Use the extraction of titanium and aluminium to illustrate your answer. (6)

6) Many metal extraction processes produce unwanted by-products that could cause pollution problems. Name the gaseous pollutants produced in the extraction of the following metals, and in each case explain the environmental consequences of each pollutant that you name.

a) Iron/steel. (6)

b) Aluminium. (2)

c) Titanium. (4)

7) For each of the following metal extractions:

- State the method of reduction. (1 each)
- Name the metal ore used in the extraction. (1 each)
- State whether the reaction is a batch process or a continuous process. (1 each)
- State the purity of the metal obtained. (1 each)

a) Iron.

b) Aluminium.

c) Titanium.

8) a) The two most abundant metals in the Earth's crust are iron and aluminium. For each of these two metals, give *two* reasons, other than their abundance, why the metals are very widely used. (4)

b) The recycling of these two metals is becoming increasingly important. Give *two* reasons why recycling these metals is desirable. (2)

Module 2 Questions

1 a) Define the *standard enthalpy of formation* of a compound. (2)

b) The standard enthalpy of formation of benzene (C_6H_6) is +49.0 kJ mol^{-1}, and the conversion of benzene into cyclohexane (C_6H_{12}) is given by the equation:

$$C_6H_6(l) + 3H_2(g) \rightarrow C_6H_{12}(l) \quad \Delta H^{\ominus} = -208.1\,\text{kJ mol}^{-1}$$

Use these data to calculate the standard enthalpy of formation of cyclohexane. (3)

c) Define the *standard enthalpy of combustion* of a compound. (2)

d) Ethanol (C_2H_6O) burns in air and can be used as a fuel. The apparatus shown in the diagram can be used to obtain an approximate value for the enthalpy of combustion of ethanol:

(i) Write an equation to show the complete combustion of ethanol.

(ii) State all measurements which must be recorded in order to determine the enthalpy of combustion of ethanol. What is the major source of error in this experiment? (7)

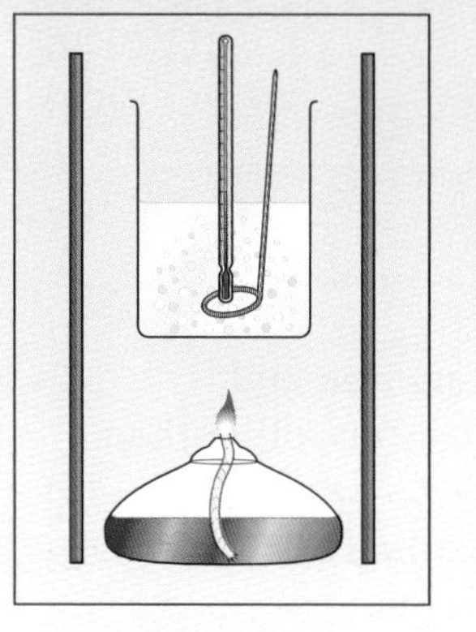

Fig. 1

2 'Reactions can only occur when collisions take place between particles having sufficient energy.'

a) What is meant by this statement? (4)

b) Use your answer to **a)** to help you explain the following observations:

(i) Powdered calcium carbonate reacts faster with 5 M hydrochloric acid than it does with 1 M hydrochloric acid.

(ii) Powdered calcium carbonate reacts faster with 1 M hydrochloric acid than if the calcium carbonate were in lumps.

(iii) Powdered copper(II) oxide reacts faster with 1 M hydrochloric acid if the solution is heated. (6)

3 Hydrogen peroxide decomposes according to the equation:

$$2H_2O_2(aq) \rightarrow 2H_2O(l) + O_2(g)$$

The reaction is very slow in the absence of a catalyst. The catalyst used in this reaction is manganese(IV) oxide.

In an experiment, carried out at room temperature, 20 cm^3 of 0.20 M hydrogen peroxide decomposed in the presence of 0.2 g of granulated manganese(IV) oxide to produce oxygen. The results of the experiment are shown below:

Oxygen produced (mol)	Time (s)
0.00020	8
0.00050	20
0.00094	40
0.00154	80
0.00184	120
0.00196	160
0.00196	200

a) By drawing a diagram of the apparatus, outline a suitable method for carrying out this experiment. (3)

b) Plot a graph of the results and use this to find the total number of moles of oxygen produced. (4)

c) Calculate the theoretical amount of oxygen that could be made from 20 cm^3 of 0.20 M hydrogen peroxide. Why would your answer to part **b)** be less than this? (3)

d) Sketch on your graph the curve you would expect to obtain if the reaction were repeated using 20 cm^3 of 0.20 M hydrogen peroxide with 0.2 g of powdered manganese(IV) oxide. Explain any differences between this sketch and the graph you plotted from the experimental results. (4)

4 The following reaction is very slow at room temperature:

$$CH_3COOCH_3(aq) + H_2O(aq) \rightleftharpoons CH_3COOH(aq) + CH_3OH(aq)$$

a) With reference to a sketch of the distribution of molecular energies (Maxwell-Boltzmann curves), explain why:

(i) the reaction speeds up when the reaction mixture is heated

(ii) the reaction speeds up when a catalyst is present

(iii) a small increase in temperature has a much greater effect than a small increase in concentration. (12)

b) Use Le Chatelier's Principle to explain what would happen to the equilibrium position for this reaction if more water were added. (3)

c) What would be the effect of a catalyst on the equilibrium position? Explain your answer. (3)

5 Sulphuric acid is manufactured from sulphur trioxide by the Contact Process. The equation for one of the stages in this process is shown below:

$$2SO_2(g) + O_2(g) \rightleftharpoons 2SO_3(g) \quad \Delta H = -189\,kJ\,mol^{-1}$$

When heated in air in the presence of finely divided vanadium(V) oxide, a homogeneous, dynamic equilibrium is established between sulphur dioxide, oxygen and sulphur trioxide.

a) Explain the terms *homogeneous*, *dynamic* and *equilibrium*. (4)

b) This reaction is carried out at about 450°C. Explain this choice of temperature. (5)

c) The vanadium(V) oxide behaves as a catalyst in this process.

(i) Explain how a catalyst works.

(ii) Why is the catalyst 'finely divided'?

(iii) How would the catalyst affect the rate of production and the yield of sulphur trioxide? (7)

Module 2 Questions

(6) **a)** Oxidation was once considered to be the *gain of oxygen*. Explain how the concept of oxidation as electron transfer is related to this earlier definition. Use examples to illustrate your answer. (3)

b) The halogens can act as oxidising agents. Their ability to act as oxidising agents decreases down Group VII.

(i) Explain this trend.

(ii) You are provided with aqueous solutions of the halogens chlorine, bromine and iodine, plus aqueous solutions containing the corresponding halide ions, chloride, bromide and iodide. Describe a series of simple test tube reactions which would confirm this trend. You should include your observations and write equations for the reactions that occur. (13)

c) Chlorine is bubbled through a cold, dilute solution of sodium hydroxide. Write an equation for this reaction, and explain why this reaction is described as a redox reaction. (5)

(7) **a)** How can the addition of silver nitrate solution be used to distinguish between aqueous samples of potassium chloride and potassium iodide? Your answer should include the results of the tests and equations for the reactions described. (4)

b) How can the addition of concentrated sulphuric acid be used to distinguish between solid samples of potassium chloride and potassium iodide? Your answer should include the results of the tests and equations for the reactions described. (6)

c) 0.600 g of table salt (impure sodium chloride) were dissolved in dilute nitric acid. When an excess of dilute aqueous silver nitrate was added to the resulting solution, a white precipitate of silver chloride was formed. The precipitate was separated by filtration, washed with distilled water, dried and weighed. The mass of silver chloride formed was 1.435 g. Calculate the percentage by mass of chloride ion in the table salt. (5)

(8) This question is about the reactions between solid halides and concentrated sulphuric acid.

a) What observations would you make during these reactions which would help you to distinguish between chloride, bromide and iodide ions? (5)

b) Explain, by using equations, which of these reactions are redox reactions. (6)

c) Astatine (At) exists as At_2 and is a black solid. Use your understanding of the redox reactions of Group VII to suggest the likely products formed, and write an equation for the reaction that occurs when concentrated sulphuric acid is added to NaAt(s). (3)

9 a) The halide ions can act as reducing agents. The ability to act as reducing agents increases down Group VII of the Periodic Table.
(i) Explain this trend.
(ii) How do the various reduction products formed when the hydrogen halides react with concentrated sulphuric acid support the theory that hydrogen iodide is a stronger reducing agent than hydrogen bromide? (9)

b) Define the term *electronegativity*. State and explain the trend in electronegativity of the halogens down Group VII. (5)

10 What type of reaction is involved in the extraction of metals from their purified ores? Illustrate your answer by reference to the metals named below. Name any necessary reagents and conditions for the extraction. Include equations to show the formation of the metal.

a) Iron from Fe_2O_3. (5)
b) Aluminium from Al_2O_3. (5)
c) Titanium from $TiCl_4$. (5)

11 a) Discuss the factors which determine the methods used to extract metals from their oxides on an industrial scale, giving examples of *two* different methods of reduction other than by carbon. (10)

b) Iron and titanium both react with carbon to form carbides. Explain why iron can be extracted from its oxide using carbon but titanium cannot. What pollution problems can arise in the manufacture of these two metals? (10)

c) An equation to represent the industrial extraction of iron is shown below:

$$Fe_2O_3(s) + 3CO(g) \rightarrow 2Fe(l) + 3CO_2(g)$$

(i) Explain the function of carbon monoxide in this reaction.
(ii) What mass of iron would be obtained from 32 kg of the oxide? (5)

12 $10.0\,cm^3$ of bleach were made up to $250\,cm^3$ in a volumetric flask. $25.0\,cm^3$ of this diluted bleach were treated with an excess of potassium iodide and acidified with sulphuric acid. The iodine produced was then titrated against 0.100 M sodium thiosulphate solution; $15.4\,cm^3$ were required.
The equations for the reactions are:

$$ClO^-(aq) + 2I^-(aq) + 2H^+(aq) \rightarrow I_2(aq) + Cl^-(aq) + H_2O(l)$$

$$I_2(aq) + 2S_2O_3^{2-}(aq) \rightarrow 2I^-(aq) + S_4O_6^{2-}(aq)$$

a) Explain why it is necessary to use excess potassium iodide, and describe how the end-point of the titration is determined. (4)
b) Calculate the mass of sodium chlorate(I) in $1.0\,dm^3$ of the bleach. (8)

Introduction to Module 3

Organic chemistry is the study of the compounds of carbon. There is such a vast number and variety of these compounds that a whole branch of chemistry is allocated to their study. Ranging from drugs to medicines to plastics, our modern world is totally dependent on organic compounds. In fact the cells in our bodies are miniature organic chemistry laboratories!

Fig. 1 *Many fuels are organic compounds*

Originally the word 'organic' was used because it was thought that these compounds could only be made by living organisms. During the eighteenth century, organic compounds such as drugs, dyes and perfumes were extracted from plant and animal tissues. One of the first scientists to study organic compounds was a Swedish chemist called Carl Scheele. He isolated lactic acid from milk, and showed that this acid was the cause of milk turning sour. In 1828 the first organic compound, urea, was synthesised in the laboratory by Friedrich Wohler. He made urea by heating a solution of ammonium cyanate, an inorganic compound:

$$NH_4CNO \rightarrow (NH_2)_2CO$$

Millions of organic compounds have since been synthesised, with thousands of new ones being produced annually. Nowadays most organic compounds are obtained directly or indirectly from petroleum.

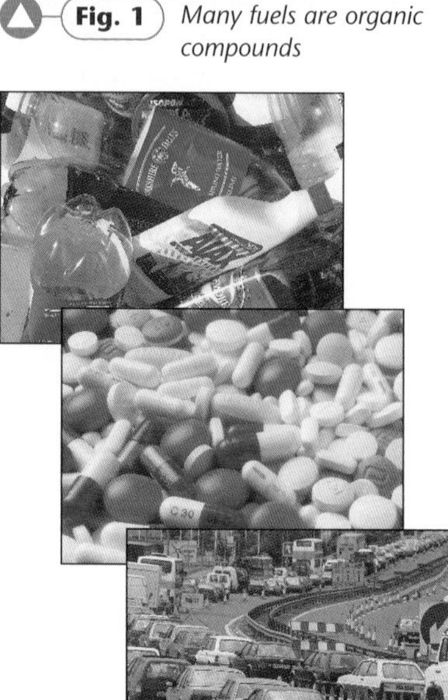

Fig. 2 *Products from the petrochemical industry*

The distinction between organic and inorganic chemistry is now arbitrary, but organic chemistry is still considered as the chemistry of carbon compounds other than its oxides and carbonates.

This module revises some of the ideas you will have met already at GCSE, and develops them further for AS level. Many of the properties of organic compounds depend on a particular atom, or group of atoms, within the molecule: these are called the functional groups. Indeed, organic chemistry can be considered to be the study of functional groups. This module looks at:

- some of the specific reactions of molecules with the same functional group, and the conditions needed for these reactions to occur
- some of the mechanisms of the reactions – that is, how the reactions proceed
- how these specific reactions link together to form synthetic routes which are particularly important in industry
- the uses of organic compounds in everyday life.

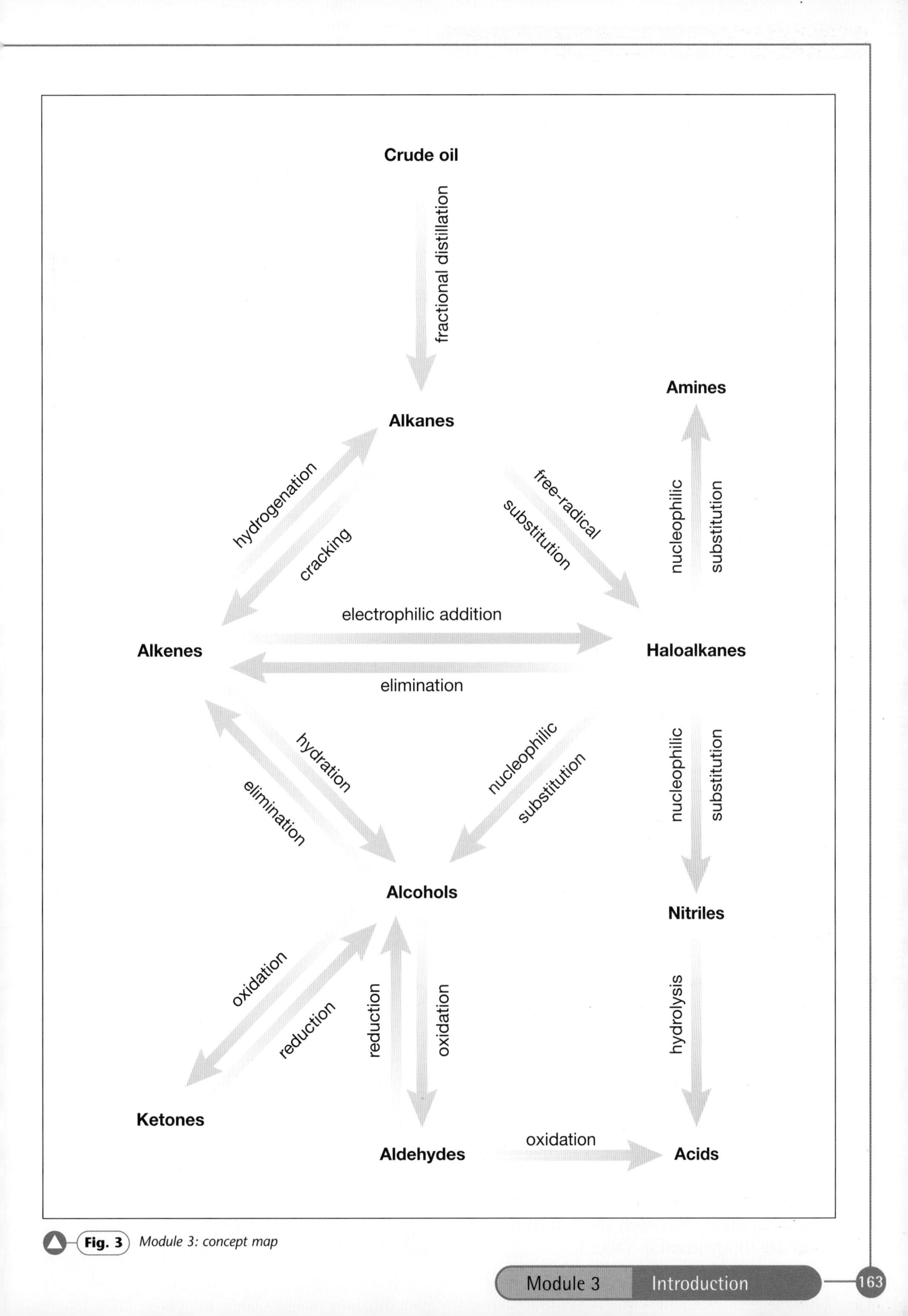

Fig. 3 *Module 3: concept map*

1 Organic chemistry is everywhere

There are a number of key words and basic ideas used in organic chemistry that need to be revised from GCSE or introduced for AS level before we can start to look at the details of organic reactions.

Introducing organic molecules

Organic compounds may be as simple as a molecule of methane or as complex as a dye molecule. Despite the fact that these molecules look very different, there are some common features. What do organic compounds have in common?

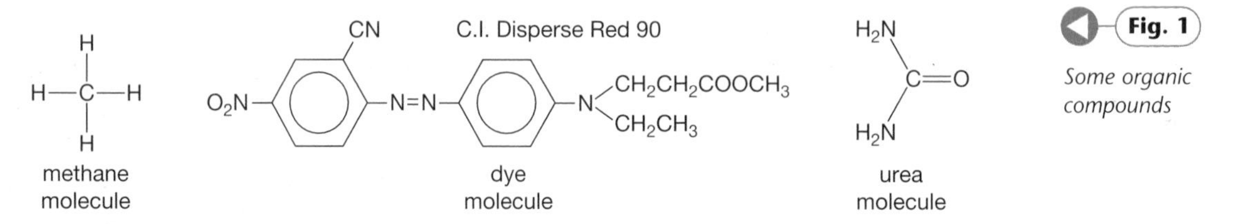

Fig. 1 *Some organic compounds*

Organic molecules contain carbon and hydrogen

You will know from GCSE that molecules which contain *only* the elements carbon and hydrogen are called **hydrocarbons**. Methane is an example of a hydrocarbon. Over eight million hydrocarbons exist. This is more than all the known compounds involving the rest of the elements put together!

Other elements may be present in addition to carbon and hydrogen. These include oxygen, nitrogen and the halogens. Figure 1 **b** and **c** illustrate two organic molecules that contain nitrogen and oxygen.

Bonding in organic molecules is predominantly covalent

Carbon forms covalent bonds with hydrogen, nitrogen, oxygen and the halogens. The bonds formed between carbon and nitrogen, oxygen and the halogens are significantly **polar** (see page 29). The polarity of the bonds is important in dictating the chemistry of organic compounds, as you will see in subsequent sections.

Notice from the examples given in Figure 1 the number of bonds that each element forms. These are summarised in Table 1.

Carbon can bond to itself or to other elements by single or double bonds. Molecules containing only single bonds are called **saturated molecules**. Molecules containing at least one double bond are called **unsaturated molecules**.

At GCSE you will have studied saturated hydrocarbons (**alkanes**) and unsaturated hydrocarbons (**alkenes**).

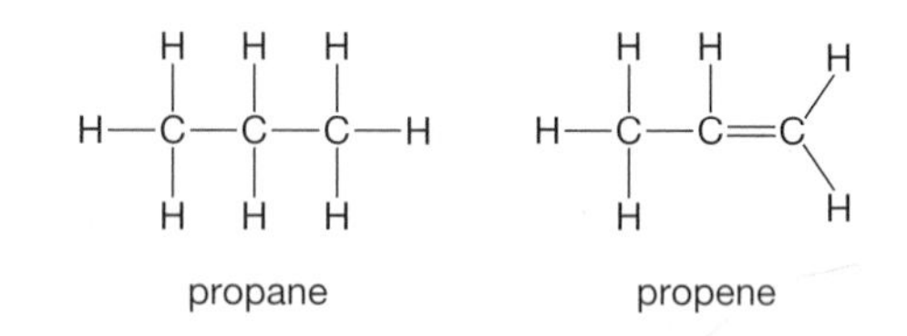

Fig. 2 *All the bonds in saturated molecules are single bonds; unsaturated molecules may contain one or more double bonds*

Element	Number of covalent bonds
C	4
H	1
O	2
N	3
Cl, Br, I	1

Table 1 *Numbers of bonds formed*

1 Draw dot-cross diagrams to show the arrangement of outer electrons of the carbon and hydrogen atoms in one molecule of:

a) methane b) propane c) propene.

2 Which bond will be most polar: C–H, C–Cl or C–Br?

Carbon atoms may be arranged in a chain or in a ring

The structures of propane and propene show the carbon atoms arranged in a chain. These are called **aliphatic molecules**.

The dye structure in Figure 1 shows carbon atoms arranged in a ring. There are many different ring structures, which are generally referred to as **cyclic structures**, but the one shown here is a special case. It is called an **aromatic molecule**. Aromatic molecules contain at least one benzene ring. The word 'aromatic' was first used because these compounds have a distinctive smell (aroma). The chemistry of aromatic compounds is studied in detail in A2 Chemistry.

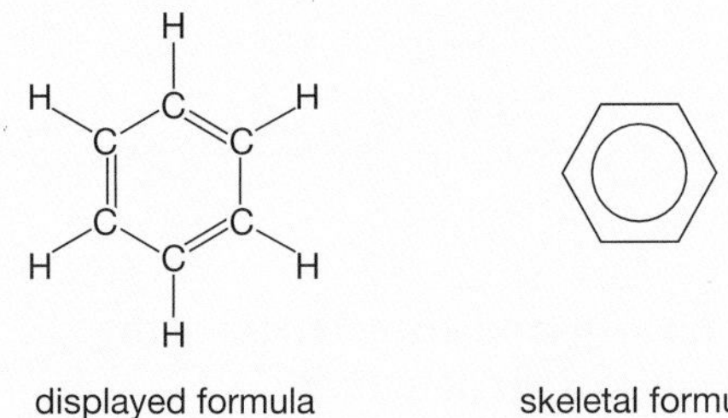

Fig. 3 *Benzene, showing full and skeletal formulae*

Organic molecules can be represented by molecular or structural formulae

The **molecular formula** of a compound gives the number of atoms of each element present in the molecule – for example, the molecular formula of propane is C_3H_8 (see page 167).

The **structural formula** shows how the atoms are bonded together in the molecule. The diagrams of the compounds in Figure 1 show the structural formulae. The displayed structural formula shows all the atoms and all the bonds. In more complicated molecules these are abbreviated to skeletal formulae, in which the carbon atoms are not shown and the hydrogen atoms are not included. Skeletal formulae are commonly used to represent cyclic and aromatic molecules. The dye structure in Figure 1 shows a combination of a displayed and a skeletal formula.

3 Give the molecular formulae for:

a) methane b) urea c) propene d) benzene.

4 Explain the meaning of the terms, and give an example to illustrate your answer, for:

a) hydrocarbon b) unsaturated c) saturated d) aliphatic.

Key Ideas 164 – 165

- **Many organic molecules contain only carbon and hydrogen. These are called hydrocarbons.**
- **The bonding in organic molecules is mainly covalent.**
- **A saturated molecule contains only single covalent bonds. An unsaturated molecule contains at least one double bond.**
- **If the carbon atoms are arranged in chains, the molecules are said to be aliphatic. If the molecule contains at least one benzene ring, the molecule is said to be aromatic.**

2 Shape up!

Too many British people are overweight because their diet contains too much fat. The biggest concern is *saturated* fats, which are linked to heart disease. Saturated molecules contain only single covalent bonds. The alkanes are a series of saturated hydrocarbons.

The structure of alkanes

Methane is the simplest alkane. The structure is usually shown as in Figure 1.

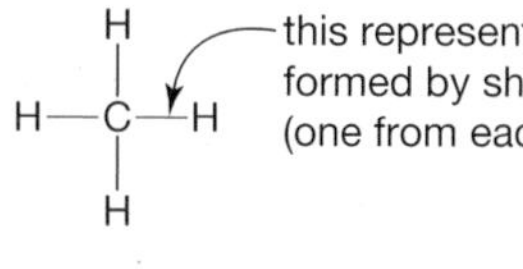

Fig. 1 *The structure of methane*

Figure 1 shows the structure or **structural formula** for methane. In organic chemistry it is often most useful to use structural formulae, as these show how the atoms are bonded together in a molecule.

This is a two-dimensional representation of a three-dimensional structure. The arrangement of the bonds around the carbon atom is actually tetrahedral. This can be seen when using models (Figure 2).

The tetrahedral arrangement of bonds around the carbon atom occurs in all saturated molecules. In each case the bond angle is 109.5° (see page 42).

Fig. 2 *Model of methane*

The atoms can rotate freely about the carbon–carbon single bonds.

It is important to bear in mind the actual three-dimensional shape of these molecules, but in practice it is only necessary to draw the structure of the molecules in two dimensions. Some examples are shown in Table 1.

You can use either the full or the abbreviated form of the structure, or a combination of the two, to represent the structural formula.

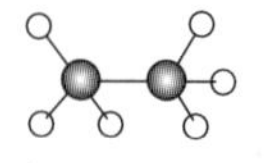

ethane, C_2H_6

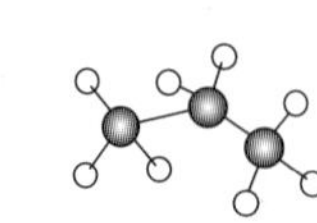

propane, C_3H_8

butane, C_4H_{10}

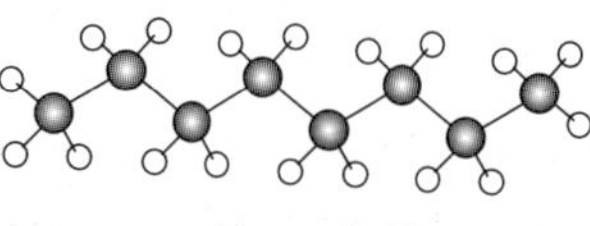

octane, C_8H_{18}

Fig. 3 *All saturated molecules have a tetrahedral arrangement of bonds around each carbon atom*

Name	Molecular formula	Empirical formula	Structural formula (full)	Structural formula (abbreviated)
Methane	CH_4	CH_4	H H–C–H H	CH_4
Ethane	C_2H_6	CH_3	H H H–C–C–H H H	CH_3CH_3
Propane	C_3H_8	C_3H_8	H H H H–C–C–C–H H H H	$CH_3CH_2CH_3$
Butane	C_4H_{10}	C_2H_5	H H H H H–C–C–C–C–H H H H H	$CH_3CH_2CH_2CH_3$

Table 1 *The first four members of the homologous series of alkanes*

Using formulae

Empirical and molecular formulae were introduced in Module 1.

- The **empirical formula** of a compound gives the *simplest ratio of atoms* of each element present in one molecule.
- The **molecular formula** of a compound gives the *actual number of atoms* of each element present in one molecule.

Here is a reminder of how to calculate a formula.

A gaseous hydrocarbon X contains 80% by mass of carbon. If the relative molecular mass of X is 30, find the molecular formula and draw out its structure.

Step 1: Calculate the empirical formula:

$\frac{80}{12}$ mol of carbon combines with $\frac{20}{1}$ mol of hydrogen

= 6.667 mol of carbon : 20 mol of hydrogen

Step 2: To obtain a simple ratio, divide by the smallest number:

$$\frac{6.667}{6.667} : \frac{20}{6.667} = 1:3$$

The empirical formula of **X** is therefore CH_3.

Step 3: The empirical formula mass = 15, but the molecular mass = 30

so the molecular formula = $(CH_3)_2 = C_2H_6$

Step 4: The structure of this molecule is:

```
   H  H
   |  |
H—C—C—H
   |  |
   H  H
```

Q

1 a) Deduce the molecular formula of an alkane which contains 6 carbon atoms.
 b) What is the empirical formula of this molecule?

2 Compound A (M_r = 58) has a percentage composition by mass of:
 C = 62.04%, H = 10.41%, O = 27.55%.
 Calculate the molecular formula of A.

Key Ideas 166 – 167

- **The arrangement of bonds around a saturated carbon atom is tetrahedral.**
- **Organic molecules can be represented by empirical, molecular and structural formulae.**

3 Series of compounds

There is such an enormous number and variety of organic compounds – over 90% of known compounds are organic, with new ones continually being produced – that we cannot study each one separately. Instead, those with similar properties are grouped together, rather as the elements are grouped in the Periodic Table.

Homologous series

A group of molecules with similar properties is called a **homologous series**. The properties of a homologous series can be studied rather than studying the individual molecules. You will be familiar with three homologous series from GCSE – alkanes, alkenes and alcohols. Some others which you will meet in this module are shown in Table 1.

What do members of a homologous series have in common?

1 They can be represented by a **general formula**. The formulae of *all* the aliphatic members of a particular homologous series fit this general formula (see Table 1).

2 The compounds in a particular series have the same **functional group**. This is an atom or a group of atoms that determines the chemical properties of the molecule. As members of the same homologous series have the same functional group, they must have the same, or very similar, chemical properties.

3 Down the series there is a trend in the physical properties. The most important physical property to consider is boiling point: the longer the hydrocarbon chain of an organic molecule, the higher the boiling point.

Table 1 *Homologous series*

Homologous series	General formula	Functional group	Example	
Alkanes	C_nH_{2n+2}	$-\overset{\vert}{\underset{\vert}{C}}-H$	CH_3CH_3	ethane
Alkenes	C_nH_{2n}	$>C=C<$	CH_3CHCH_2	propene
Haloalkanes	$C_nH_{2n+1}X$ X is a halogen	$-X$	CH_3Cl	chloromethane
Alcohols	$C_nH_{2n+2}O$	$-OH$	CH_3CH_2OH	ethanol
Carboxylic acids	$C_nH_{2n}O_2$	$-C(=O)OH$	CH_3CH_2COOH	propanoic acid
Aldehydes	$C_nH_{2n}O$	$-C(=O)H$	CH_3CHO	ethanal
Ketones	$C_nH_{2n}O$	$>C=O$	CH_3COCH_3	propanone

- The general formulae given above do *not* apply to *cyclic* molecules.
- Notice that aldehydes and ketones have the same general formula. This is because they are functional group isomers (see page 222).
- C_nH_{2n+1} is a hydrocarbon chain called an **alkyl group** (see page 172). If only the functional group of the molecule is being considered, the alkyl group can be represented by **R**.

Trends in boiling points

The trend in boiling points can be illustrated by looking at the first six hydrocarbons in the homologous series of alkanes. The boiling points of the straight-chain alkanes are shown in Table 2.

> **Q**
> 1 Draw a graph of the data in Table 2 by plotting boiling point (vertical axis) against number of carbon atoms.

Name	Molecular formula	Boiling point/°C
Methane	CH_4	–164
Ethane	C_2H_6	–89
Propane	C_3H_8	–42
Butane	C_4H_{10}	–1
Pentane	C_5H_{12}	36
Hexane	C_6H_{14}	69

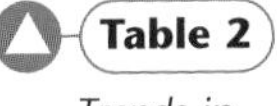

Trends in boiling points

Alkanes are non-polar molecules, so the only type of intermolecular force between the molecules is van der Waals forces. All alkanes, therefore, have relatively low boiling points (see page 33). But as the size of the molecule increases, the strength of the van der Waals forces increases too, so the boiling point increases.

Petroleum, the source of many organic compounds, is a mixture of mainly alkanes. Petroleum can be separated into fractions (see page 184) by **fractional distillation**, a method that takes advantage of this increase in boiling points as the number of carbon atoms increases.

> **Q**
> 2 a) Place these molecules in the correct homologous series:
> (i) C_2H_4 (ii) C_4H_{10} (iii) C_2H_5Br (iv) C_2H_5OH (v) $C_3H_6O_2$
> b) Write out the full structural formulae for these molecules. Table 1 (page 164) will help you to remember how many bonds the different atoms make.
>
> 3 Give the molecular formula of a molecule containing six carbon atoms given that it is:
> a) an alkene b) a ketone c) a carboxylic acid.
> Use the general formulae in Table 1 to help you.

Key Ideas 168 – 169

- **Organic compounds are grouped into homologous series.**
- **Members of a homologous series can be represented by a general formula.**
- **Members of a homologous series have the same functional group and therefore very similar chemical properties.**
- **Members of a homologous series show a trend in boiling points.**

4 Naming series

With such a variety of different molecules, we need rules of nomenclature so that we can give compounds systematic names. These rules are laid down by IUPAC (the International Union of Pure and Applied Chemistry). Once you have practised applying these rules you should be able to name any organic molecule from this module; or, given the name, you should be able to write out its structure.

Naming organic compounds

Look back at Table 1 on page 168. What do all compounds starting with *eth-* or *prop-* have in common?

The answer is that all molecules whose names start with **eth-** contain 2 carbon atoms, while those which start with **prop-** have 3 carbon atoms.

The *prefix* of the name indicates the number of carbon atoms present in the molecule.

This applies to straight-chain molecules. In this module you will study examples of molecules containing up to 6 carbon atoms.

The functional group, and hence the homologous series to which a compound belongs, is usually indicated by the *suffix* of the name.

Prefix	No. of carbon atoms
meth-	1
eth-	2
prop-	3
but-	4
pent-	5
hex-	6

Suffix	Homologous series
-ane	alk**ane**
-ene	alk**ene**
-ol	alcoh**ol**
-oic	carboxy**lic** acid
-al	**al**dehyde
-one	ket**one**

Table 1 *Prefixes and suffixes*

Some examples are shown in Table 1.

We can use this information, together with the general formulae given in Table 1 on page 168, to name some organic molecules from the formula and to write some formulae from the names. Some examples are given below.

1 What is the molecular formula of butane?

Step 1: The suffix is *-ane*, so this molecule is an alkane with the general formula C_nH_{2n+2}.

Step 2: The prefix is *but-*, so the molecule contains 4 carbon atoms ($n = 4$).

Step 3: Applying the general formula gives $\mathbf{C_4H_{10}}$ as the molecular formula for butane.

2 What is the name of the molecule with the formula C_2H_4?

Step 1: There are 2 carbon atoms in this molecule, so its name starts with *eth-*.

Step 2: The formula shows that the molecule is a hydrocarbon which fits the general formula for the alkenes. Alkene molecules have the suffix *–ene*.

Step 3: This molecule is therefore called **ethene**.

Names are based on the corresponding alkane

All homologous series are named from the corresponding alkane. For example, an alcohol containing 6 carbon atoms is called hex**an**ol, a carboxylic acid containing 5 carbon atoms is called pent**an**oic acid, and an aldehyde containing 3 carbon atoms is called prop**an**al.

3 What is the formula of butanoic acid?

Step 1: This molecule is a carboxylic acid, and contains 4 carbon atoms (*but-*).

Step 2: The functional group is —COOH. This leaves 3 carbon atoms in the hydrocarbon chain.

Step 3: The formula is therefore C_3H_7COOH, or more simply $C_4H_8O_2$.

Q

1 Complete the following table:

Name	Formula	Homologous series
Ethane		
	C_3H_6	
	CH_3COOH	
	CH_3CH_2CHO	aldehydes
Butanol		

Finding the structure from the formula

$C_4H_8O_2$ is actually the molecular formula of butanoic acid. C_3H_7COOH is somewhere between a molecular and a structural formula. It should be obvious that the molecule is an acid from the —COOH group but how are the other atoms arranged? There are two possibilities, as shown in Figure 1.

These molecules have the same molecular formula and contain the same functional group (—COOH), but the atoms in the hydrocarbon chain (C_3H_7) can be arranged in two different ways. They are called **structural isomers**. One is known as the **straight-chain isomer**, the other as the **branched-chain isomer**.

You can learn more about structural isomers in the next section. Before starting the next section, check that you can answer the following questions, which review some important key words and ideas used in organic chemistry.

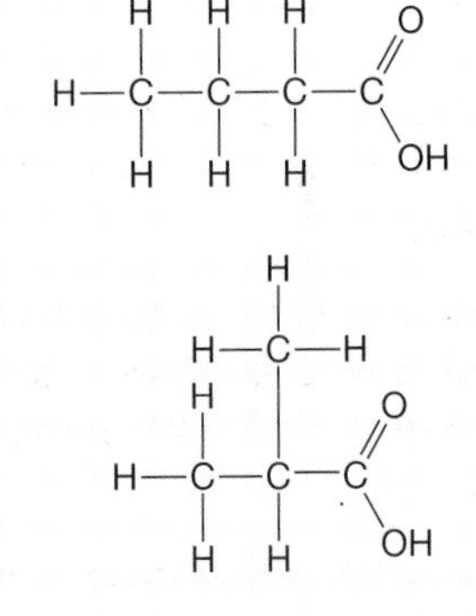

Fig. 1 *Two structures for butanoic acid*

Q

2 What is meant by:

a) molecular formula? b) structural formula? c) homologous series? d) functional group?

Illustrate your answer with *one* example in each case.

Key Ideas 170 – 171

- **Organic compounds are named according to how many carbon atoms they contain and which homologous series they belong to.**
- **The functional group indicates the homologous series to which the compound belongs.**

5 Arranging the same atoms differently

As the number of carbon atoms increases, it becomes possible to arrange the atoms in more and more ways to make different structures. Below we consider *structural isomerism.*

Structural isomers

If you make models of a butane molecule, you will see that there are two different ways to fix the atoms together:

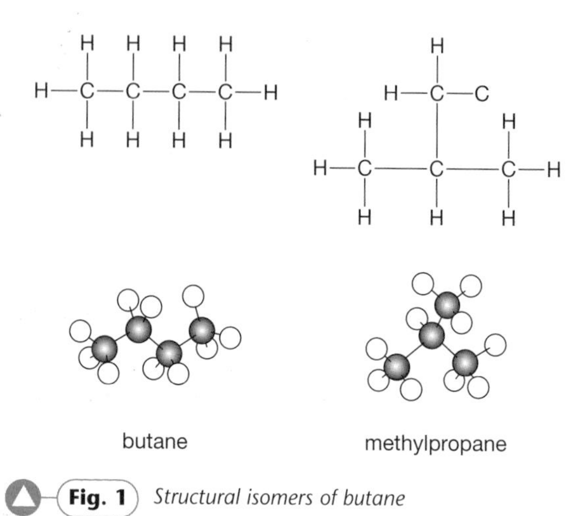

Fig. 1 *Structural isomers of butane*

When dealing with butane and subsequent members of the alkanes, you need to consider the structural isomers that exist.

Structural isomers have the *same molecular formula* but *different structural formulae*. This means that the atoms in the molecules are arranged differently. The arrangement can affect the chemical or physical properties of the molecules. For example, the two isomers of butane shown in Figure 1 have different boiling points: the straight-chain isomer boils at –1 °C, and the branched-chain isomer boils at –12 °C (see page 33).

As the number of carbon atoms increases, so the different ways of arranging them in a molecule also increases. For $C_{30}H_{62}$ there are over 4 million different ways of arranging the atoms!

Name	Molecular formula	Number of isomers
Butane	C_4H_{10}	2
Pentane	C_5H_{12}	3
Hexane	C_6H_{14}	5

Table 1 *Numbers of structural isomers*

These structural isomers are called **chain isomers**: the hydrocarbon chain can either be straight or branched.

Q

1 How many isomers of heptane, C_7H_{16}, can you find?

Naming the side chains

The branch or side chain is an **alkyl group**. Alkyl groups have the general formula $—C_nH_{2n+1}$. You can see that the alkyl group is part of an alkane molecule with one hydrogen removed: the group can therefore bond to other atoms. In the case of alkanes, the alkyl group bonds to another carbon atom. In the case of alcohols, the alkyl group bonds to an —OH group.

Some examples of alkyl groups are given below:

$—CH_3$ is called a **methyl group**

$—CH_2CH_3$ is called an **ethyl group**

$—CH_2CH_2CH_3$ is called a **propyl group**

When dealing with homologous series other than alkanes, it is sometimes convenient to use **R** to represent an alkyl group, or **R** and **R′** if there are two different alkyl groups present in one molecule. For instance, R—OH can be used to represent an alcohol.

The most common branch is the methyl group.

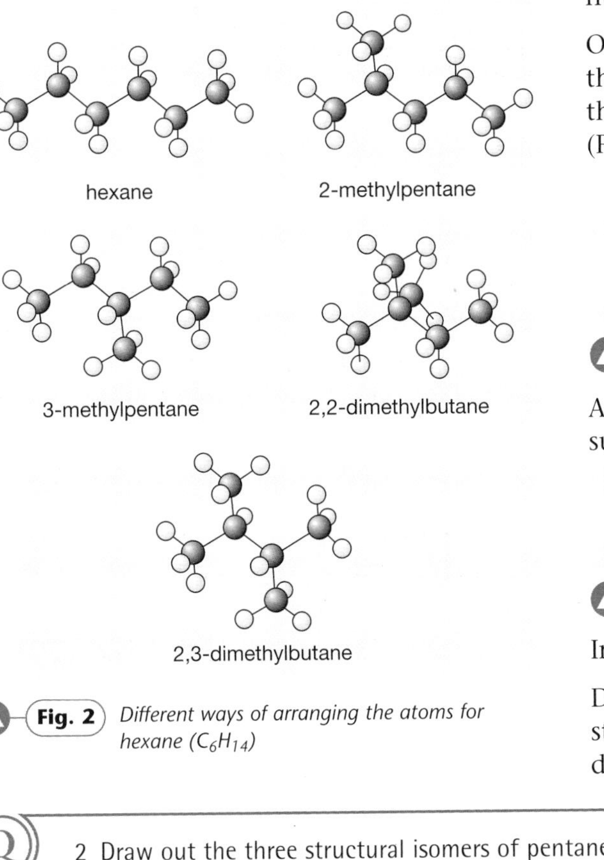

Fig. 2 *Different ways of arranging the atoms for hexane (C_6H_{14})*

This can be seen in the isomers of hexane shown in Figure 2.

The branched-chain isomers of hexane have either one or two methyl group side chains in the molecules. The position of the side chains is numbered (see page 174).

One common mistake is to draw branches on the carbon atoms at the ends of a chain and think that this has produced a different isomer (Figure 3).

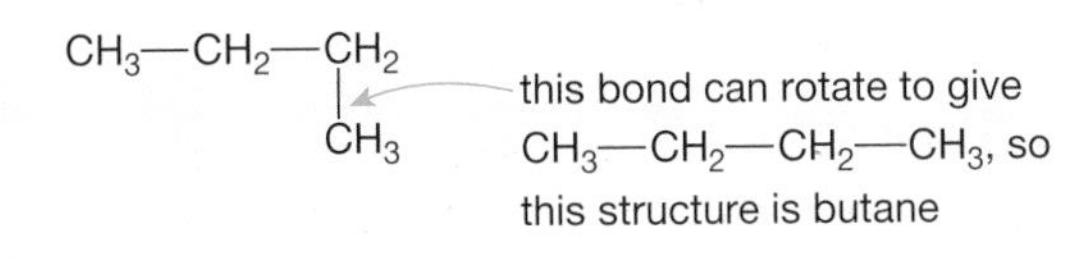

Fig. 3 *Not a side chain ...*

Another mistake is to think that two molecules such as those in Figure 4 are *different*:

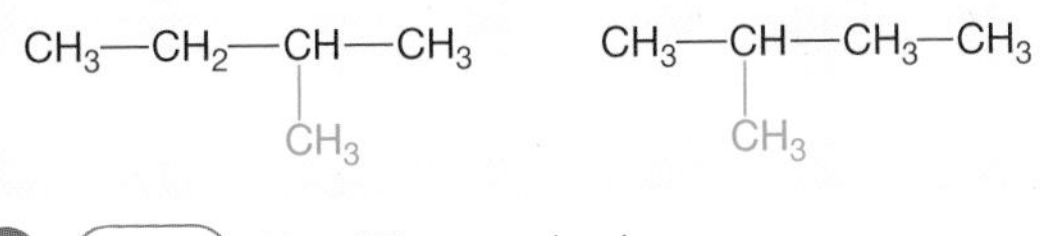

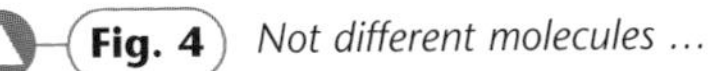

Fig. 4 *Not different molecules ...*

In fact they are the same, but turned through 180°.

Drawing out isomers becomes easier when we start to name them. The naming of isomers is dealt with in the next unit.

Q

2 Draw out the three structural isomers of pentane.

3 Which one of the following is a branched isomer of hexane?

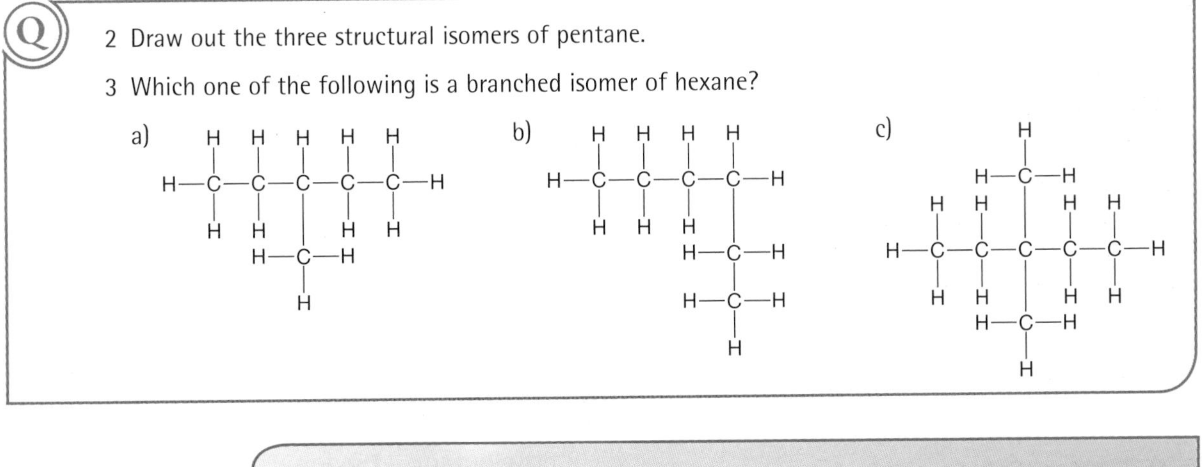

Key Ideas 172 – 173

- **Alkanes that contain four or more carbon atoms have structural isomers.**
- **Structural isomers of a molecule have the same *molecular* formula but different *structural* formulae.**
- **One type of structural isomerism is called chain isomerism. In this, the hydrocarbon chain can be straight or branched.**

6 What's in a name?

To distinguish between chain isomers, we need to use systematic names which precisely describe the structure of the molecule.

Naming alkanes

When naming alkanes you need to apply the following rules.

Suffix

The suffix is always **–ane**.

Straight-chain alkanes

For a *straight-chain alkane*, the prefix denotes the number of carbon atoms.

Example:

$CH_3—CH_2—CH_2—CH_2—CH_2—CH_3$

This molecule, $CH_3CH_2CH_2CH_2CH_2CH_3$, is called **hex**ane (C_6H_{14}).

Branched-chain alkanes

For a branched-chain alkane:

1 First look for the longest straight chain. Name this chain as though it were a straight-chain alkane.

2 Then add the name of the alkyl group branch to the front of this.

Example:

$CH_3—CH—CH_3$ ← the longest straight chain is propane (3 carbon atoms)

CH_3 (side chain on the middle CH) ← the side chain is a methyl group

This molecule is called methylpropane. It is an isomer of butane (C_4H_{10}). It can also be written as $CH_3CH(CH_3)CH_3$, with the side chain shown in brackets.

3 If there is a *choice* of positions for the alkyl group:

- Number the carbon atoms in the *longest chain*.
- Give the side chain the *lowest number possible*.

Example: Look back at the isomers of hexane (page 173).

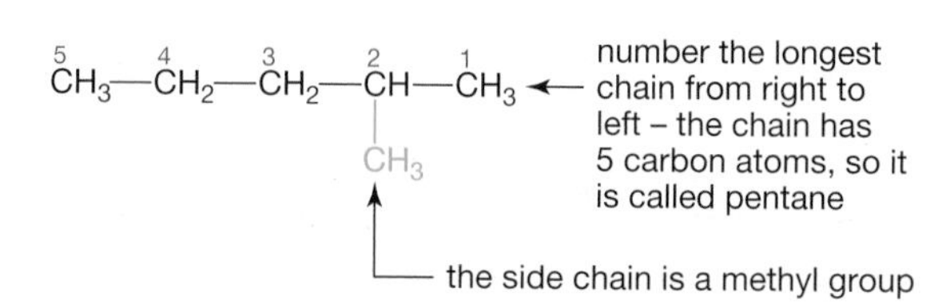

This molecule is called 2-methylpentane, $CH_3CH_2CH_2CH(CH_3)CH_3$ or $CH_3CH(CH_3)CH_2CH_2CH_3$.

Why not call it 4-methylpentane? Because that would break the rule of giving the side chain the lowest possible number.

Example:

$\overset{1}{CH_3}—\overset{2}{CH_2}—\overset{3}{CH}—CH_3$ ← the methyl group side chain is on C–3

(below C-3:) $4\,CH_2$ — $5\,CH_3$

Don't be tempted to call this molecule 2-ethylbutane – the longest chain contains 5 carbon atoms: this molecule is 3-methylpentane.

The carbon atoms can rotate about a single bond so this can also be shown as:

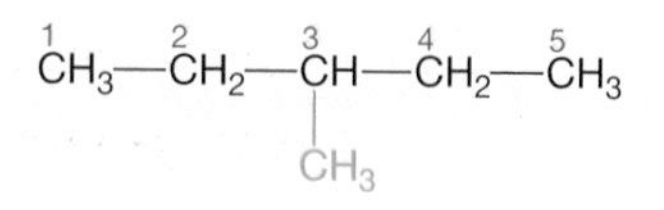

4 If there is *more than one* alkyl side chain, then *each* one carries the number of the carbon atom to which it is attached: the alkyl group is prefixed by **di-**, **tri-** or **tetra-** depending on how many side chains there are.

Example:

$$CH_2{-}C(CH_3)_2{-}CH_2{-}CH_3$$

This molecule is called 2,**2-di**methylbutane: the longest chain has 4 carbon atoms and *both* the methyl groups are on C-2. The abbreviated structure is written as $CH_3C(CH_3)_2CH_2CH_3$.

Example:

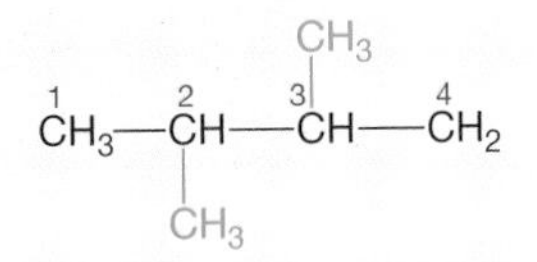

This molecule is called 2,**3**-dimethylbutane – here one methyl group is on C-2 and the other on C-3. The abbreviated structure is written as $CH_3CH(CH_3)CH(CH_3)CH_3$.

Q

1 Name, and write out the full structural formulae for, the following compounds:

a) $CH_3C(CH_3)_2CH_3$

b) $CH_3C(CH_3)_2CH(CH_3)CH_3$

2 Deduce the structural formulae of:

a) 2-methylbutane

b) 3-ethylpentane

Cyclic alkanes

Cyclic alkanes are named from the corresponding straight-chain alkane by adding the prefix **cyclo-**. The most common example is cyclohexane:

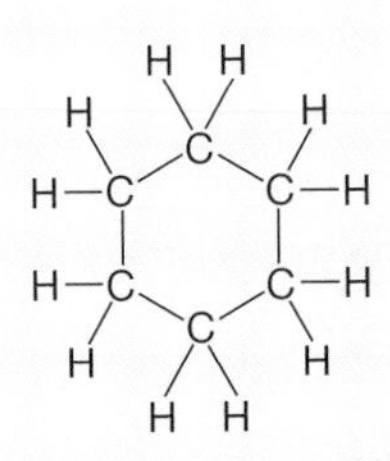

a displayed formula

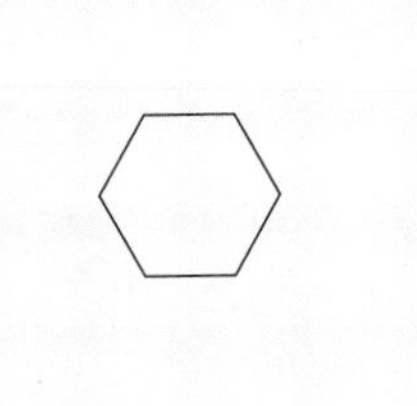

b skeletal formula

Diagram **a** shows the full structural formula of cyclohexane, with all the atoms and all the bonds.

Diagram **b** shows the skeletal formula: the carbon atoms are not labelled, and the hydrogen atoms are not included. It is common when drawing out cyclic structures to use skeletal formulae.

Remember that the formulae of cyclic molecules do not fit the general formula. The molecular formula of cyclohexane is C_6H_{12}.

Fig. 1 *Compounds containing a ring of carbon atoms are prefixed by 'cyclo-'*

Q

3 Give the molecular formula and structure of:

a) cyclohexanol

b) cyclohexene.

Key Ideas 174 – 175

- **When naming branched-chain alkanes, systematic naming avoids ambiguity.**
- **Each side chain is numbered to indicate its position on the main chain.**
- **Cyclic alkanes are prefixed by 'cyclo-'.**

Unit 11 Questions

1. The structural formula of aspirin is shown in Figure 1.

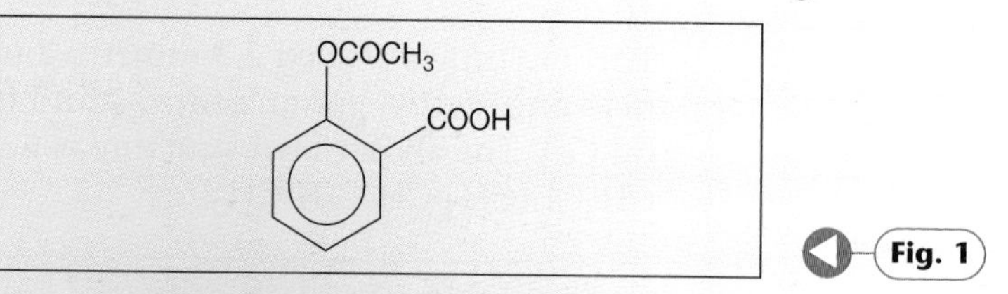

Fig. 1

a) (i) Write the molecular formula for aspirin.
(ii) One of the functional groups present in the aspirin molecule is the carboxyl group, which makes aspirin a carboxylic acid. Identify this group in the given structure. (2)

b) It was found that on average a commercial brand of aspirin tablets contains 1.67×10^{-3} mol of aspirin per tablet. If each tablet on average weighs 0.400 g, what is the percentage of aspirin by mass in this commercial brand of tablet? (3)

2. The diagrams in Figure 2 show models of two structural isomers of C_2H_6O.

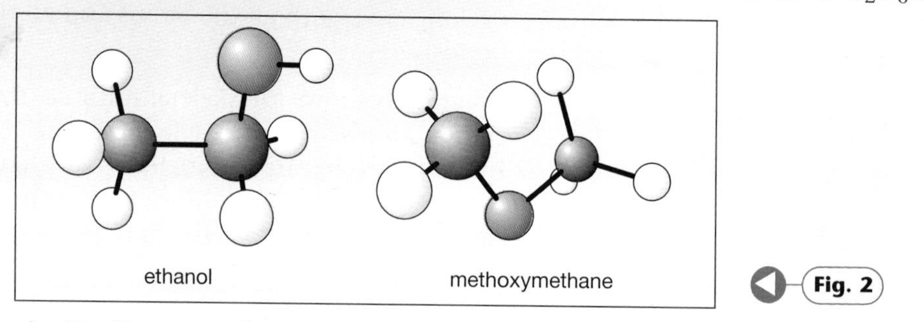

Fig. 2

a) (i) Draw out the structural formula for each molecule.
(ii) Why are these molecules described as structural isomers? (3)

b) (i) On your diagram, circle the functional group of the ethanol molecule.
(ii) Why is it important to recognise the functional group of a molecule? (2)

3. a) To which homologous series does the molecule C_5H_{12} belong? Explain how you arrived at your answer. (2)

b) (i) Draw out the structures of, and name, the two branched-chain isomers of C_5H_{12}.
(ii) What are the bond angles in these molecules? (5)

c) Explain why the three structures in Figure 3 represent only one isomer of pentane. (2)

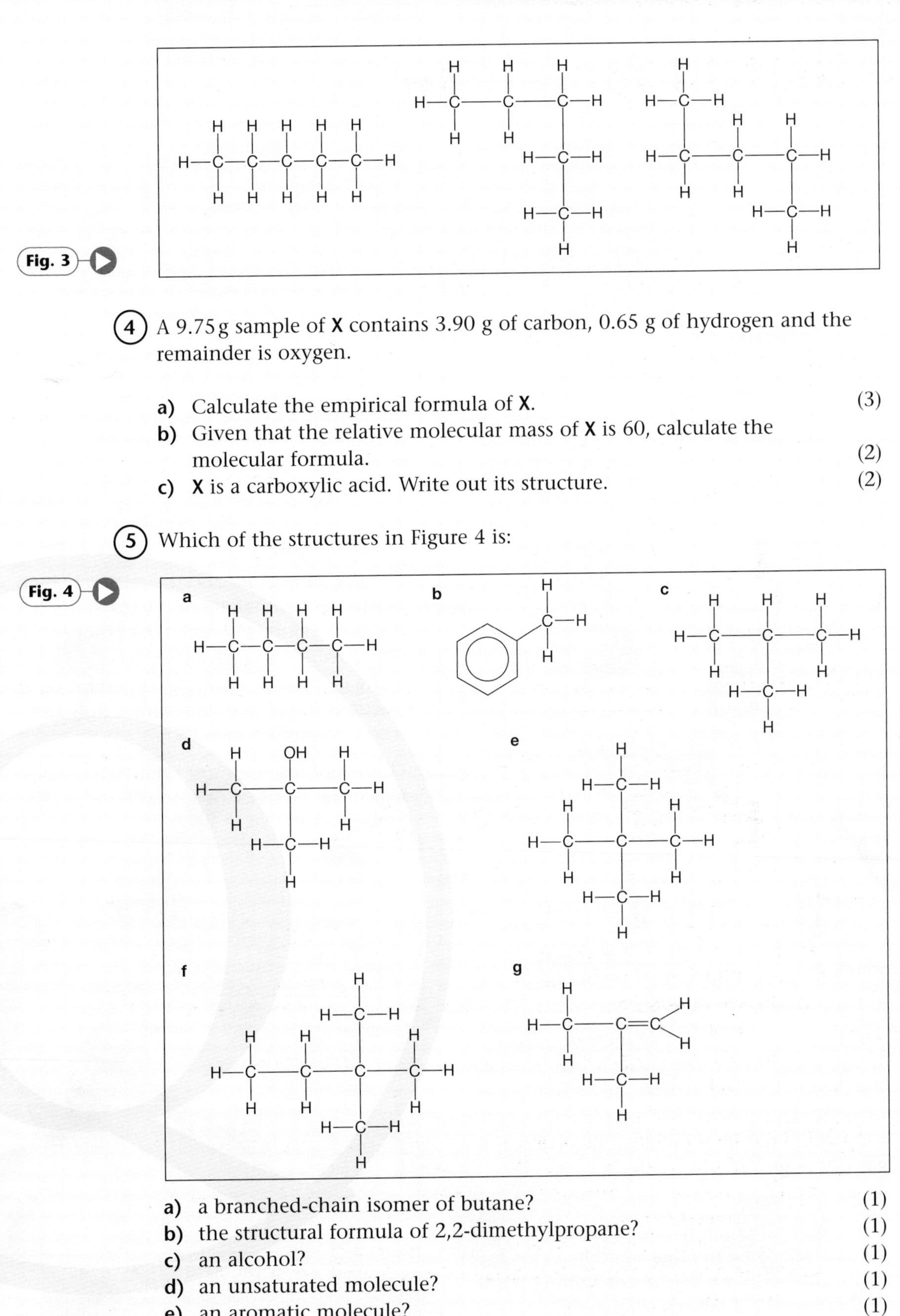

4) A 9.75 g sample of **X** contains 3.90 g of carbon, 0.65 g of hydrogen and the remainder is oxygen.

a) Calculate the empirical formula of **X**. (3)

b) Given that the relative molecular mass of **X** is 60, calculate the molecular formula. (2)

c) **X** is a carboxylic acid. Write out its structure. (2)

5) Which of the structures in Figure 4 is:

a) a branched-chain isomer of butane? (1)

b) the structural formula of 2,2-dimethylpropane? (1)

c) an alcohol? (1)

d) an unsaturated molecule? (1)

e) an aromatic molecule? (1)

1 The burning issue

One of the characteristics of a homologous series is that successive members show a trend in physical properties. This is illustrated by looking at the boiling points of the alkanes. Alkanes are non-polar saturated molecules. This makes them chemically unreactive; they do, however, have important uses as fuels.

The alkanes

We need first to recap what you have already met about alkanes. The alkanes are saturated hydrocarbons. The atoms around each carbon atom are tetrahedrally arranged, with a bond angle of 109.5°.

Fig. 1 *Model of ethane*

The alkane molecules are non-polar. The individual C—H bonds are weakly polar, due to the difference in electronegativity between carbon and hydrogen. But since the molecules are symmetrical, they have no *overall* polarity (see page 29).

The lack of polarity and the relatively high mean bond enthalpies of the C—H bond and the C—C bond (412 and 348 kJ mol^{-1} respectively) make alkanes unreactive towards many chemicals. The only important chemical reactions to consider are combustion and chlorination (see page 180).

This is not true for most homologous series, in which the presence of a polar bond makes the molecule reactive.

Trends in boiling points

The boiling points of the straight-chain alkanes are shown in the Figure 2.

As the size of the molecule increases, the strength of the van der Waals forces between the molecules increases, so the boiling points increase. This means that more energy must be supplied to release individual molecules as a gas (see page 32).

Structural isomers of the same alkane have different boiling points, even though they contain the same number of atoms. This can be illustrated by looking at the boiling points of the isomers of pentane (Table 1).

Branched-chain alkanes have a lower boiling point than their straight-chain isomers; and as

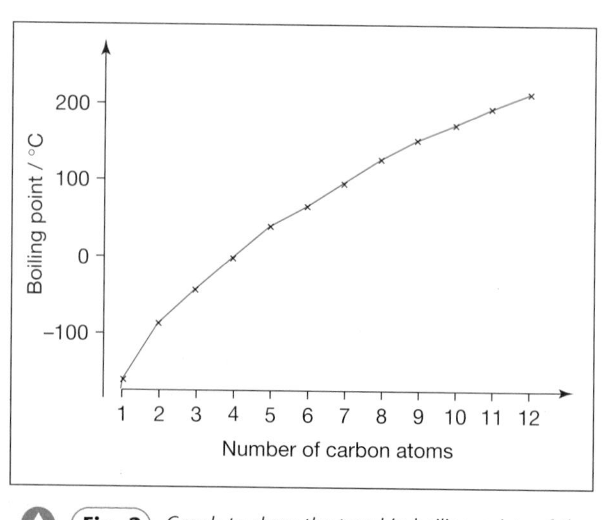

Fig. 2 *Graph to show the trend in boiling points of the straight-chain alkanes*

Name	Structural formula	Boiling point/°C
pentane	$CH_3CH_2CH_2CH_2CH_3$	36
2-methylbutane	$CH_3CH(CH_3)CH_2CH_3$	28
2,2-dimethylpropane	$CH_3C(CH_3)_2CH_3$	10

Table 1 *Boiling points of pentane isomers*

the branching increases, the boiling point decreases even further. This is because as the molecules become more branched, their shape becomes more spherical. This reduces the points of contact between neighbouring molecules and thus reduces the van der Waals forces present between the molecules.

Q

1 Explain why:

a) hexane has a higher boiling point than ethane

b) methylpropane has a lower boiling point than butane.

Combustion

Alkanes are used as fuels. You studied fuels at GCSE and you will know that when a fuel burns in air, energy is released as heat. This means that the reaction between an alkane and oxygen must be exothermic.

Octane (C_8H_{18}) is one of the components of petrol. It burns according to the following equation:

$$C_8H_{18}(l) + 12\tfrac{1}{2}O_2(g) \rightarrow 8CO_2(g) + 9H_2O(l)$$

$$\Delta H_c = -5470 \text{ kJ mol}^{-1}$$

This reaction has a high activation energy, due to the strength of the C—C and C—H bonds which must be broken during the reaction. This energy is supplied from a spark in the internal combustion engine. The bond-making involved, however, is more than enough to compensate for the activation energy, and the overall reaction is highly exothermic.

All alkanes burn in a similar way to octane. Given a plentiful supply of oxygen, the products of the reaction are carbon dioxide and water, and reaction continues to **complete combustion**. In a *limited* supply of oxygen, the reaction is still exothermic but the products are carbon monoxide, or carbon (soot) and water:

$$C_8H_{18}(l) + 8\tfrac{1}{2}O_2(g) \rightarrow 8CO(g) + 9H_2O(l)$$

This is called **incomplete combustion**. The carbon monoxide produced is one of the **pollutants** which is released into the atmosphere from the exhausts of vehicles with petrol engines and from gas fires and boilers in poorly ventilated rooms. Carbon monoxide is toxic because it combines with the haemoglobin in the blood and reduces its oxygen-carrying capacity. Death can occur by suffocation (see page 190).

Q

2 Write equations for

a) the complete and

b) the incomplete combustion of butane.

Key Ideas 178 – 179

- **In alkanes, the longer the carbon chain, the higher the boiling point.**
- **Branched-chain alkanes boil at lower temperatures than their straight-chain isomers.**
- **Alkanes are used as fuels. The complete combustion of alkanes produces carbon dioxide and water as waste products. The incomplete combustion of alkanes produces carbon monoxide and water.**
- **Carbon monoxide is a pollutant. It is toxic to humans because it reduces the oxygen-carrying capacity of the blood.**

2 Introducing reaction mechanisms

You are now familiar with using an equation to represent a chemical reaction. The equation tells us the reactants and products, but gives no indication how the reaction occurs or in how many steps. To study how the reaction occurs and the steps involved, we need to look at the reaction *mechanism*.

The chlorination of methane

One chemical with which the alkanes will react is chlorine. When chlorine and methane are mixed together in the presence of ultra-violet light, the following reaction occurs:

$$CH_4 + Cl_2 \rightarrow CH_3Cl + HCl$$

chloromethane

You can see from the equation that this is a **substitution reaction** – one of the chlorine atoms replaces one of the hydrogen atoms in methane. The product of the reaction is called chloromethane. This molecule belongs to the homologous series called the **haloalkanes** (see page 210).

Once the reaction has started the process may continue until all the hydrogen atoms are replaced by chlorine. This is called a **chain reaction**.

$$CH_3Cl + Cl_2 \rightarrow CH_2Cl_2 + HCl$$

dichloromethane

$$CH_2Cl_2 + Cl_2 \rightarrow CHCl_3 + HCl$$

trichloromethane

$$CHCl_3 + Cl_2 \rightarrow CCl_4 + HCl$$

tetrachloromethane

Further substitution produces by-products, as shown in the above equations.

These chlorinated methane compounds were once important solvents used in degreasing and dry cleaning. Their use is now very limited and their production is being phased out because they are some of the chemicals responsible for the depletion of the **ozone layer**. Destruction of the ozone layer results in our exposure to harmful UV light from the sun and a greater risk of skin cancer.

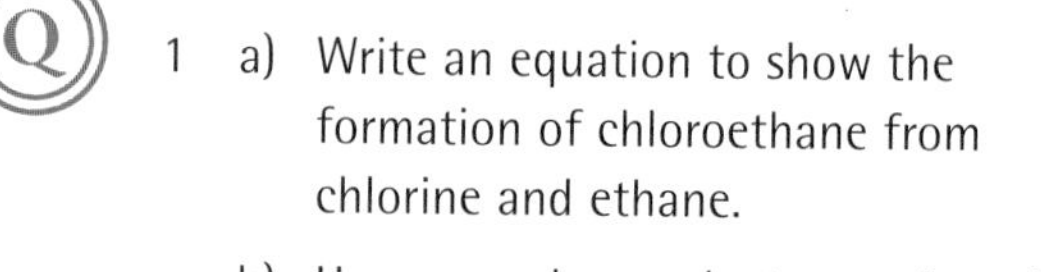

1 a) Write an equation to show the formation of chloroethane from chlorine and ethane.

b) How many by-products may form in this reaction?

The main product of the reaction described above is chloromethane. What happens during the reaction?

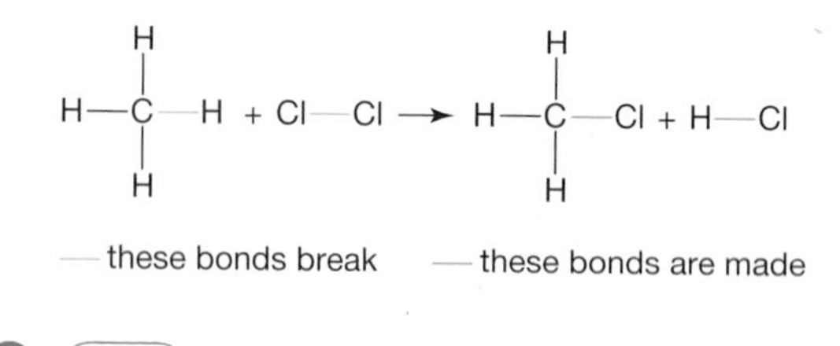

Fig. 1 *The reaction between methane and chlorine*

You can see from the equation in Figure 1 that for the reaction to occur, a C—H bond and a Cl—Cl must break and an H—Cl and a C—Cl bond must be made. This is the **mechanism** of the reaction: the series of steps that occur during the reaction.

Compounds in different homologous series have different reaction mechanisms, depending on the functional group present.

Different ways of breaking bonds

The first step in any reaction mechanism has to involve breaking bonds. A covalent bond is a shared pair of electrons. When the bond is broken, one of two things can happen: *homolytic fission* or *heterolytic fission*.

Homolytic fission

In **homolytic fission**, the electrons from the bond are shared between the two atoms involved. See Figure 2.

$$H{:}Cl \rightarrow H^{\bullet} + Cl^{\bullet}$$

• represents an unpaired electron

Fig. 2 *Homolytic fission*

In this example the covalent bond between the hydrogen atom and chlorine atom in hydrogen chloride is broken, leaving each atom with one unpaired electron. These atoms are very reactive. In organic chemistry they are called *free radicals*. The chlorination of methane involves free-radical intermediates.

- **Free radicals** are high-energy intermediates in many reactions.
  The unpaired electron is available for bonding.
- **Intermediates** are species that are formed in one step of a reaction mechanism but used up in another step. They do not appear in the overall equation either as reactants or as products.

Heterolytic fission

In **heterolytic fission**, the electrons from the bond are both transferred to the more electronegative atom. See Figure 3: the curly arrow represents the movement of a pair of electrons.

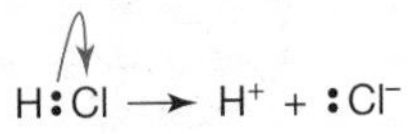

$$H{:}Cl \rightarrow H^{+} + {:}Cl^{-}$$

Fig. 3 *Heterolytic fission*

The hydrogen ion formed is electron-deficient. In organic chemistry electron-deficient species are called **electrophiles**. Alkenes react with electrophiles (see page 198).

The chloride ion formed has a lone pair of electrons which it can donate to another species. In organic chemistry an electron-pair donor is called a **nucleophile**. Haloalkanes react with nucleophiles (see page 211). 'Electrophile' and 'nucleophile' are important terms in organic chemistry.

During heterolytic fission the positively charged species may contain a carbon atom. Such species are called **carbocations**. They are very important intermediates in organic reactions – for instance, the alkenes (page 199) and the alcohols (page 224) react via carbocations.

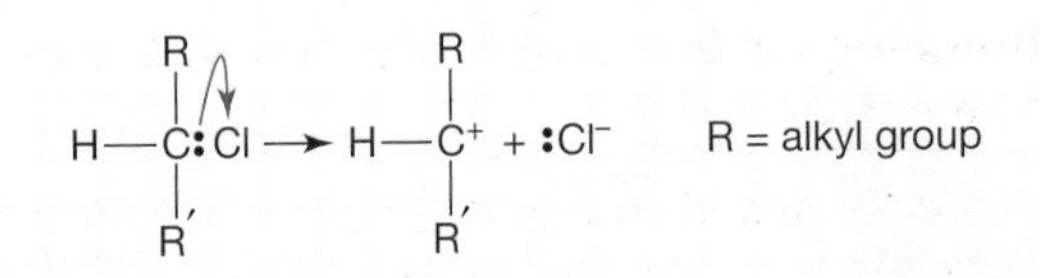

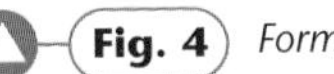
Fig. 4 *Formation of a carbocation*

2 What is the difference between homolytic and heterolytic fission?

3 What is a carbocation?

Key Ideas 180 – 181

- Methane reacts with chlorine to form chloromethane as the major product.
- Chlorination is a chain reaction, and can continue until all the hydrogen atoms in the alkane have been substituted by chlorine atoms.
- There are two ways a bond can break: by homolytic fission or by heterolytic fission.

3 Introducing free radicals

The chlorination of methane occurs via a free-radical mechanism. There are three main stages to this process which we will now consider.

A free-radical substitution mechanism

The mechanism for the formation of chloromethane is called a **free-radical substitution reaction**. It occurs in three main stages.

Initiation

We have already said that to make chloromethane, chlorine and methane are reacted together in the presence of ultra-violet light. There is no reaction in the dark.

The light provides the energy to break the covalent bonds in some of the chlorine molecules to produce **free radicals**. This is an example of homolytic fission:

$$Cl{:}Cl \rightarrow Cl\cdot + Cl\cdot$$

or $Cl_2 \rightarrow 2Cl\cdot$ *Equation 1*

The Cl—Cl bond is weaker than the C—H bonds in methane and is therefore broken first. We can see this by considering the mean bond enthalpies: the value for the Cl—Cl bond is 242 kJ mol^{-1}, whilst the value for the C—H bond is 412 kJ mol^{-1}.

> **Q** 1 Explain why Equation 1 above is an example of homolytic fission.

Propagation

The highly reactive chlorine radicals may collide with the methane molecules. C—H bonds break, resulting in the formation of stable hydrogen chloride molecules but unstable **methyl radicals**:

$$\begin{array}{c}H\\|\\H{-}C{:}H\\|\\H\end{array} + Cl\cdot \rightarrow H{:}Cl + \begin{array}{c}H\\|\\H{-}C\cdot\\|\\H\end{array}$$

or $CH_4 + Cl\cdot \rightarrow HCl + CH_3\cdot$ *Equation 2*

Then the methyl radicals can collide with the chlorine molecules to produce more chlorine free radicals:

$$\begin{array}{c}H\\|\\H{-}C\cdot\\|\\H\end{array} + Cl{:}Cl \rightarrow \begin{array}{c}H\\|\\H{-}C{:}Cl\\|\\H\end{array} + Cl\cdot$$

or $CH_3\cdot + Cl_2 \rightarrow CH_3Cl + Cl\cdot$ *Equation 3*

Notice in the propagation steps that although free radicals are used up, they are also produced. The chlorine radicals are made and used until no more chlorine is left to react.

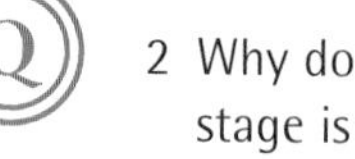

> **Q** 2 Why do you think the propagation stage is called a chain reaction?

Termination

If two free radicals collide, the product will be a molecule. Once all the intermediates have been used up the reaction is finished. For example:

$CH_3\cdot + Cl\cdot \rightarrow CH_3Cl$ *Equation 4*

$CH_3\cdot + CH_3\cdot \rightarrow CH_3CH_3$ (ethane)

All alkanes react with chlorine in this way. The greater the number of hydrogen atoms to be substituted, the greater the variety of products which can be made. Remember that during a reaction involving gases there will be millions of collisions every second. The above mechanism illustrates some of the reactions resulting from successful collisions.

Equations 1–4 show how you would be expected to write out the mechanism for the formation of chloromethane in an examination:

- Step 1: Initiation – the formation of free radicals:

$$Cl_2 \rightarrow 2Cl\cdot$$

- Step 2: Propagation – the regeneration of free radicals:

$$CH_4 + Cl\cdot \rightarrow HCl + CH_3\cdot$$

$$CH_3\cdot + Cl_2 \rightarrow CH_3Cl + Cl\cdot$$

- Step 3: Termination – the removal of free radicals:

$$CH_3\cdot + Cl\cdot \rightarrow CH_3Cl$$

The bromination of alkanes also occurs via a free-radical substitution reaction, but as bromine is less reactive than chlorine the reaction is much slower.

Although chloromethane is the main product of the reaction between chlorine and methane, this would not be a good method of preparation – there are too many by-products, and their removal would be too costly. The number of by-products can be reduced by carrying out the reaction in the presence of excess methane. Alternatively, haloalkanes can be prepared from alkenes (see page 198).

Q

3 The major product in the chlorination of ethane is chloroethane. Write one equation in each case to show:

a) the initiation step in this reaction

b) a possible propagation step

c) a termination step showing the formation of chloroethane.

Health and free radicals

Recent concerns about the environment have reduced the uses of some organic chloro-compounds, such as 1,1,1-trichoroethane and chlorofluorocarbons (CFCs), because they escape into the atmosphere and eventually decompose to form chlorine free radicals. These react with **ozone** and reduce the protective ozone layer in the upper atmosphere.

More and more people are taking **antioxidants** as dietary supplements. These are thought to protect the body from free radicals which may cause cell damage.

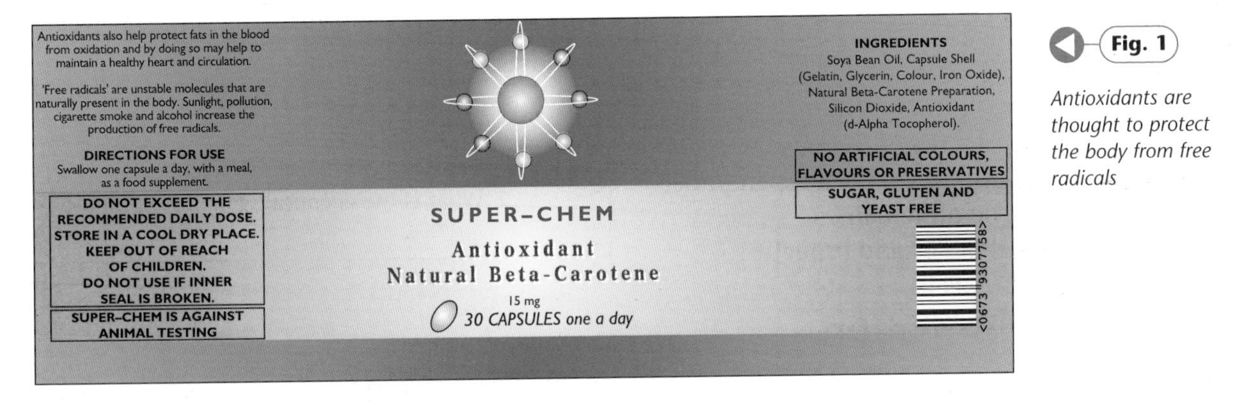

Fig. 1

Antioxidants are thought to protect the body from free radicals

Key Ideas 182 – 183

- **The mechanism of a reaction shows how the reaction takes place and which species are involved in each step.**
- **Alkanes undergo free-radical substitution reactions with chlorine.**
- **Free radicals are reactive species which have an unpaired electron available for bonding.**

4 The fractions separate

At GCSE you studied how oil was formed and how it was refined using fractional distillation. At AS level we will look in more detail at this process and at the uses of the fractions. **Crude oil** (petroleum) is a complex mixture of hydrocarbons. It is of little economic value until it has been refined. The first stage is to separate this complex mixture into its components; this is called **fractional distillation.** This process separates the hydrocarbons from petroleum into smaller mixtures, or **fractions**, according to their boiling-point ranges.

What is petroleum?

Petroleum or crude oil is a fossil fuel comprising a complex mixture of hydrocarbons. Its composition varies, depending on where in the world it is found, but generally it contains alkanes, cycloalkanes and aromatic hydrocarbons.

Crude oil is classed as a **finite** or non-renewable fuel as it cannot be replaced. The mixture was formed in the Earth's crust over 300 million years ago. It was formed from the remains of tiny marine organisms which sank to the bottom of the oceans and were covered by layers of sedimentary rock. As the layers increased, so did the temperature and the pressure; both helped the decay process.

As it was formed, the oil was absorbed by porous rock. Being less dense than water, it tended to migrate upwards until trapped under non-porous rock. The oil remains in these 'traps' until it is drilled for and brought to the surface. **Natural gas** is often found with oil found with oil deposits.

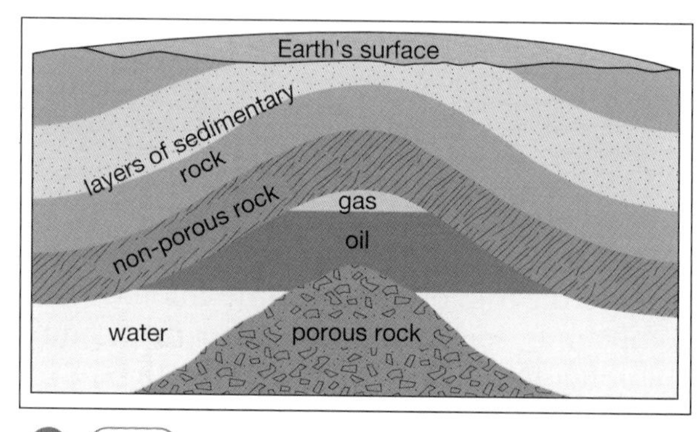

Fig. 1 *An oil trap*

> **Q**
> 1 What is a *finite energy resource?*
> 2 What is the chemical name for natural gas?

Fractional distillation

The process of fractional distillation depends on the fact that different components (**fractions**) of petroleum have different boiling points. The crude oil is heated until it vaporises. The vapour then passes up the **fractionating tower**, which is kept very hot at the bottom (about 350 °C) and cooler towards the top. The fraction containing the smallest molecules has the lowest boiling point range, so this boils off first and rises to the top of the tower.

The fractions containing the highest boiling-point hydrocarbons condense first. The liquid formed is collected in a series of trays containing bubble caps. The bubble caps force the rising vapours to pass through the liquid in the trays. The vapours condense to a liquid when they arrive at a tray that is sufficiently cool. The liquids are then piped off separately.

The fraction containing the largest molecules does not vaporise but falls to the bottom of the tower and is removed as the **residue**. The intermediate fractions are collected according to their boiling-point ranges at different levels up the tower, as shown in Figure 3.

Q

3 What property of the fractions allows them to be separated in the fractionating tower?

4 The gases produced by fractional distillation include butane, which is used as bottled fuel by campers. Write an equation for the complete combustion of butane.

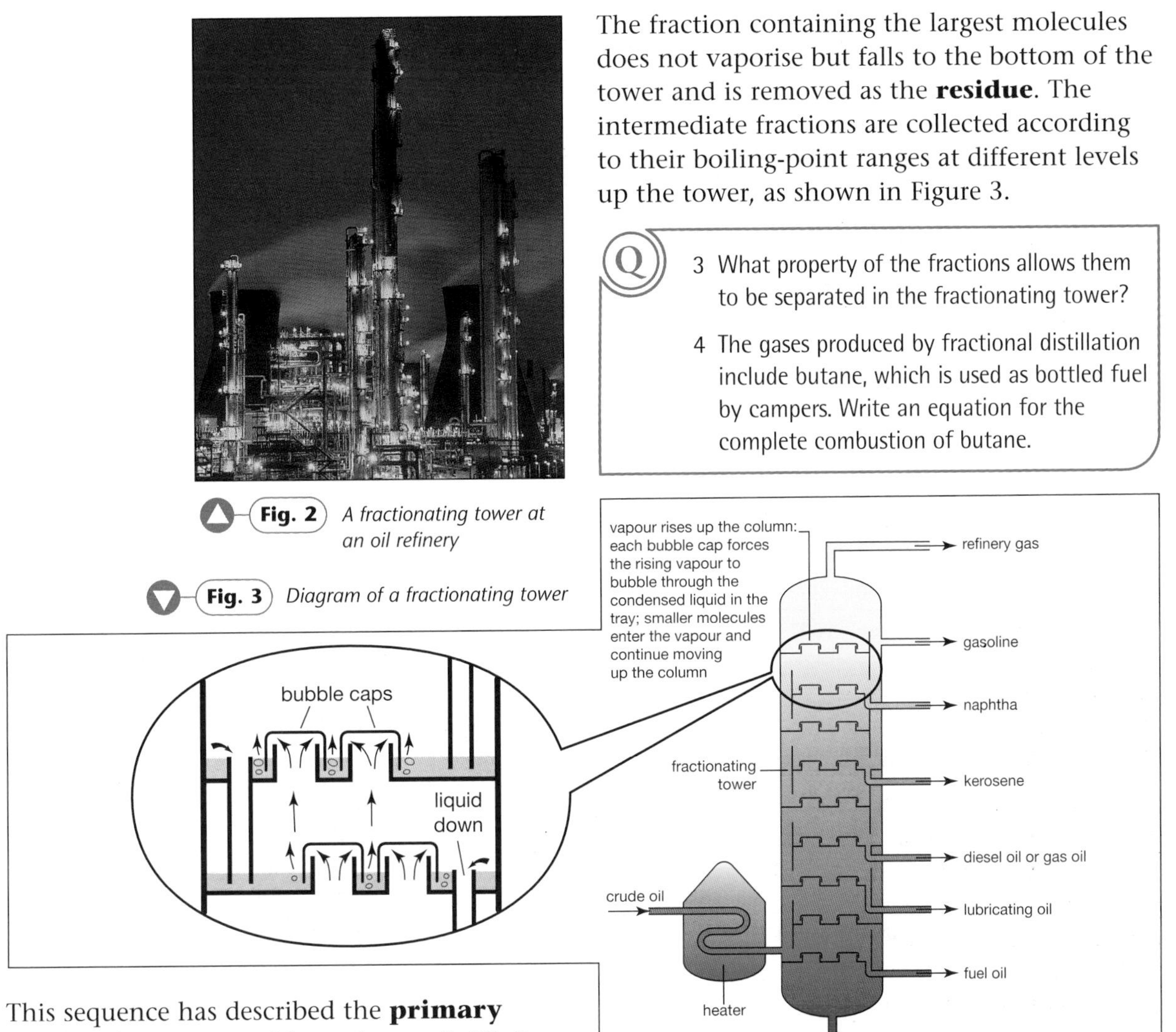

Fig. 2 *A fractionating tower at an oil refinery*

Fig. 3 *Diagram of a fractionating tower*

This sequence has described the **primary distillation** process. After primary distillation the fractions are treated in several different ways:

- Some of the sulphur is removed, to reduce the amount of pollution produced when the fractions used as fuels are burned in motor vehicle engines (see page 190).
- Some of the fractions are **'cracked'** to produce more useful chemicals. (See page 188.)
- The residue is further distilled. (This is dealt with in the next section.)

Key Ideas 184 – 185

- **Petroleum is a mixture consisting mainly of alkane hydrocarbons.**
- **The different components of this mixture can be separated at different levels in a fractionating column because of the temperature gradient in the column.**

5 Making use of the fractions

About 90% of the oil is burnt for fuels. The remaining 10% is used in the petrochemical industry to produce plastics, dyes, detergents, medicines, paints and insecticides. The naphtha fraction is the most important starting material for these products.

Uses of petroleum

Name of fraction	Boiling-point range/°C (approx.)	Number of carbon atoms present (approx.)	Uses
Refinery gas	−160 → 25	1 → 4	fuel for gas cookers, gas fires, camping gas
Petrol (gasoline)	30 → 100	4 → 10	petrol to fuel cars
Naphtha	100 → 180	6 → 14	feedstock for petrochemical industry
Kerosene (paraffin)	150 → 250	10 → 16	fuel for jets, feedstock for petrochemicals
Diesel (gas) oil	220 → 375	13 → 25	fuel for transport and central heating
Residue	over 350	over 25	(see Table 2)

Table 1 *The fractions obtained from petroleum and their uses*

The fractions obtained from the primary distillation of petroleum are shown in Table 1, together with their uses. Notice that:

- As the carbon chain gets longer, the boiling point increases (see page 169).
- Most of the fractions are used as fuels – yet petroleum is a finite energy resource, and as the supplies become more limited these fractions may become too valuable to burn.
- The naptha and kerosene fractions are further refined, mainly by cracking, to produce smaller molecules which may be branched or unsaturated. Aromatic compounds are also produced. Alkanes and aromatics are useful starting materials (**feedstock**) for the petrochemical industry. They are converted into a variety of important products, including plastics, dyes, detergents, medicines, paints and insecticides.

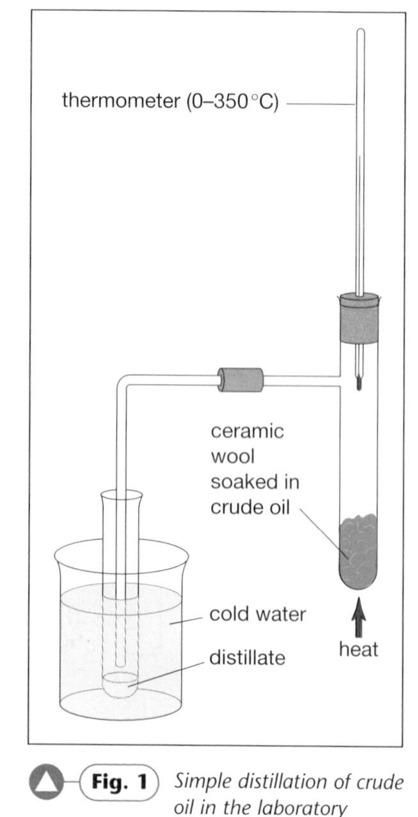

Fig. 1 *Simple distillation of crude oil in the laboratory*

Q

1. Which fraction is used to fuel jet engines?
2. What is the boiling-point range of diesel oil?
3. Give the formula of one of the hydrocarbons present in naptha.

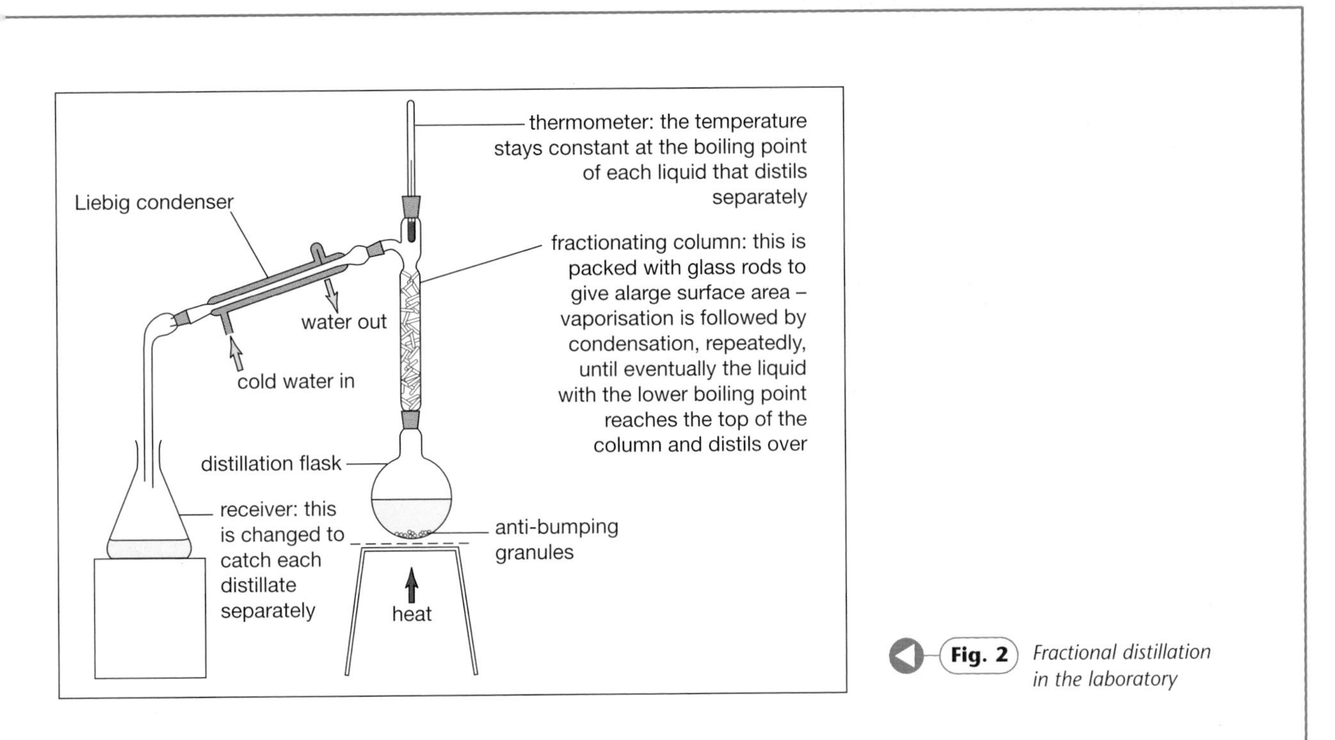

Fig. 2 *Fractional distillation in the laboratory*

Separating the residue

Name of fraction	Number of carbon atoms (approx.)	Uses
Lubricating (mineral) oil	20 → 30	lubricating oils, feedstock for petrochemicals
Fuel oil	30 → 40	fuel for ships and power stations
Paraffin waxes	40 → 50	polishing waxes, candles, lubricating grease
Bitumen (tar)	over 50	surfacing roads, roofing

Table 2 *Fractions obtained by vacuum distillation*

The residue, containing the non-volatile hydrocarbons which only vaporise above 350 °C at atmospheric pressure, is further distilled, but this time under reduced pressure. This process, called **vacuum distillation**, allows the mixture to be separated at a temperature below its normal boiling point, and prevents the components from decomposing. Lubricating oils and waxes are obtained by this method. This leaves a thick, black tar which is turned into bitumen for surfacing roads and pavements.

The fractions obtained from the residue are shown in Table 2.

Q

4 Why cannot the residue be further separated into paraffin waxes and tar by strong heating?

Key Ideas 186 – 187

- **The fractions from the petroleum industry are mainly used as fuels.**
- **The residue from primary distillation is further separated to produce a range of useful oils and waxes.**

6 A cracking solution

From distillation alone there is never enough petrol to match the demand. Other fractions containing larger, surplus molecules are further processed into petrol in an attempt to meet this demand. The process used is called **cracking**.

Supply and demand

In general the amount of each fraction obtained from primary distillation does not match demand. The largest demands are for the petrol and naphtha fractions.

To try to match demand chemists process some of the surplus fractions from the distillation process. One method of processing is called *cracking*.

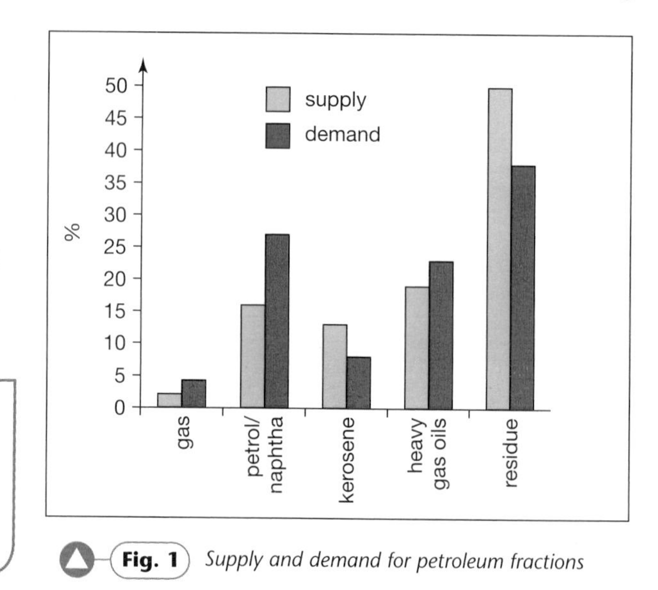

Fig. 1 *Supply and demand for petroleum fractions*

> **Q** 1 The kerosene fraction from the distillation of petroleum contains molecules in the range C_{10}–C_{16}. Why are molecules of this size chosen for cracking?

Cracking

Cracking is the process by which large alkane molecules are broken down into smaller, more useful molecules. Cracking not only produces molecules that can be used in petrol, but also produces alkenes which form the basis of the plastics industry (see page 204).

There are two types of cracking. Both involve breaking C—C bonds and C—H bonds.

Thermal cracking

The alkane is heated under pressure to a temperature between 450 and 800°C. This provides enough energy to break the C—C bonds. The process occurs via a **free-radical mechanism**. It produces a high percentage of alkenes.

When a molecule undergoes cracking it may break up in a variety of ways, forming a mixture of products. The products will always contain fewer carbon atoms than the original molecule. For example:

$$C_{12}H_{26} \rightarrow \underset{\text{octane}}{C_8H_{18}} + \underset{\text{ethene}}{2C_2H_4}$$

Octane is used in petrol, ethene is used to make plastics and antifreeze.

The mixture of hydrocarbons produced by cracking can be separated by fractional distillation.

> **Q** 2 Write an equation to show the formation of ethene and propene by the cracking of $C_{12}H_{26}$. Name the other product formed.

Catalytic cracking

The presence of a catalyst means that the process can be carried out at a lower temperature, as the catalyst reduces the activation energy required for the reaction (see page 106). This means that less energy is required.

In the catalytic cracker the vaporised hydrocarbons are mixed with the catalyst. Modern catalysts are synthetic **zeolites**, based on a type of clay. They contain atoms of silicon, oxygen and aluminium: these provide a series of pores and interconnecting channels which create a large surface area. The pore size can be altered to accommodate hydrocarbon molecules which are cracked inside the structure. The mechanism of this reaction involves a **carbocation intermediate** (see page 181).

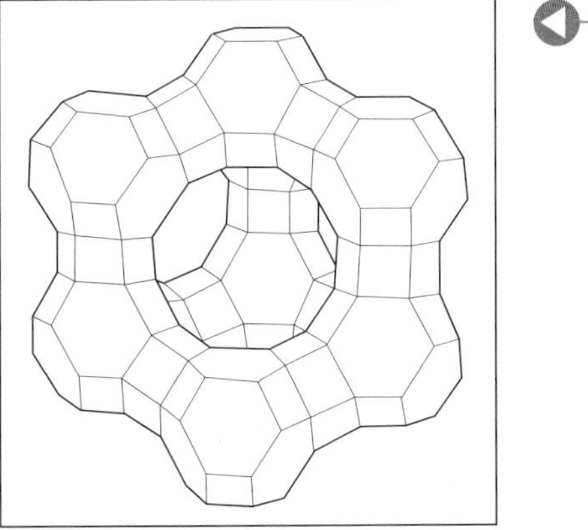

Fig. 3 *A computer graphic of zeolite Y used in the catalytic cracker*

Like thermal cracking, this process produces a mixture of shorter-chain alkanes and alkenes, but the molecules are more likely to be branched or aromatic (see Figure 2).

Small branched-chain alkanes and aromatic hydrocarbons are important components in petrol. They help the fuel to burn more efficiently and make it unnecessary to add lead (see page 190). The straight and branched-chain hydrocarbons can be separated as they come into contact with a zeolite. This acts as a molecular sieve: it has pores small enough to allow the straight-chain molecules to pass through, but the branched-chain molecules are too bulky and are therefore separated.

3 Why are the products of catalytic cracking more useful as motor fuels than the products of thermal cracking?

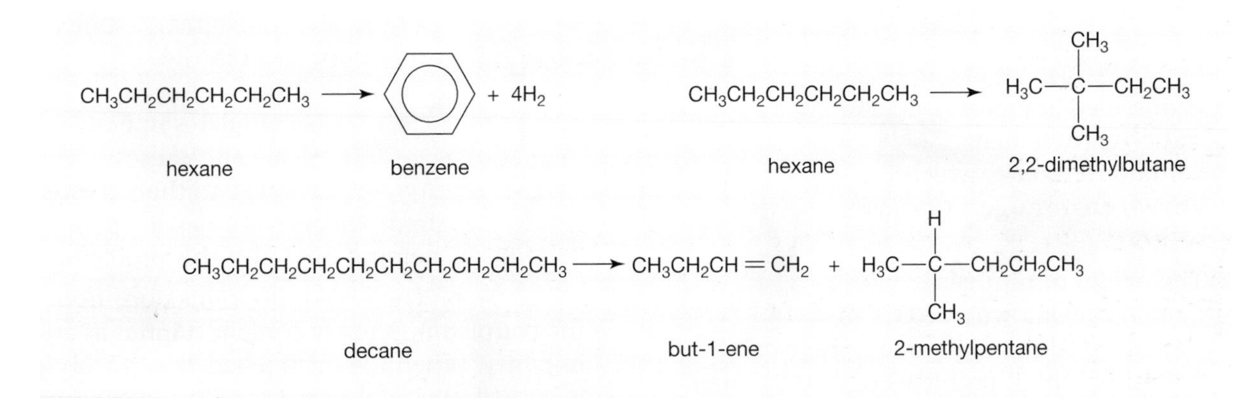

Fig. 2 *Equations to show formation of branched and aromatic molecules*

Key Ideas 188 – 189

- **Cracking involves the breaking of C—C bonds in alkanes.**
- **Thermal cracking takes place at high temperatures and pressures to produce a high percentage of alkenes. These alkenes are used to make plastics.**
- **Catalytic cracking uses a zeolite catalyst and is used mainly to produce motor fuels and aromatic hydrocarbons.**

7 Pollution is a problem

At GCSE you studied fuels and the gases produced when they burn, so you will know that some of these gases are responsible for polluting our atmosphere. Acid rain in particular damages buildings and reduces the pH in rivers and lakes to such an extent that animals and plants cannot survive.

Petrol and pollution

There is much concern about the use of petrol and diesel to fuel our transport systems.

1 The incomplete combustion of petrol vapour in a car engine produces **carbon monoxide**:

$$C_8H_{18}(l) + 8\frac{1}{2}O_2(g) \rightarrow 8CO(g) + 9H_2O(l)$$

The incomplete combustion of diesel fuel deposits carbon (soot) into the environment. Carbon monoxide is toxic (see page 179).

2 Unburnt hydrocarbons escape in exhaust fumes, and from petrol tanks when the tanks are being filled at petrol stations. These may contain **benzene**, which is a known **carcinogen** (it causes cancer).

3 The oxides of nitrogen (commonly known as $\mathbf{NO_x}$) are present in the exhaust gases. These cause several pollution problems once they enter the atmosphere. One of these is **acid rain**.

4 In the presence of sunlight the nitrogen dioxide present in exhaust gases also reacts with the unburnt hydrocarbons from petrol to form **photochemical smog**. This is caused by a complex series of reactions involving free radicals. The smog contains a variety of chemicals which are irritants, and may be responsible for the increase in asthma attacks during the summer.

5 **Lead compounds**, which are still added to some petrol to improve the engine's performance, are a major source of airborne lead in the atmosphere. Lead accumulates in body tissue and causes damage to the nervous system. The use of unleaded petrol is helping to reduce this problem.

6 The hydrocarbon fractions from the petroleum industry contain sulphur as an impurity. When the fraction (such as petrol) is burned, the sulphur reacts with the oxygen in the air to produce oxides of sulphur which mix with rainwater, causing **acid rain**.

The formation of acid rain

Acid rain is caused by the presence of non-metal oxides in the atmosphere. The main culprits are the oxides of nitrogen and sulphur.

Sulphur is an impurity found in petroleum fractions. It was present when the petroleum was formed in the Earth's crust. Petroleum, as you may remember from GCSE, was formed from the remains of the living organisms. Sulphur is a constituent of protein. If the sulphur is not removed from the fraction and the fraction is burned, sulphur dioxide forms:

$$S + O_2(g) \rightarrow SO_2(g)$$

This may be oxidised to sulphur trioxide:

$$2SO_2(g) + O_2(g) \rightarrow 2SO_3(g)$$

These are both acidic oxides which dissolve in water to form acids:

$$SO_2(g) + H_2O(l) \rightarrow H_2SO_3(aq)$$

$$SO_3(g) + H_2O(l) \rightarrow H_2SO_4(aq)$$

The burning of petrol and diesel accounts for about 2% of sulphur dioxide emissions. By far the greatest source of sulphur dioxide entering the atmosphere is the burning of coal in power stations.

Nitrogen is not present in petroleum, so where do the oxides of nitrogen come from? The oxygen required for combustion of petrol comes from the air, which is about 20%

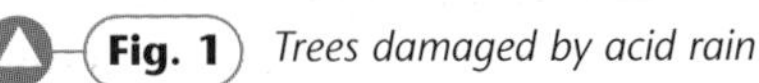

Fig. 1 *Trees damaged by acid rain*

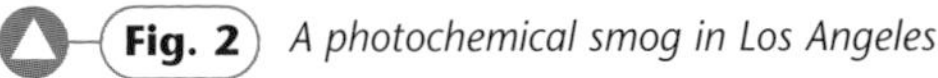

Fig. 2 *A photochemical smog in Los Angeles*

oxygen and 80% nitrogen. The combustion mixture must be sparked to overcome the activation energy of the reaction; the temperature can be as high as 2500 °C. This is enough to overcome not only the activation energy for the combustion reaction but the activation energy of the reaction between oxygen and nitrogen to form nitrogen monoxide:

$$N_2(g) + O_2(g) \rightarrow 2NO(g)$$

Nitrogen monoxide readily combines with oxygen in the air or engine to form nitrogen dioxide:

$$2NO(g) + O_2(g) \rightarrow 2NO_2(g)$$

In the atmosphere nitrogen dioxide reacts with rainwater to form nitrous acid, and this in turn is oxidised by oxygen to form nitric acid:

$$4NO_2(g) + O_2(g) + 2H_2O(l) \rightarrow 4HNO_3(aq)$$

Q

1 What are the names of $H_2SO_3(aq)$ and $H_2SO_4(aq)$?

2 Explain, with equations, how the oxides of nitrogen are formed.

3 What type of intermediates are involved in the formation of a photochemical smog?

Key Ideas 190 – 191

- **The internal combustion engine produces a number of pollutants, including NO_x, CO and unburnt hydrocarbons.**
- **Sulphur-containing impurities are found in petroleum fractions. The combustion of these impurities produces oxides of sulphur which are toxic and cause acid rain.**

8 Convert the problem ...

Is there an answer to modern-day pollution problems? The increased use of catalytic converters is helping to reduce the amount of pollutants which enter the atmosphere from vehicle exhausts. Another solution may be to change the composition of the fuel, or to develop a better public transport system.

Catalytic converters

Fig. 1 *The air over Britain contains about 8 million tonnes of carbon monoxide*

Catalytic converters fitted to the exhaust systems of modern cars are helping to reduce the levels of carbon monoxide, unburnt hydrocarbons, and oxides of nitrogen in our atmosphere.

A **catalytic converter** consists of a stainless steel cylinder with an opening at each end to allow the exhaust gases to pass through. Inside there is a ceramic block with a honeycomb structure which is thinly coated in metal catalysts, usually a mixture of platinum, palladium and rhodium. The honeycomb structure provides a large surface area, increasing the efficiency of the catalyst and reducing the amount of catalyst needed and hence the cost. Only about one gram of each metal is needed to produce a surface area the size of two football pitches!

1 In which block of the Periodic Table do platinum, palladium and rhodium appear?

2 Exhausts are normally made from mild steel. Why do you think stainless steel is used for catalytic converters?

How the catalytic converter works

Inside the converter a series of redox reactions take place. Air is drawn into the converter. The carbon monoxide and hydrocarbons are oxidised:

$$2CO + O_2 \rightarrow 2CO_2$$

$$C_8H_{18} + 12\tfrac{1}{2}O_2 \rightarrow 8CO_2 + 9H_2O$$

whilst the oxides of nitrogen are reduced:

$$2NO_2 \rightarrow N_2 + 2O_2$$

It is important to get the amount of oxygen in the fuel–air mixture exactly right, otherwise the catalytic converter cannot work efficiently. The car is fitted with a sensor to monitor the amount of oxygen passing through the exhaust system.

Other reactions that may occur inside the converter include:

$$2CO + 2NO \rightarrow 2CO_2 + N_2$$

$$C_8H_{18} + 25NO \rightarrow 8CO_2 + 12\tfrac{1}{2}N_2 + 9H_2O$$

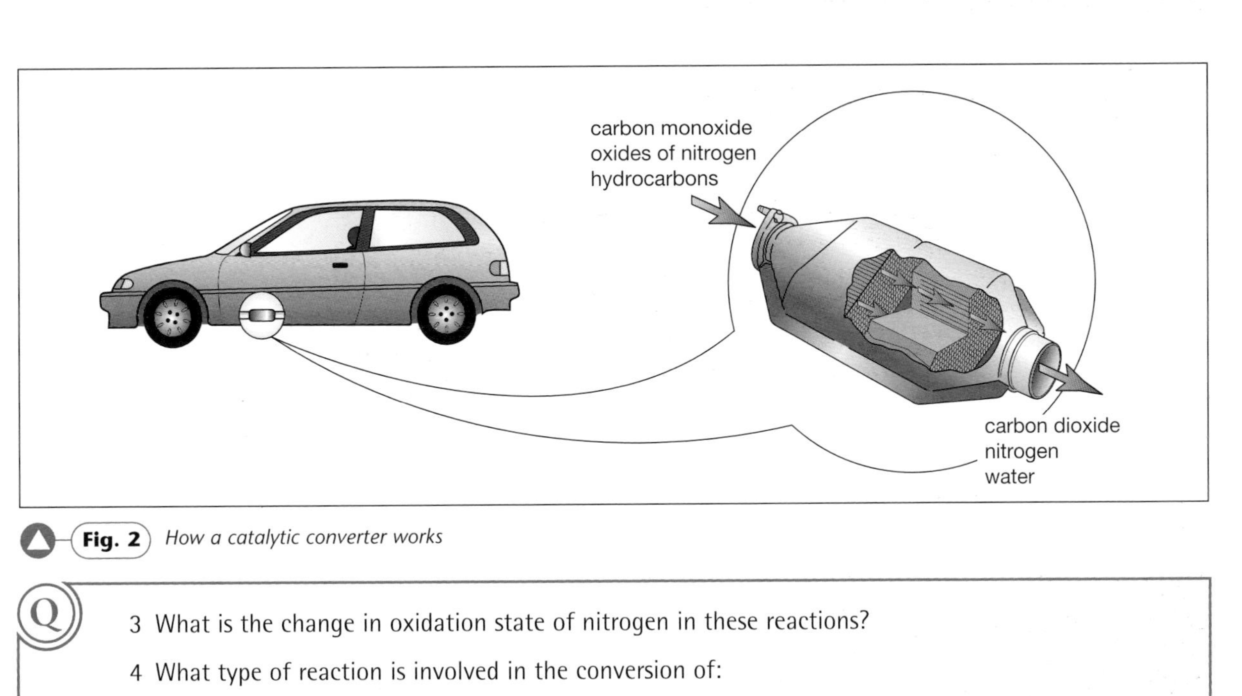

Fig. 2 *How a catalytic converter works*

3 What is the change in oxidation state of nitrogen in these reactions?

4 What type of reaction is involved in the conversion of:

a) nitrogen monoxide into nitrogen? b) carbon monoxide into carbon dioxide?

Problems with the converter

In summary, the changes taking place in the converter are shown in Table 1.

In	Out
Carbon monoxide	Carbon dioxide
Oxides of nitrogen	Nitrogen
Unburnt hydrocarbons	Carbon dioxide and water

Table 1 *Reactions within the converter*

The nitrogen is harmless. Carbon dioxide is much less toxic to humans than carbon monoxide, but its release causes another problem. Because carbon dioxide is a **'greenhouse' gas**, it is one of the main contributors to global warming.

Catalytic converters only work with lead-free petrol. The presence of lead would reduce the efficiency of the catalyst.

The metals in the converter work more efficiently when they are warm: this is why the catalytic converter is positioned near the engine. The problem is that on short journeys, particularly in the winter, the engine does not warm up sufficiently for the catalysts to work properly.

5 Why must lead-free petrol be used in a vehicle fitted with a catalytic converter?

6 Which of the products formed by the converter is still undesirable? Give a reason for your answer.

Key Ideas 192 – 193

- **The internal combustion engine produces a number of pollutants including NO_x, CO and unburnt hydrocarbons.**
- **These pollutants can be removed by catalytic converters.**
- **The catalyst is a mixture of expensive metals.**
- **The conversion of the pollutants into less harmful products involves redox reactions.**

Unit 12 Questions

9

(1) The mechanism for the chlorination of ethane to form chloroethane is given below:

Step 1: $Cl_2 \rightarrow 2Cl\cdot$
Step 2: $Cl\cdot + CH_3CH_3 \rightarrow CH_3CH_2\cdot + HCl$
Step 3: $CH_3CH_2\cdot + Cl_2 \rightarrow CH_3CH_2Cl + Cl\cdot$
Step 4: $CH_3CH_2\cdot + Cl\cdot \rightarrow CH_3CH_2Cl$

a) (i) What name is given to this overall mechanism?
(ii) Name the individual steps. (5)

b) Why is this reaction described as a *chain reaction*? (2)

c) Why is this reaction not a good method of preparation for chloroethane? Suggest an alternative. (3)

d) Explain why chloroethane has no isomers. (2)

(2) Figure 1 represents the fractional distillation of petroleum. Petroleum is a mixture of saturated hydrocarbons.

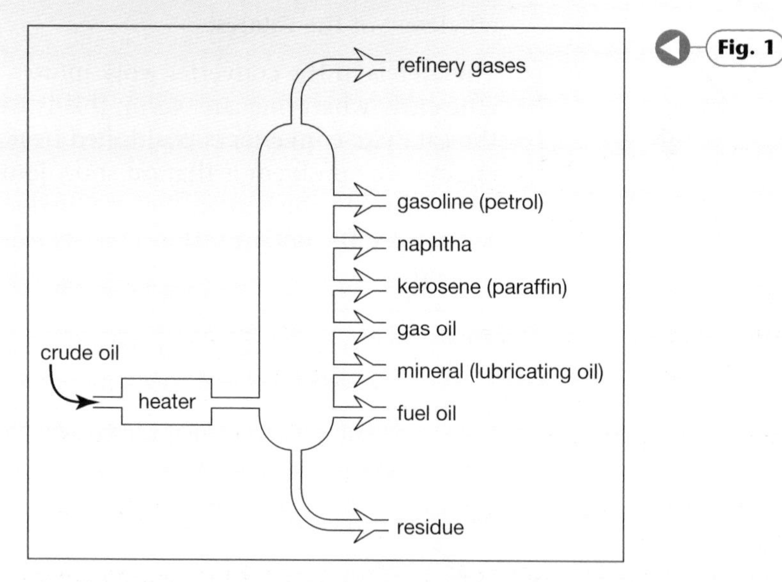

Fig. 1

a) (i) Explain the terms *saturated* and *hydrocarbon*.
(ii) To which homologous series do saturated hydrocarbons belong? (3)

b) (i) Give an alternative name for the fraction labelled *gas oil*.
(ii) Why would this fraction be unsuitable for use as camping gas?
(iii) Which fraction would be a better alternative for use as camping gas? (4)

c) What is the fraction labelled *fuel oil* used to fuel? (1)

d) Name the solid present in the residue which is used for surfacing roads. (1)

3 The process of cracking of hydrocarbons in the petroleum industry involves the breaking of the weakest C—C bonds to form smaller molecules.

a) (i) Name the type of mechanism involved in thermal cracking.
(ii) Complete the following equation to show cracking in a molecule of $C_{12}H_{26}$:
$C_{12}H_{26} \rightarrow$ .... + $2C_2H_4$ (2)

b) (i) Name the type of mechanism involved in catalytic cracking.
(ii) Name the type of catalyst used in this process.
(iii) Give one advantage of catalytic cracking over thermal cracking. (3)

c) Another way of cracking $C_{12}H_{26}$ would be that shown in Figure 2.

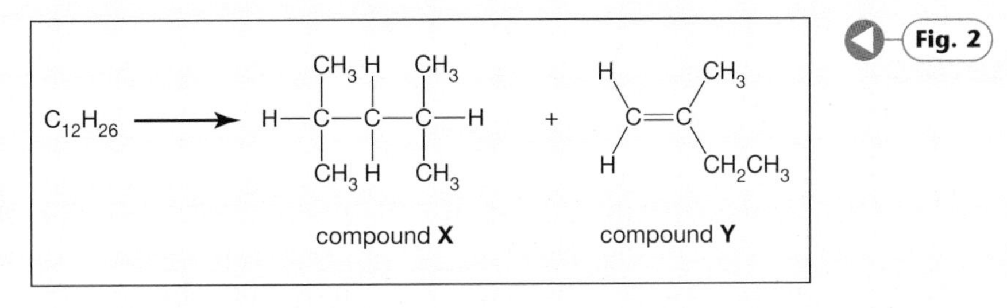

Fig. 2

(i) Give the systematic name for compound **X**.
(ii) To which homologous series does compound **Y** belong? (2)

4 Figure 3 shows a ball-and-stick model of an isomer of octane (C_8H_{18}) which is present in petrol.

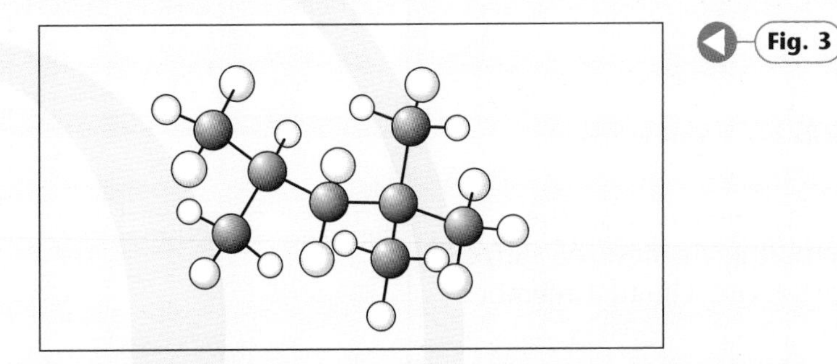

Fig. 3

a) (i) Draw out the structural formula of this molecule.
(ii) What is the systematic name for this molecule?
(iii) Give one advantage of the addition of this compound to petrol. (5)

b) (i) Write an equation to show the complete combustion of this compound.
(ii) What volume of oxygen, measured in dm^3, would be needed to react completely with $1\,dm^3$ of C_8H_{18}? (3)

c) The incomplete combustion process that occurs in the engine of a car produces carbon monoxide, which is a pollutant.
(i) Write an equation for the incomplete combustion of C_8H_{18}.
(ii) Explain why carbon monoxide is described as a *pollutant*.
(iii) Explain briefly how the presence of a catalytic converter reduces carbon monoxide emissions into the atmosphere. (6)

1 Unsaturated hydrocarbons ...

At GCSE you will have studied the hydrocarbons, and you will have learned that some hydrocarbons are unsaturated because they have a carbon–carbon double bond. Hydrocarbons that contain a carbon–carbon double bond are called alkenes. In this unit you will study the reactions and properties of the alkenes.

Alkenes

The alkenes are a homologous series of hydrocarbons that contain a carbon—carbon double bond. The presence of the double bond makes the alkenes **unsaturated**. The general formula for the alkenes is C_nH_{2n}. The carbon—carbon double bond is planar, with bond angles of 120°.

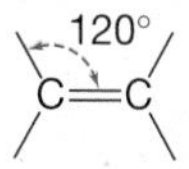

120°
C=C

Fig. 1 *Structure of the alkene functional group*

The names of the alkenes are based on the names of the corresponding alkanes: the suffix is changed from *–ane* to *–ene*. The first member of the series is ethene.

Name	Structural formula	Structure
Ethene	$CH_2{=}CH_2$	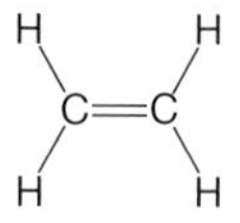
Propene	$CH_3CH{=}CH_2$	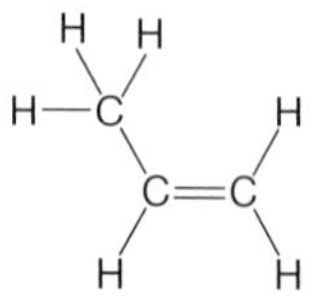

Table 1 *Naming alkenes*

The higher alkenes have **position isomers**. Position isomers occur when the functional group can appear in different places in the molecule. The position of the functional group is included in the name of the isomer.

Table 2 *Naming position isomers*

Name	Structural formula	Structure
But-1-ene	$CH_3CH_2CH{=}CH_2$	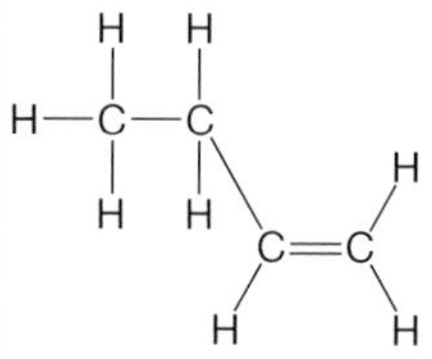
But-2-ene	$CH_3CH{=}CHCH_3$	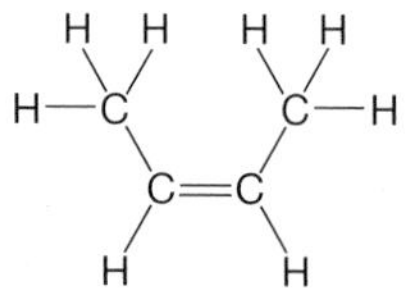
Methylpropene	$(CH_3)_2C{=}CH_2$	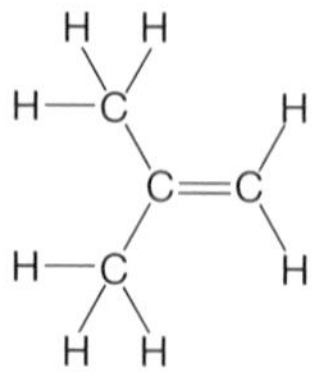

Q

1 Draw the structures of the following alkenes.

a) pent-1-ene
b) pent-2-ene
c) 2-methylpent-2-ene.

Geometrical isomerism

The presence of the double bond in the alkenes can result in the existence of **geometrical isomers**. The carbon—carbon double bond is resistant to rotation, and this produces ***cis*** and ***trans*** isomers.

Fig. 2 *Geometrical isomers of but-2-ene*

Viewed along the bond, in the *cis* form, both $—CH_3$ groups are on the *same* side of the double bond. In the *trans* form, the $—CH_3$ groups are on *opposite* sides.

Geometrical isomerism is a form of **stereoisomerism**. The two isomers have the same molecular and structural formulae, but differ in the orientation of some atoms.

The necessary condition for an alkene to have geometrical isomers is that each of the carbon atoms either side of the double bond must have two different groups of atoms attached (Figure 3).

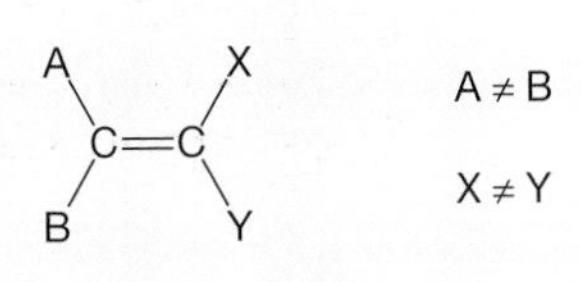

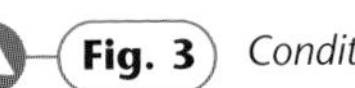

Fig. 3 *Conditions for geometrical isomerism*

But-2-ene has geometrical isomers because each of the two carbon atoms in the carbon—carbon double bond has two different groups attached. But-1-ene does *not* have geometrical isomers because one of the carbons in the double bond has two H atoms attached.

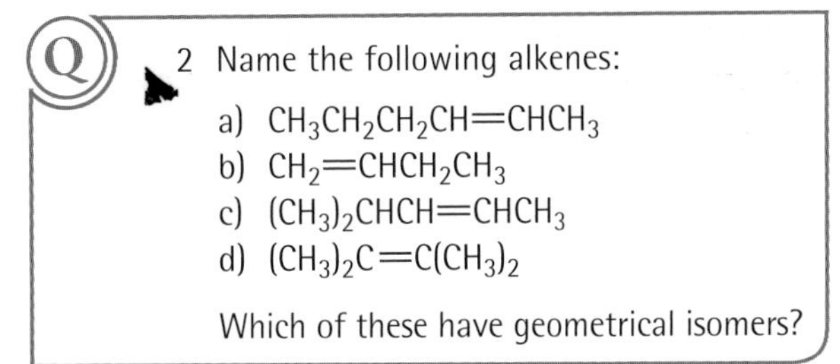

Q

2 Name the following alkenes:

a) $CH_3CH_2CH_2CH{=}CHCH_3$
b) $CH_2{=}CHCH_2CH_3$
c) $(CH_3)_2CHCH{=}CHCH_3$
d) $(CH_3)_2C{=}C(CH_3)_2$

Which of these have geometrical isomers?

Combustion reactions

The alkenes are hydrocarbons and will undergo combustion. In excess air or oxygen, the carbon is oxidised to form carbon dioxide and the hydrogen is oxidised to form water.

$$CH_2{=}CH_2 + 3O_2 \rightarrow 2CO_2 + 2H_2O$$

$$CH_3CH{=}CH_2 + 4\tfrac{1}{2}O_2 \rightarrow 3CO_2 + 3H_2O$$

In a limited supply of air or oxygen, incomplete combustion can occur, leading to the formation of carbon monoxide.

Q

3 Write equations for the complete combustion of:

a) butene
b) pentene
c) methylpropene.

Key Ideas 196 – 197

- **The alkenes are unsaturated hydrocarbons containing a carbon–carbon double bond.**
- **The carbon–carbon double bond is resistant to rotation, and this results in geometrical isomers.**
- **Alkenes undergo combustion reactions, forming carbon dioxide and water.**

2 Breaking the double bond ...

The alkenes are far more reactive than the alkanes. It is the carbon–carbon double bond that is responsible for the greater reactivity of the alkenes.

When the carbon–carbon double bond breaks the alkenes undergo addition reactions to form a single bond. This allows each carbon atom to form a bond with other atoms or groups of atoms, thereby forming a saturated product.

Electrophilic addition

There is a region of high **electron density** around the carbon—carbon double bond. This area is readily attacked by **electron-deficient** species known as **electrophiles**. An electrophile has a partial positive charge (δ^+) on one of its atoms. An electrophile accepts a pair of electrons.

Alkenes undergo addition reactions with electrophiles, producing a saturated product. We can use X—Y, where X has a δ^+ charge, to represent an electrophile in such a reaction:

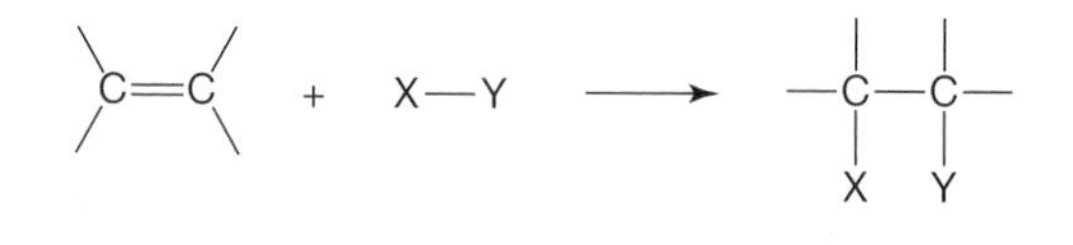

Fig. 1 *Electrophilic addition*

The reaction takes place in two steps.

Step 1: A pair of electrons from the carbon—carbon double bond in the alkene form a covalent bond with the δ^+ X atom. The X—Y bond breaks, forming a Y^- ion. The addition of the X atom onto the alkene results in the formation of an intermediate species with a positively charged carbon atom (a **carbocation**).

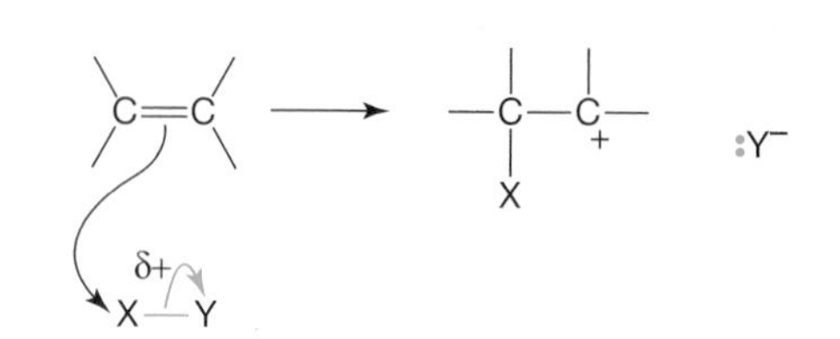

Fig. 2 *First step of mechanism*

Step 2: The lone pair of electrons on the Y^- ion form a covalent bond with the positively charged carbon atom.

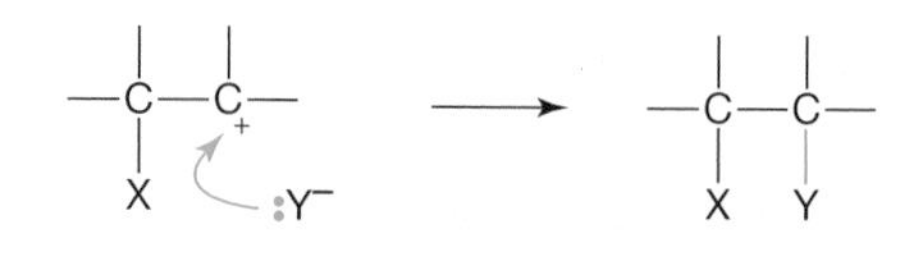

Fig. 3 *Second step of mechanism*

The full mechanism of the reaction is shown in Figure 4.

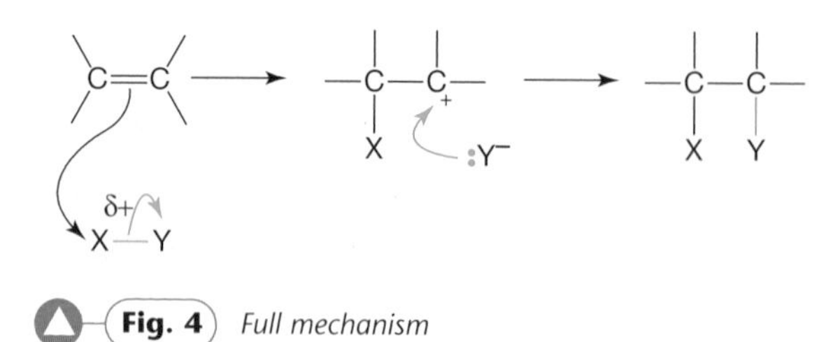

Fig. 4 *Full mechanism*

Reaction with hydrogen bromide

Ethene gas is passed through concentrated HBr, or through HBr dissolved in an inert solvent such as tetrachloromethane, CCl_4:

$$CH_2{=}CH_2 + HBr \rightarrow CH_3CH_2Br$$

ethene bromoethane

The reaction occurs in two steps.

Step 1: HBr is a polar molecule with a δ^+ charge on the hydrogen atom. In the reaction of HBr with an alkene, a pair of electrons from the carbon—carbon double bond forms a bond

with the δ^+ hydrogen atom. The hydrogen—carbon bond breaks, leaving a bromide ion (Figure 5).

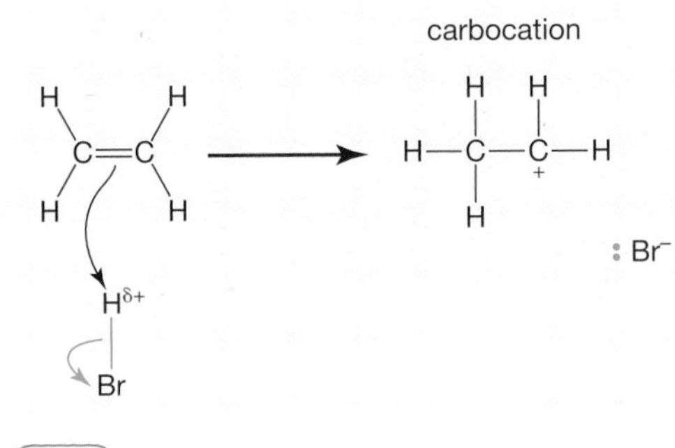

Fig. 5 *First step of mechanism*

The addition of a hydrogen atom to the alkene forms a carbocation (sometimes called a **carbonium ion**) intermediate, which has a positive charge on a carbon atom.

Step 2: Carbocations are very reactive species and are readily attacked by negative ions (and by species that have a lone pair of electrons). The carbocation formed in Step 1 is attacked by the bromide ion (Figure 6).

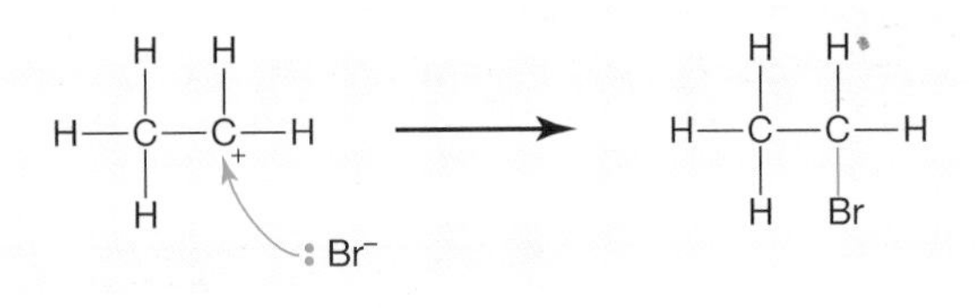

Fig. 6 *Second step of mechanism*

The full mechanism of the reaction is shown in Figure 7.

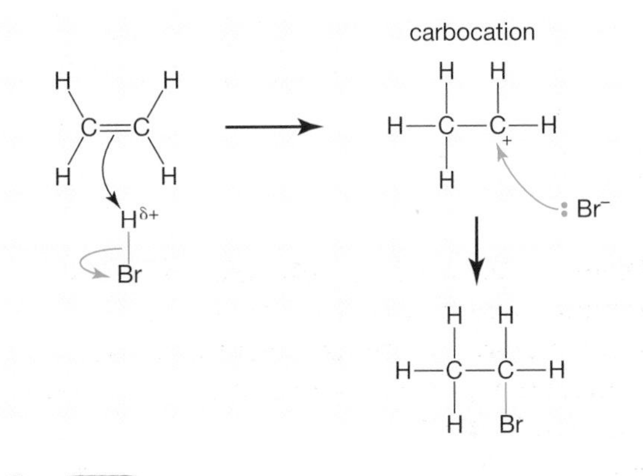

Fig. 7 *Full mechanism*

The curly arrow in the mechanism shows the movement of the pair of electrons when a bond is formed and when a bond is broken.

Q

1 Explain why HBr is a polar molecule.

2 Explain why alkenes are attacked by electrophiles.

3 What does a curly arrow represent in the mechanism of a reaction?

Key Ideas 198 – 199

- The double bond in alkenes is a region of high electron density.
- Alkenes are readily attacked by electrophiles.
- Alkenes undergo electrophilic addition reactions.
- The mechanism of a reaction shows how the bonds are formed and broken in a reaction.
- In mechanisms a curly arrow is used to show the movement of a pair of electrons involved in bond-forming and bond-breaking.

3 Breaking the double bond ... (2)

Two other electrophiles that undergo addition reactions with alkenes are concentrated sulphuric acid and bromine. The reactions with concentrated sulphuric acid and bromine each involve a two-step reaction.

Reaction with concentrated sulphuric acid

Ethene gas is passed through cold concentrated sulphuric acid.

$$CH_2{=}CH_2 + H_2SO_4 \rightarrow CH_3CH_2OSO_3H$$

ethene ethyl hydrogensulphate

The hydrogen atoms in H_2SO_4 carry a δ^+ charge and this makes H_2SO_4 an electrophile.

Step 1: The first step of the mechanism involves the addition of a hydrogen from the sulphuric acid to form a carbocation (Figure 1).

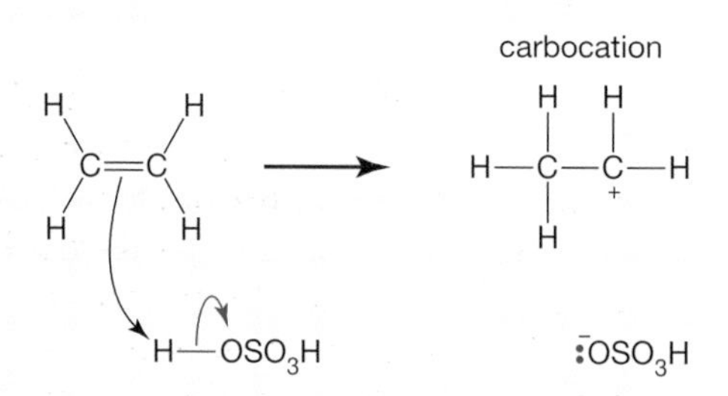

Fig. 1 *First step of mechanism*

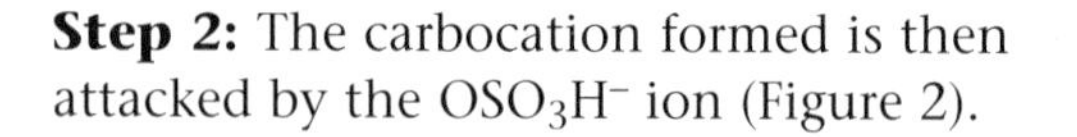

Step 2: The carbocation formed is then attacked by the OSO_3H^- ion (Figure 2).

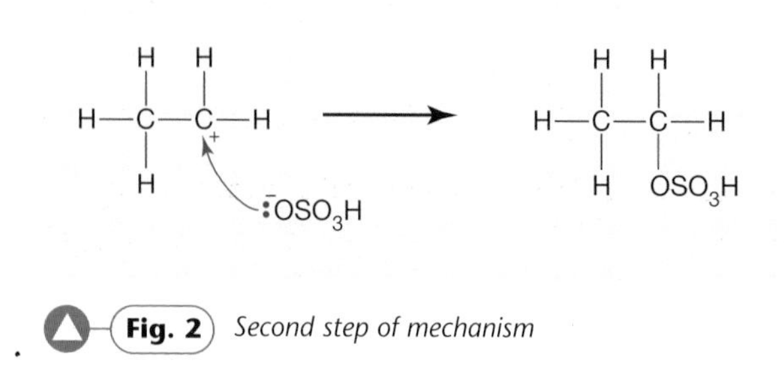

Fig. 2 *Second step of mechanism*

The full mechanism for the reaction is shown in Figure 3.

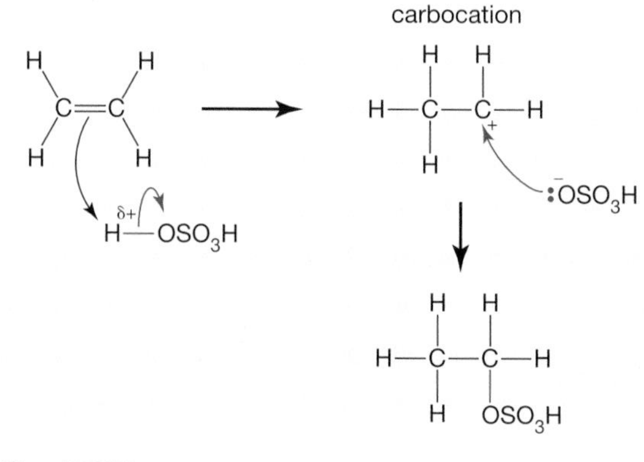

Fig. 3 *Full mechanism*

Ethyl hydrogensulphate is readily hydrolysed when warmed with water forming an alcohol:

$$CH_3CH_2OSO_3H + H_2O \rightarrow CH_3CH_2OH + H_2SO_4$$

ethanol

The overall reaction of an alkene with concentrated sulphuric acid followed by hydrolysis can be represented by a single equation. (Sulphuric acid is not included in this equation because it acts as a catalyst in the overall reaction.)

$$CH_2{=}CH_2 + H_2O \rightarrow CH_3CH_2OH$$

Q

1 Give the structural formula of the carbocation formed when but-2-ene reacts with concentrated sulphuric acid.

Reaction with bromine

Ethene gas is passed through bromine.

Step 1: The bromine molecule is non-polar, but when a bromine molecule approaches the carbon—carbon double bond in an alkene, the region of high electron density induces a temporary polar bond in the bromine molecule.

A pair of electrons from the carbon–carbon double bond in the alkene forms a bond with the δ^+ bromine atom in the bromine molecule, forming a carbocation intermediate and a bromide ion (Figure 4).

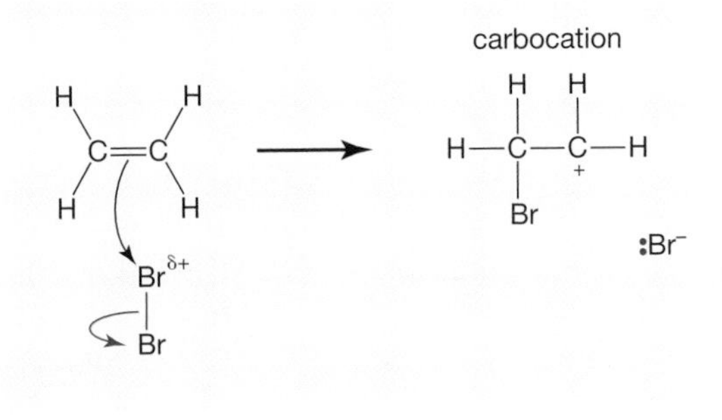

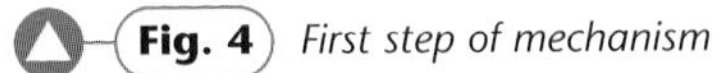
Fig. 4 *First step of mechanism*

Step 2: The carbocation formed is then attacked by the bromide ion as shown in Figure 5.

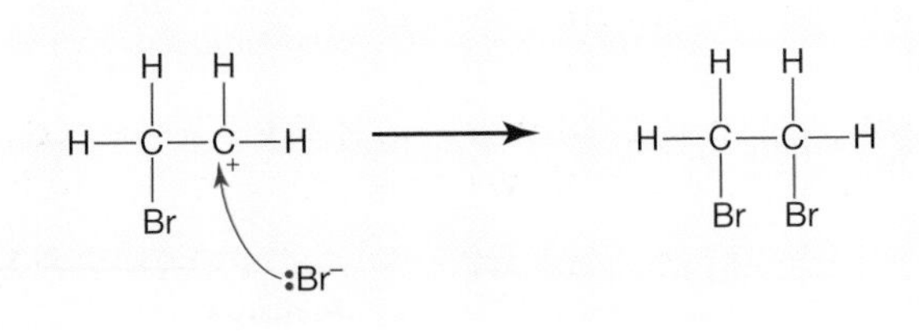

Fig. 5 *Second step of mechanism*

The full mechanism for the reaction is as shown in Figure 6.

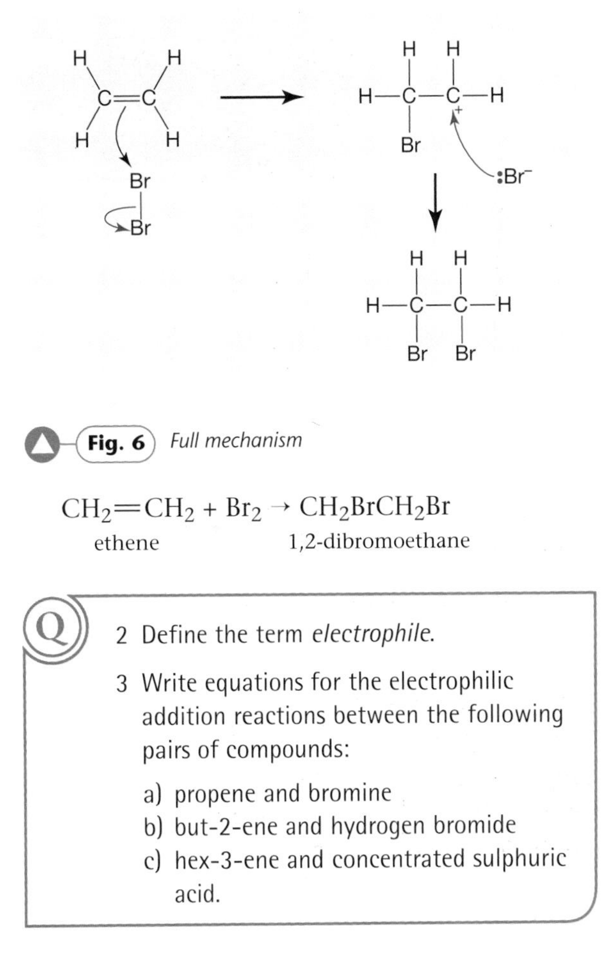

Fig. 6 *Full mechanism*

$$CH_2{=}CH_2 + Br_2 \rightarrow CH_2BrCH_2Br$$

ethene 1,2-dibromoethane

> **Q**
>
> 2 Define the term *electrophile*.
>
> 3 Write equations for the electrophilic addition reactions between the following pairs of compounds:
>
> a) propene and bromine
> b) but-2-ene and hydrogen bromide
> c) hex-3-ene and concentrated sulphuric acid.

The reaction of alkenes with bromine provides the basis of a test for the alkenes. Bromine is a very hazardous chemical to use, and therefore a dilute solution of bromine in water (**bromine water**) is used for the test. Bromine water is an orange-coloured solution, and undergoes an addition reaction with alkenes to form a colourless saturated product. Addition of bromine water to *any* alkene will result in the bromine water changing from an orange solution to a colourless solution.

Key Ideas 200 – 201

- **Bromine water is used as a test for alkenes.**
- **Alcohols are formed when alkenes react with concentrated sulphuric acid and the products are hydrolysed.**

4 Major and minor products ...

In all the examples so far, the alkenes used have been symmetrical alkenes. When the alkene is asymmetrical, the addition of HBr and H_2SO_4 leads to the formation of two products.

Reaction with hydrogen bromide

When HBr adds onto propene, there are two products, a major product and a minor product.

$$CH_3CH{=}CH_2 + HBr \rightarrow CH_3CHBrCH_3$$
2-bromopropane

$$CH_3CH{=}CH_2 + HBr \rightarrow CH_3CH_2CH_2Br$$
1-bromopropane

2–bromopropane is the major product; 1–bromopropane is the minor product.

To understand why this reaction produces two products, and why they are not formed in equal quantities, we need to consider the mechanism of the reaction.

In the initial step of the mechanism, the hydrogen from the HBr can bond to either of the carbon atoms in the carbon—carbon double bond. This results in the formation of two different carbocations.

In the reaction of HBr and propene a **primary carbocation** and a **secondary carbocation** are both formed. The secondary carbocation is *more stable* than the primary carbocation, and exists for longer. Therefore a bromide ion is *more likely* to react with a secondary carbocation. This results in 2-bromopropane being formed in greater quantities than 1-bromopropane.

The major product is always formed from the more stable carbocation.

Carbocations

Carbocations can be primary, secondary or tertiary, depending upon the number of carbon atoms directly bonded to the positively charged carbon atom (see Figure 2).

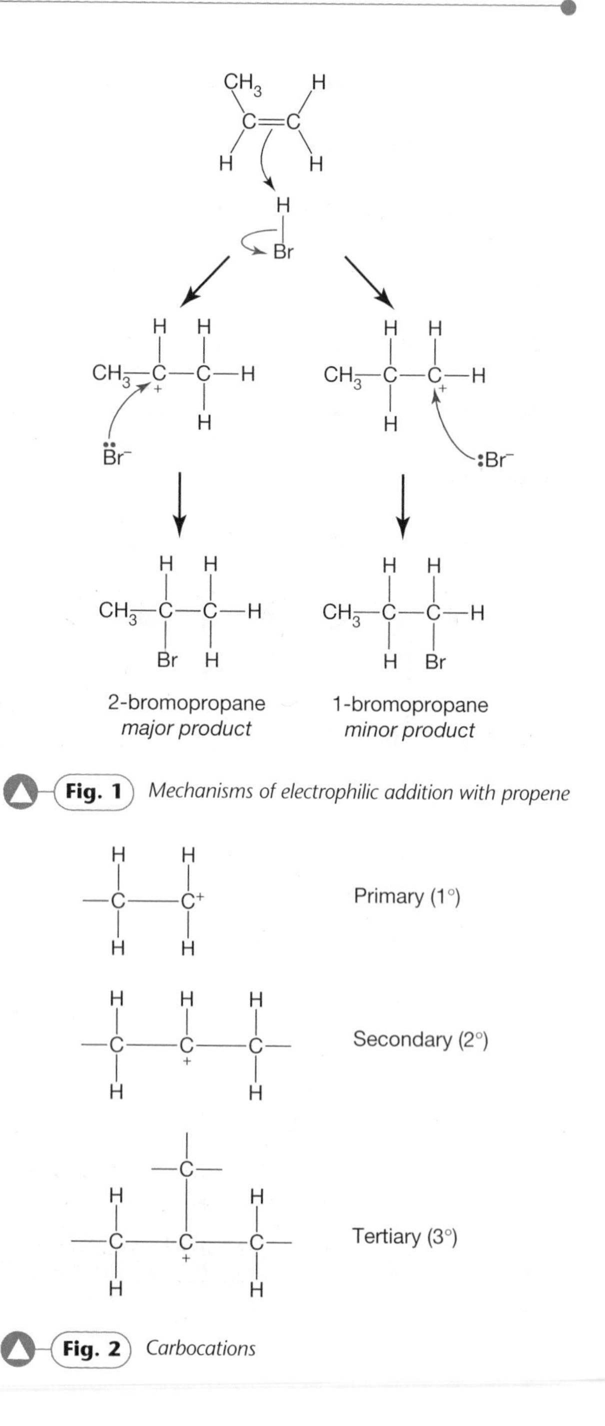

Fig. 1 Mechanisms of electrophilic addition with propene

Fig. 2 Carbocations

These carbocations have different stabilities. The stability is increased by the presence of alkyl groups. These groups are **electron-releasing** relative to hydrogen, and help to stabilise the positive charge on the carbocation. As the number of alkyl groups increases, so the stability of the carbocation increases.

The order of stability of carbocations is thus:

increasing stability ↑
- tertiary
- secondary
- primary

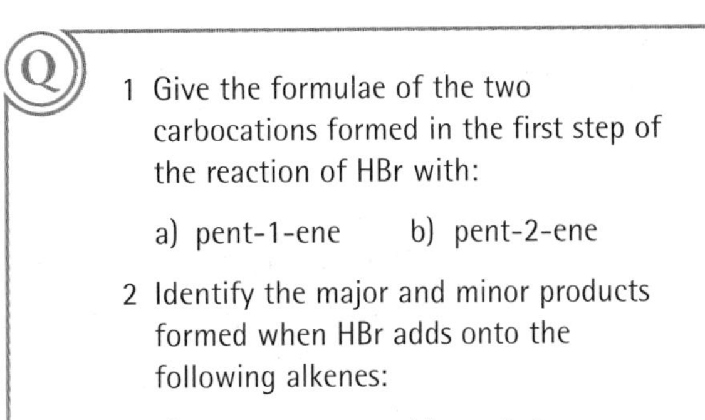

1 Give the formulae of the two carbocations formed in the first step of the reaction of HBr with:

a) pent-1-ene b) pent-2-ene

2 Identify the major and minor products formed when HBr adds onto the following alkenes:

a) but-1-ene b) methylpropene

Reaction with concentrated sulphuric acid

The mechanism of the reaction with concentrated sulphuric acid follows the same pattern as the reaction with HBr. In the first step of the mechanism a hydrogen from the sulphuric acid adds onto the alkene forming two different carbocations. Each of these carbocations then reacts with OSO_3H^- ions to give different products. When these products are hydrolysed by water, two different alcohols are formed.

3 Give the formulae of the two alcohols formed when each of the following alkenes reacts with concentrated sulphuric acid and the products are hydrolysed by water. In each case state which is the major product.

a) pent-1-ene b) 2-methylbut-2-ene

4 Give the names and formulae of the alkenes from which the following alcohols could be prepared:

a) $CH_3CH_2CH(OH)CH_3$ b) $(CH_3)_3COH$

$CH_3CH_2CH=CH_2$ + $H^{\delta+}$—OSO_3H

secondary carbocation: $CH_3CH_2CH^+CH_3$ + $:OSO_3H^-$ → $CH_3CH_2CH(OSO_3H)CH_3$ → hydrolysis → $CH_3CH_2CH(OH)CH_3$
butan-2-ol
major product

primary carbocation: $CH_3CH_2CH_2CH_2^+$ + $:OSO_3H^-$ → $CH_3CH_2CH_2CH_2OSO_3H$ → hydrolysis → $CH_3CH_2CH_2CH_2OH$
butan-1-ol
minor product

Mechanisms of electrophilic addition with but-1-ene **Fig. 3**

Key Ideas 202 – 203

- **The addition of HBr or H_2SO_4 to an asymmetrical alkene leads to the formation of two products.**
- **The initial addition of a hydrogen leads to the formation of two different carbocations.**
- **The major product is formed from the more stable carbocation.**

5 Industry matters ...

The alkenes are very important in the production of a large range of commercial chemicals. In this section you will study the hydrogenation reactions of alkenes and the formation of addition polymers.

Hydrogenation reactions

A reaction in which hydrogen molecules are added to an alkene is called a **hydrogenation reaction** (it is a *reduction* reaction).

Alkenes will undergo an addition reaction with hydrogen gas in the presence of a catalyst to form alkanes. The catalyst often used is nickel, at 150°C. The reaction occurs on the surface of the catalyst.

Examples of hydrogenation reactions:

$CH_2{=}CH_2 + H_2 \rightarrow CH_3CH_3$
ethene → ethane

$CH_3CH{=}CH_2 + H_2 \rightarrow CH_3CH_2CH_3$
propene → propane

$CH_2{=}CHCH{=}CH_2 + 2H_2 \rightarrow CH_3CH_2CH_2CH_3$
buta-1,3-diene → butane

A hydrogenation reaction is used to convert vegetable oils into margarine. Vegetable oils are liquids with long carbon chains containing many carbon—carbon double bonds – they are **polyunsaturated**. The polyunsaturated oils have lower melting points than the saturated versions. When *some* of the double bonds are hydrogenated, the melting point increases and the vegetable oil becomes a solid at room temperature. This process is called **hardening**.

Butter is made from saturated animal fats. Unsaturated vegetable oils are a healthier alternative to these saturated fats.

Q

1 Write an equation for the hydrogenation of each of the following alkenes:

a) methylpropene
b) pent-2-ene
c) 2-methylbut-2-ene

2 Name the two alkenes that form butane when they react with hydrogen gas.

Addition polymerisation

A **polymerisation** reaction involves joining together a large number of small molecules, **monomers**, to form a very large molecule called a **polymer**.

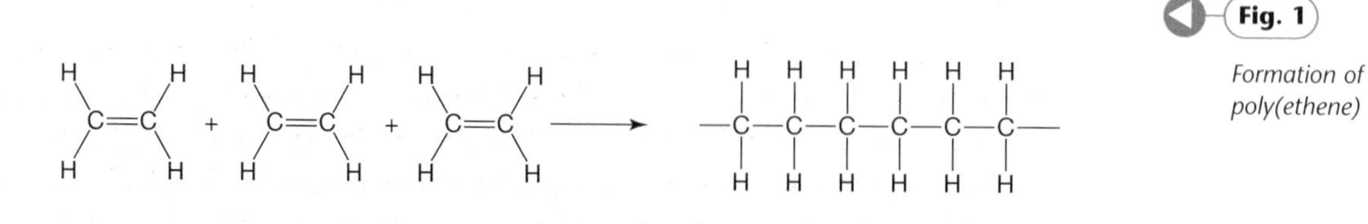

Fig. 1 *Formation of poly(ethene)*

Alkenes undergo addition polymerisation in which thousands of alkene monomers join together to form a long-chain polymer. Addition polymers are usually formed by a free-radical reaction. Addition polymerisation reactions are carried out at high pressure, in the presence of an initiator which provides the free radicals needed to start a chain reaction.

The simplest addition polymer, poly(ethene), is formed from ethene. Its tradename is 'polythene'. For other examples see Figure 3.

- Poly(ethene) is used to make a wide range of products. Low-density poly(ethene) is used to make washing-up bowls, squeezy bottles and plastic bags. High-density poly(ethene) is more rigid than low-density poly(ethene), and is used to make trays and crates.
- Poly(propene) is used to make plastic pipes, chairs and wrapping film.
- Poly(chloroethene) is used to make records, imitation leather, plastic pipes and floor tiles.
- Poly(phenylethene) is used to make toys. Expanded poly(phenylethene) is used to make insulation tiles, packaging materials and plastic cups for hot drinks.
- Poly(methyl-2-methylpropenoate), or perspex, is used as a glass substitute in windows and lenses.
- Poly(tetrafluoroethene) is used as a low-friction material in curtain rails, and as a non-stick surface on cooking utensils.

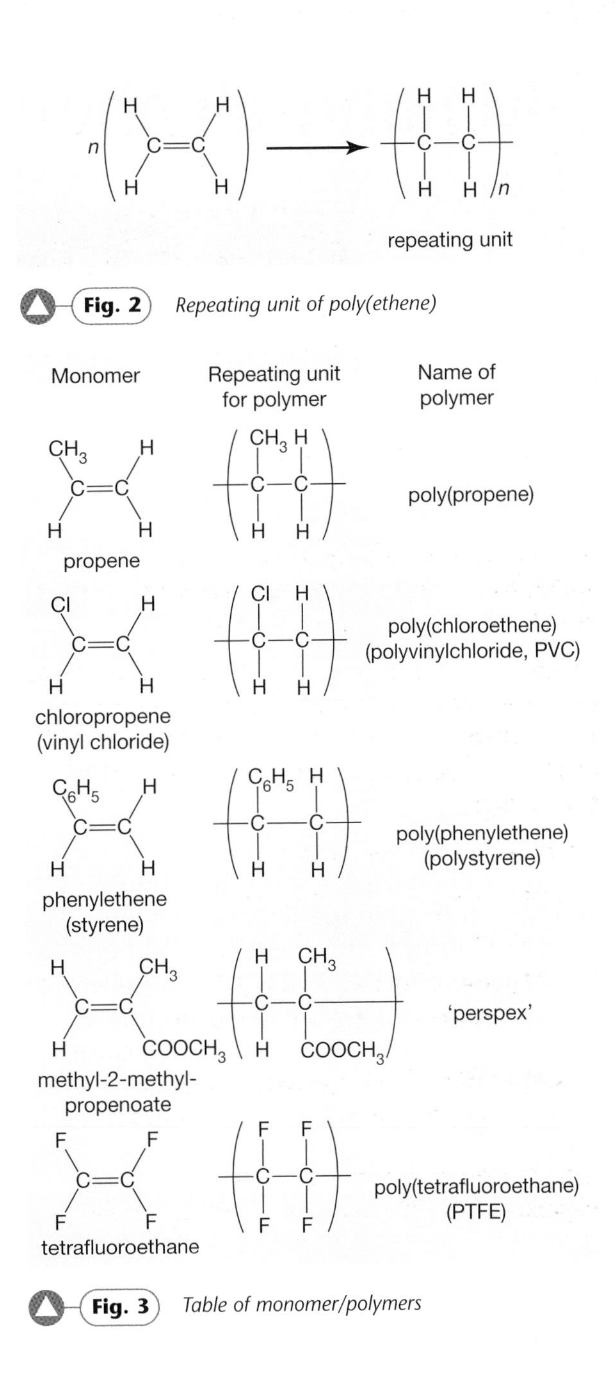

Fig. 2 *Repeating unit of poly(ethene)*

Fig. 3 *Table of monomer/polymers*

Q

3 Draw the repeating unit of the polymer formed from:

a) $CH_3CH{=}CHCH_3$ b) $(CH_3)_2C{=}CH_2$

Key Ideas 204 – 205

- **Alkenes will undergo an addition reaction with hydrogen to form an alkane. This reaction is the basis of the manufacture of margarine.**
- **Alkenes undergo addition polymerisation reactions to form long-chain polymers. Many different addition polymers can be made from substituted ethene compounds. Each polymer has its own physical properties, so the different polymers can be used to make a range of commercial products.**

6 Making alcohols ...

Alkenes can be reacted to form alcohols in two different industrial reactions. One method involves a direct hydration reaction with water. In another reaction, alkenes react with oxygen to form cyclic ethers, which then react with water to form products with two alcohol groups.

Direct hydration

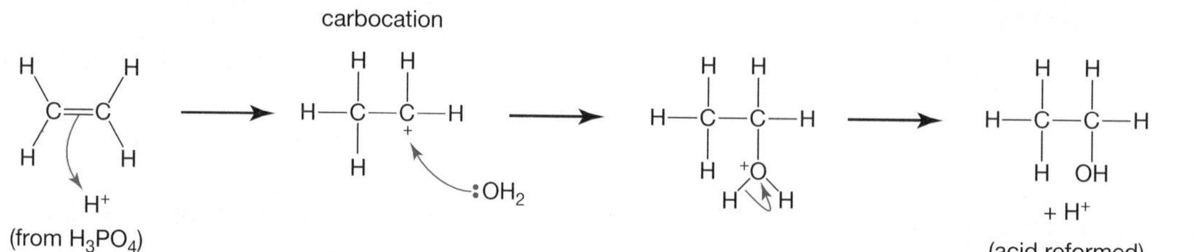

Fig. 1 *Mechanism of direct hydration*

Alkenes will undergo a direct hydration reaction with steam in the presence of a catalyst to form an alcohol. The conditions used are a pressure of 7 MPa (70 atmospheres), a temperature of 300 °C, and a concentrated phosphoric acid (H_3PO_4) catalyst on a silica support. For example, ethanol is made industrially in a **continuous process** from ethene:

$$\underset{\text{ethene}}{CH_2{=}CH_2(g)} + H_2O(g) \rightleftharpoons \underset{\text{ethanol}}{CH_3CH_2OH(l)}$$

A continuous process is one in which more reactants are added as the products are removed.

The reaction is an electrophilic addition reaction. Phosphoric acid is a strong acid and reacts with the ethene to form a carbocation. The carbocation is then attacked by a water molecule, forming ethanol.

The pressure in the reaction is carefully controlled, as very high pressures could result in a polymerisation reaction.

Other alcohols are also made by this industrial method:

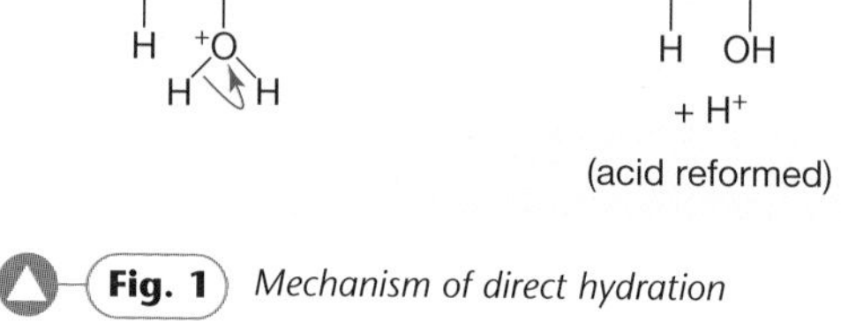

$$\underset{\text{propene}}{CH_3CH{=}CH_2(g)} + H_2O(g) \rightleftharpoons \underset{\text{propan-2-ol}}{CH_3CH(OH)CH_3(l)}$$

Propan-2-ol is the major product formed from a secondary carbocation.

Q

1 Use Le Chatelier's Principle to explain why a pressure of 7 MPa is used in the direct hydration reaction of alkenes.

2 Give the structure of the carbocation formed during the formation of propan-2-ol.

Epoxyethane

Epoxyethane is manufactured by the reaction of ethene with oxygen, using a silver catalyst at a temperature of 180 °C.

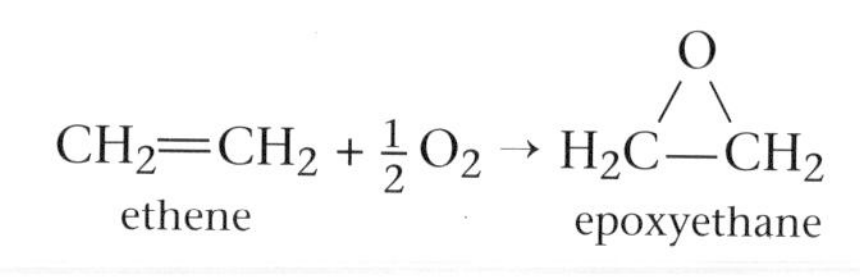

$$\underset{\text{ethene}}{CH_2{=}CH_2} + \tfrac{1}{2}O_2 \rightarrow \underset{\text{epoxyethane}}{H_2C{-}CH_2 \text{ (with O bridging)}}$$

Fig. 2 *Structure of epoxyethane*

The industrial manufacture is very carefully controlled because epoxyethane is very reactive and there is a risk of violent reactions or explosions.

Epoxyethane is a cyclic **ether** and has a 3-membered cyclic structure (Figure 2). This is a strained structure and the carbon—oxygen bonds are very easily broken. This makes epoxyethane very reactive.

Epoxyethane is readily hydrolysed by water to form an alcohol:

$$\underset{\text{epoxyethane}}{H_2C\overset{O}{\diagup\diagdown}CH_2} + H_2O \rightarrow \underset{\text{ethane-1,2-diol}}{HOCH_2CH_2OH}$$

Ethane-1,2-diol contains two alcohol groups. The old name of ethane-1,2-diol is ethylene glycol (or glycol). It is used as an **antifreeze**, and added to the water in the cooling systems of the engines of motor vehicles. Ethane-1,2-diol lowers the freezing point of water and stops the water in cooling systems from freezing in cold weather.

Ethane-1,2-diol is also used to make condensation polymers. Polyesters such as Terylene™ are made using ethane-1,2-diol. Terylene is used in clothing and furnishings.

Epoxyethane will also react with alcohols (ROH):

$$H_2C\overset{O}{\diagup\diagdown}CH_2 + ROH \rightarrow ROCH_2CH_2OH$$

The product is an alkoxyalcohol and can react with more epoxyethane to produce poly(alkoxyalcohols):

$$H_2C\overset{O}{\diagup\diagdown}CH_2 + ROCH_2CH_2OH \rightarrow ROCH_2CH_2OCH_2CH_2OH$$

The reaction of 1 mole of an alcohol (ROH) with *n* moles of epoxyethane can be represented by the following equation:

$$nH_2C\overset{O}{\diagup\diagdown}CH_2 + ROH \rightarrow R(OCH_2CH_2)_nOH$$

Poly(alkoxyalcohols) are used to make solvents, plasticisers and non-ionic surfactants.

3 Explain why epoxyethane is a very reactive compound.

4 Write an equation for the reaction of 3 moles of epoxyethane with 1 mole of ethanol.

Key Ideas 206 – 207

- **Alcohols are manufactured by direct hydration reactions of alkenes with steam. The reaction is an electrophilic addition reaction.**
- **Ethene reacts with oxygen to form the cyclic ether epoxyethane.**
- **Epoxyethane has a strained cyclic structure and is very reactive.**
- **Epoxyethane is reacted with water or alcohols to form a large range of useful chemicals, including ethane-1,2-diol and alkoxyalcohols.**

Unit 13 Questions

7

1. Name each of the following alkenes:

   a) $CH_2{=}CHCH_2CH_2CH_3$ (1)
   b) $CH_3CH{=}C(CH_3)CH_2CH_3$ (1)
   c) $(CH_3)_2C{=}C(CH_3)_2$ (1)

2. Write equations for the complete combustion of the following alkenes:

   a) $CH_3CH{=}CHCH_3$ (1)
   b) $CH_3CH_2CH_2CH{=}CHCH_3$ (1)

3. a) Define the term *electrophile.* (1)
   b) Explain why alkenes are readily attacked by electrophiles. (1)
   c) Outline the mechanism for the electrophilic addition of HBr with ethene. (4)

4. When but-1-ene reacts with HBr, a mixture of 1-bromobutane and 2-bromobutane is formed.

   a) Outline the mechanism for the formation of the two bromobutane isomers. (5)
   b) Which of the bromobutane isomers is obtained in greater quantity? (1)
   c) Explain why equal quantities of the two bromobutanes are not formed in the reaction. (3)

5. a) Write an equation for the reaction of concentrated sulphuric acid with but-2-ene. (1)
   b) Outline the mechanism of the reaction. (4)
   c) Write an equation for the hydrolysis reaction of the product. (2)

6. When concentrated sulphuric acid reacts with methylpropene and the product is hydrolysed, the alcohol methylpropan-2-ol is formed.

   a) Outline the mechanism for the reaction of methylpropene and concentrated sulphuric acid. (4)
   b) Explain why methylpropan-1-ol is also formed as a minor product. (3)

7. a) Ethanol, CH_3CH_2OH, can be prepared industrially by the reaction of ethene with steam.
      (i) Give the essential conditions for this reaction.
      (ii) Write an equation for the reaction. (4)
   b) Propene reacts with steam under the same conditions to form propan-2-ol. Outline a possible mechanism for the reaction. (4)

8 **a)** Name the compound in Figure 1: (1)

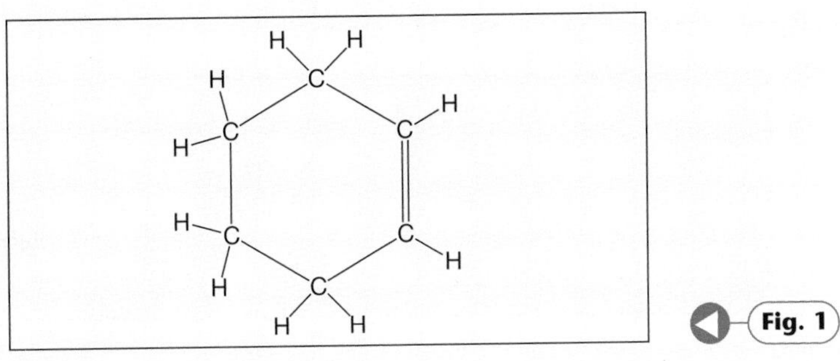

Fig. 1

b) Outline the mechanism of the reaction between this compound and bromine. (4)

9 **a)** Explain why some alkenes exhibit geometrical isomerism. (2)

b) Name and draw the structure of the two geometrical isomers of pent-2-ene. (4)

c) Explain why pent-1-ene does not have geometrical isomers. (2)

10 **a)** Write an equation and state the essential conditions for the formation of epoxyethane. (3)

b) Explain why epoxyethane is a very reactive compound. (2)

c) Write an equation for the reaction of epoxyethane with one mole of water. (1)

d) Give *two* uses of the product. (2)

e) Write an equation for the reaction of one mole of ethanol with:

(i) 2 moles of epoxyethane

(ii) *n* moles of epoxyethane. (4)

11 Give the repeating structures of the addition polymers formed from the following monomers:

a) $CH_3CH{=}CH_2$ (1)

b) $CH_3CH{=}CHCH_3$ (1)

c) $CF_2{=}CF_2$ (1)

d) $CH_3CH{=}CHCl$ (1)

12 By naming reagents and stating observations, describe a simple chemical test to differentiate between cyclohexane and cyclohexene. (3)

13 Bromine water can be considered to contain Br^+OH^-.

a) Give the structural formulae of the two products formed when propene reacts with Br^+OH^-. (2)

b) Outline the mechanism involved in the formation of the major product. (4)

1 Introducing a halogen

In the section on alkane chemistry you will have studied the free-radical substitution reaction of methane with chlorine forming chloromethane. Chloromethane is member of the haloalkane homologous series. Haloalkanes are alkane molecules in which one of the hydrogen atoms has been replaced by a halogen atom.

Haloalkanes

The haloalkanes are named as substituted alkanes with **chloro-**, **bromo-** or **iodo-** used as a prefix.

Name	Structural formula	Structure
Chloromethane	CH_3Cl	
Bromoethane	CH_3CH_2Br	
1-chloropropane	$CH_3CH_2CH_2Cl$	
2-chloropropane	$CH_3CHClCH_3$	

The higher haloalkanes show **position isomerism**. In position isomers the number indicating the position of the halogen atom is included in the name of the isomer (Figure 1).

Haloalkanes are classified as primary, secondary and tertiary. The classification depends upon the number of carbon atoms directly attached to the carbon atom carrying the halogen atom (Figure 2).

The haloalkanes contain a polar carbon—halogen bond. This results in the haloalkanes having permanent dipole–dipole forces between

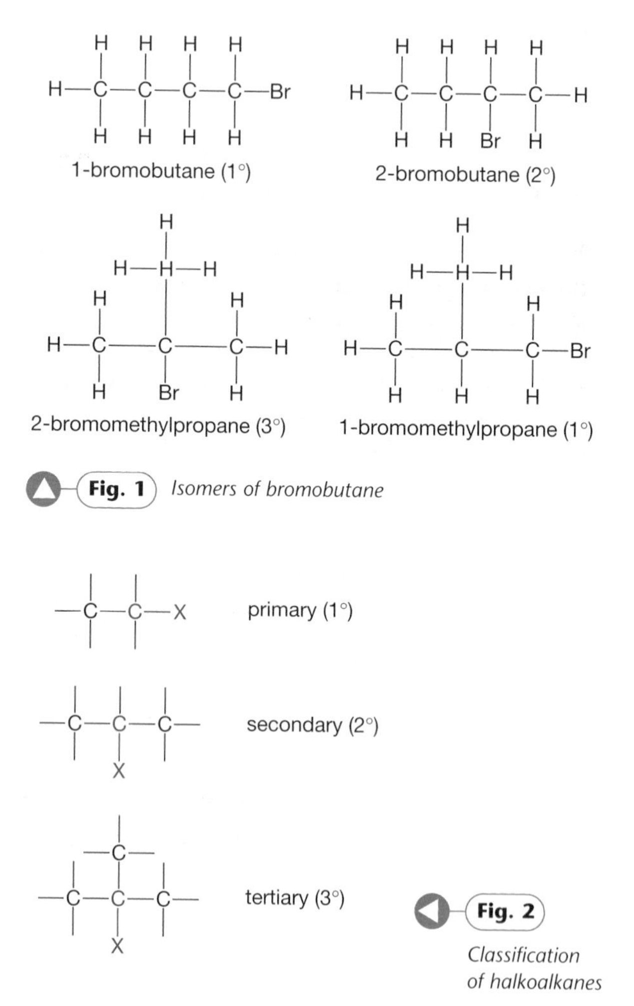

Fig. 1 *Isomers of bromobutane*

Fig. 2 *Classification of halkoalkanes*

the molecules. As a result, haloalkanes have higher melting points and boiling points than alkanes and alkenes with a similar mass. Alkanes and alkenes are non-polar, and there are van der Waals forces between the molecules.

1 Name each of the following haloalkanes:

a) CH_3CH_2I b) $CH_3CHClCH_3$ c) $CH_3CH_2CH_2CHBrCH_3$ d) $CH_3CHBrCHBrCH_3$

2 Give the structural formula for each of the following haloalkanes:

a) 3-bromopentane b) 1,1-dichloropropane c) 2-iodomethylpropane

3 Classify each of the haloalkanes in questions 1 and 2 as primary, secondary or tertiary.

Nucleophilic substitution reactions

The covalent bond between the halogen atom and the carbon atom is polar because the halogens are more electronegative than carbon. As a result the carbon atom has a δ^+ charge. It is this polar bond that influences the chemical reactions of the haloalkanes.

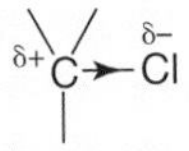

Fig. 3 *Polar bond on a haloalkane*

This **electron-deficient** carbon atom is readily attacked by species that contain a lone pair of electrons. In organic chemistry these species are called **nucleophiles** ('nucleus-loving'). A reaction in which a species with a lone pair attacks a haloalkane and replaces the halogen atom is called **nucleophilic substitution**.

Haloalkanes readily undergo substitution reactions in which the halogen atom is replaced by a different atom or group of atoms. The mechanism of nucleophilic substitution is essentially very simple. The lone pair on the nucleophile forms a covalent bond with the δ^+ carbon atom, the carbon—halogen bond breaks, and the pair of electrons from the carbon—halogen bond pass to the halogen atom, forming a halide ion (Figure 3).

Fig. 4 *Nucleophilic substitution reaction*

Key Ideas 210 – 211

- **The haloalkanes form a homologous series based on the alkanes in which a hydrogen atom has been substituted by a halogen atom.**
- **The haloalkanes contain a polar hydrogen—halogen bond with a partial positive charge on the carbon atom. The δ^+ carbon is readily attacked by nucleophiles.**
- **A nucleophile has a lone pair of electrons.**
- **Haloalkanes undergo substitution reactions with nucleophiles.**

2 Nucleophilic substitution reactions

Haloalkanes undergo nucleophilic substitution with potassium or sodium hydroxide, potassium cyanide and ammonia. The nucleophiles involved are the hydroxide ion, cyanide ion and the ammonia molecule.

Reactions with nucleophiles

In this section we consider the reactions of haloalkanes with three different nucleophiles.

Reaction with a warm aqueous solution of potassium or sodium hydroxide

These are ionic compounds containing the nucleophile $:OH^-$.

$$CH_3CH_2CH_2Br + :OH^- \rightarrow CH_3CH_2CH_2OH + :Br^-$$

1-bromopropane → propan-1-ol

The mechanism of the reaction is:

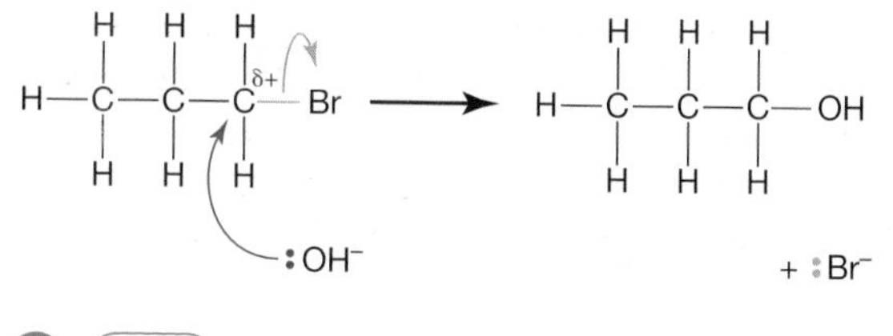

Fig. 1 Nucleophilic substitution mechanism with a hydroxide ion

The product of the reaction of hydroxide ions with a haloalkane is an alcohol.

> **Q**
>
> 1 Give the structural formula of the alcohols formed when the following haloalkanes react with KOH:
>
> a) 1-bromobutane
>
> b) 2-bromobutane
>
> c) chloroethane

Reaction with a hot ethanolic solution of potassium or sodium cyanide

$$CH_3CH_2I + CN^- \rightarrow CH_3CH_2CN + I^-$$

iodoethane → propanenitrile

Potassium and sodium cyanide are ionic compounds containing the nucleophile $:CN^-$. In a hot ethanolic solution, ethanol is used as the solvent.

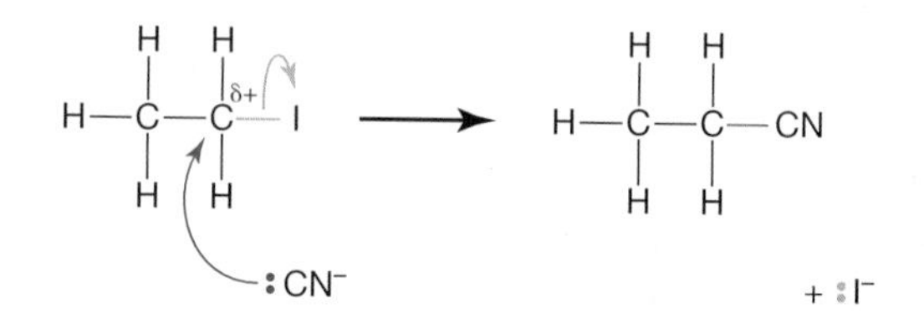

Fig. 2 Nucleophilic substitution mechanism with a cyanide ion

The product of the reaction of cyanide ions with a haloalkane is a **nitrile**. This reaction provides a useful method of increasing the length of a carbon chain.

Nitriles are readily hydrolysed in acid solution to form **carboxylic acids**:

$$CH_3CH_2CN + 2H_2O + HCl \rightarrow CH_3CH_2COOH + NH_4Cl$$

propanenitrile → propanoic acid

Reaction with excess ammonia

Ammonia has a lone pair of electrons on the nitrogen atom.

$$CH_3CH_2CH_2Cl + 2NH_3 \rightarrow CH_3CH_2CH_2NH_2 + NH_4Cl$$

1-chloropropane → 1-aminopropane

The product of the reaction of ammonia with a haloalkane is an **amine**. Amines have a lone pair of electrons on the nitrogen atom, and can act as nucleophiles. Excess ammonia is used in the reaction to prevent successive substitution.

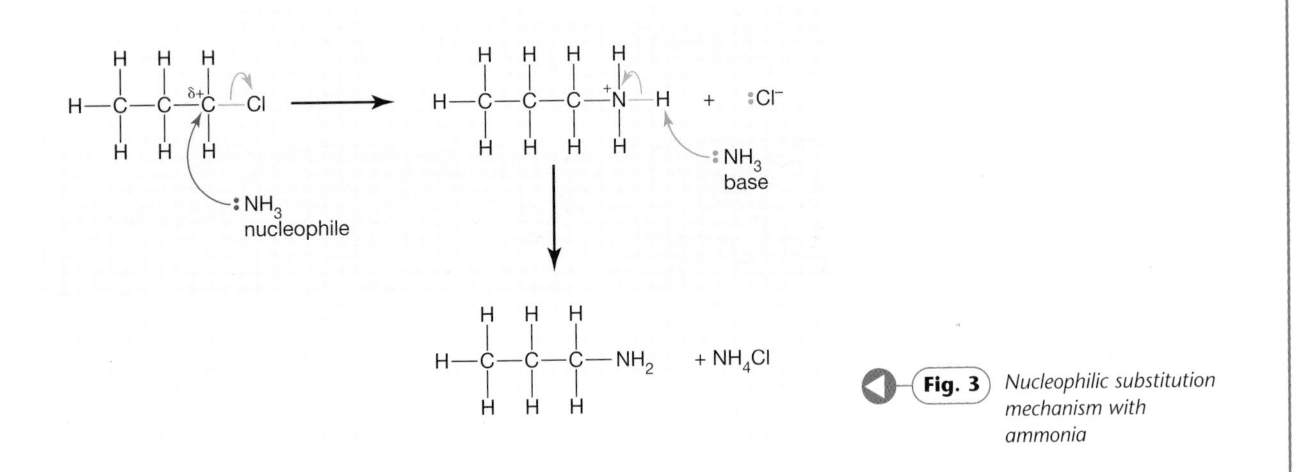

Fig. 3 *Nucleophilic substitution mechanism with ammonia*

Reactivity of haloalkanes

The reactivity of the haloalkanes depends upon the strength of the carbon—halogen bond. The weaker the bond, the more easily it is broken, and hence the more reactive is the haloalkane.

Bond	Mean bond enthalpies/kJ mol^{-1}
C—F	484
C—Cl	338
C—Br	276
C—I	238

The mean bond enthalpy of the C—X bond decreases down Group VII, and hence the rate of reaction with nucleophiles increases.

The carbon—fluorine bond has a very high mean bond enthalpy, so fluoroalkanes are not readily attacked by nucleophiles.

The haloalkanes are very significant organic compounds, and are used in the synthesis of many important organic chemicals. The reason that they are so useful is that they can be prepared from many different compounds, and they can be changed into many different homologous series.

Q

2 Write equations for the following reactions and name the organic products:

a) $CH_3CH_2CH_2Br$ and KCN b) $CH_3CH_2CH_2CH_2Cl$ and NH_3
c) $CH_3CH_2CH_2I$ and KOH(aq)

Key Ideas 212 – 213

- **Haloalkanes react with a range of nucleophiles: reaction with hydroxide ions forms alcohols; reaction with ammonia forms amines; reaction with cyanide ions forms nitriles. Nitriles can be hydrolysed to form carboxylic acids.**

3 Competition time ...

We have seen that haloalkanes react with hydroxide ions in a nucleophilic substitution reaction. However, a different type of reaction also occurs when hydroxide ions are involved in a reaction with haloalkanes.

Elimination reactions

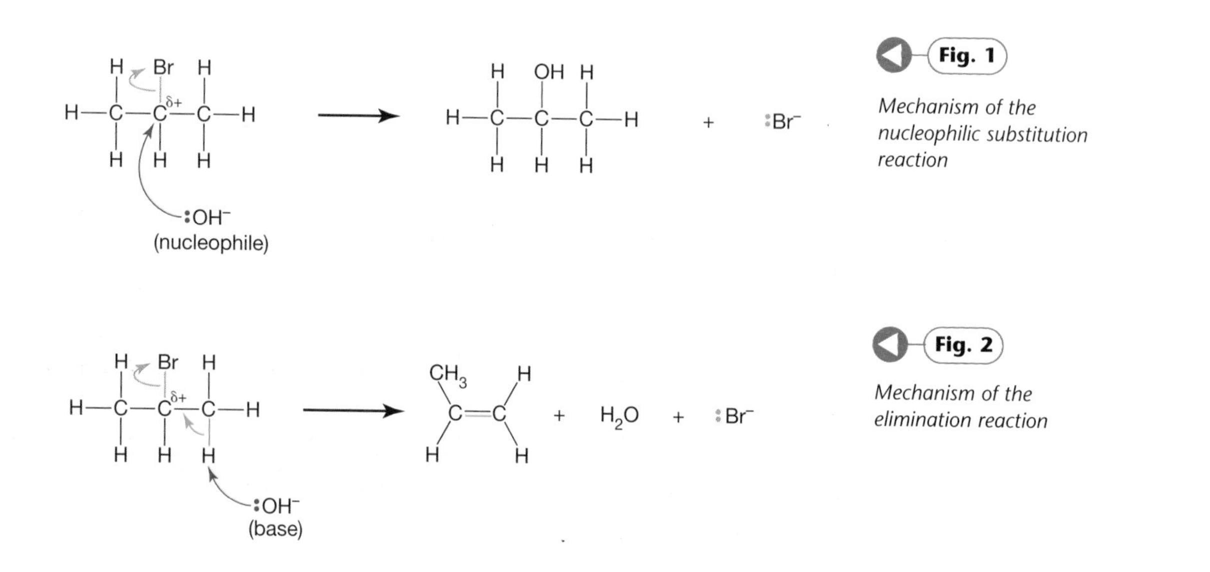

Fig. 1 *Mechanism of the nucleophilic substitution reaction*

Fig. 2 *Mechanism of the elimination reaction*

When 2-bromopropane reacts with potassium hydroxide, two products are formed. The two concurrent reactions produce an alcohol by a nucleophilic substitution reaction, and an alkene by an **elimination reaction**.

In the substitution reaction (Figure 1), the hydroxide ion acts as a *nucleophile* and attacks the electron-deficient carbon atom.

$$CH_3CHBrCH_3 + OH^- \rightarrow CH_3CH(OH)CH_3 + Br^-$$

At the same time, propene is produced by an elimination reaction.

$$\underset{\text{2-bromopropane}}{CH_3CH(Br)CH_3} + OH^- \rightarrow \underset{\text{propene}}{CH_3CH{=}CH_2} + H_2O + Br^-$$

In an elimination reaction (Figure 2), the hydroxide ion acts as a **base** attacking a hydrogen atom on a carbon atom next to the carbon atom in the carbon—halogen bond. The pair of electrons from the carbon—hydrogen bond then form the double bond. At the same time the carbon—halogen bond breaks and the pair of electrons pass to the halogen atom, forming a halide ion.

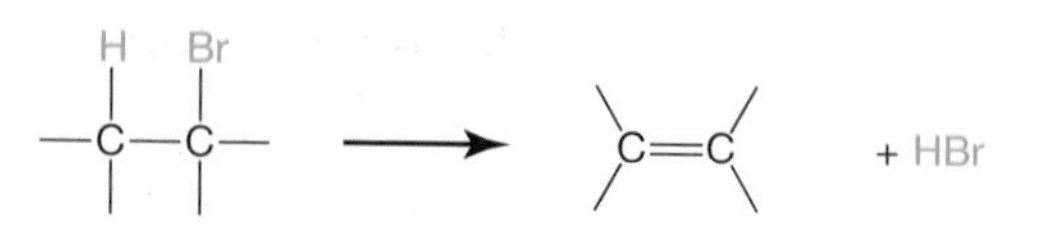

Fig. 3 *Elimination to form an alkene*

Q

1 Give the structural formulae of the products formed when KOH reacts with:

a) 1-bromobutane
b) 2-chloromethylpropane

When haloalkanes react with hydroxide ions, both substitution and elimination reactions occur at the same time (they are concurrent). The elimination reaction is favoured by the following conditions:

1 *Using ethanol as the solvent rather than water.* Using an aqueous solution increases the percentage of substitution. Using an ethanolic solution increases the percentage of elimination.
2 *Increasing the temperature.* For a given haloalkane, the percentage of the elimination reaction is increased at higher temperatures.
3 *Increased branching of the haloalkane.* There is more elimination with tertiary haloalkanes than with secondary haloalkanes, which in turn undergo more elimination than with primary haloalkanes.

When a haloalkane undergoes an elimination reaction, the haloalkane loses the halogen atom, and a hydrogen atom from an adjacent carbon atom. This can lead to the formation of two different alkenes.

When 2-bromobutane undergoes an elimination reaction, the haloalkane can lose a hydrogen atom from the carbon atom on either side of the carbon—halogen bond. This results in the formation of but-1-ene and but-2-ene.

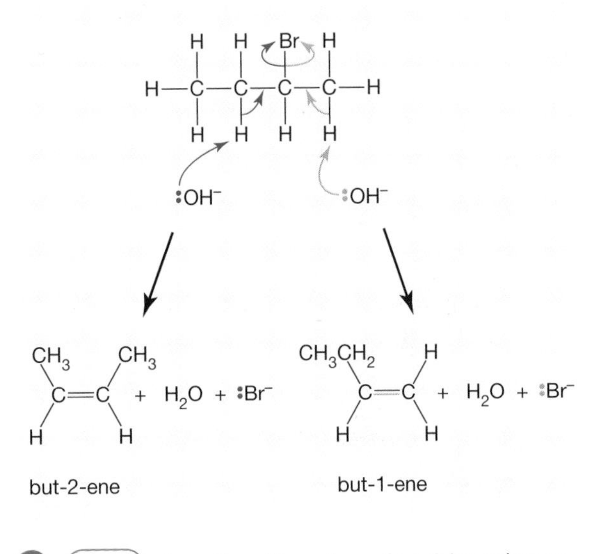

Fig. 4 *Formation of two alkenes from 2-bromobutane*

Q

2 Outline the mechanisms for the reactions involved when 2-bromopentane reacts with:

a) an ethanolic solution of potassium hydroxide
b) an aqueous solution of potassium hydroxide.

Key Ideas 214 – 215

- **Haloalkanes undergo concurrent substitution and elimination reactions with sodium hydroxide and potassium hydroxide.**
- **In the substitution reaction, the hydroxide ion acts as a nucleophile and attacks the δ^+ carbon atom.**
- **In the elimination reaction, the hydroxide ion acts as a base and attacks a hydrogen atom.**
- **The elimination reaction is favoured by using ethanol as a solvent, by increasing the temperature, and by increased branching of the haloalkane.**

Unit 14 Questions

1 Name the following haloalkanes:

a) $CH_3CH_2CH_2CH_2Br$ (1)
b) $CH_3CH_2CH_2Cl$ (1)
c) $CH_3CHBrCHBrCH_3$ (1)
d) $CH_3CH_2CH(CH_3)CH_2I$ (1)

2 Haloalkanes can be prepared from alkenes and from alkanes.

a) Name the mechanism involved in the formation of haloalkanes from alkenes. (1)
b) Write an equation for, and name the organic product formed in, the reaction between but-2-ene and HBr. (2)
c) Outline the mechanism for the reaction. (4)
d) Name the mechanism involved in the formation of haloalkanes from alkanes. (1)
e) Write an equation for the formation of chloroethane from ethane. (1)
f) Outline the mechanism of the reaction. (4)

3 Haloalkanes undergo nucleophilic substitution reactions.

a) Define the term *nucleophile*.
b) Explain why haloalkanes are readily attacked by nucleophiles.
c) For each of the following pairs of compounds:
1 Write an equation for the reaction. (1 each)
2 Name the organic product of the reaction. (1 each)
3 Outline the mechanism of the reaction. (3 each)
(i) $CH_3CH_2CH_2Br$ and KOH
(ii) $CH_3CH_2CH_2CH_2I$ and KCN
(iii) CH_3CH_2Cl and NH_3

4 The structure of bromocyclohexane is shown in Figure 1.

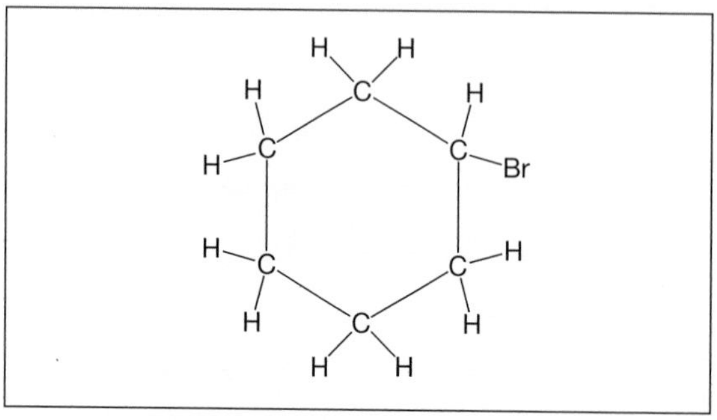

Fig. 1

a) Name the mechanism of the reaction between bromocyclohexane and KCN dissolved in ethanol. (1)

b) Outline the mechanism of the reaction. (3)

5) When 1-bromopropane reacts with KOH, two different reactions can occur.

a) For each of these reactions:
   1 Write an equation for the reaction. (1 each)
   2 Name the organic product formed. (1 each)
   3 Outline the mechanism of the reaction. (3 each)

b) Explain the different role of the KOH in the two reactions. (2)

6) a) Write an equation for the reaction of 1-chloropropane with KCN. (1)

b) Name the essential conditions for the reaction. (2)

c) Write an equation to show how the product of this reaction is converted into a carboxylic acid. (2)

7) Explain why the reaction of KOH with iodoethane occurs more readily than the reaction of KOH with chloroethane. (2)

8) Phenylmethanol (compound **B**) can be prepared in a two-step reaction from methylbenzene (compound **A**).

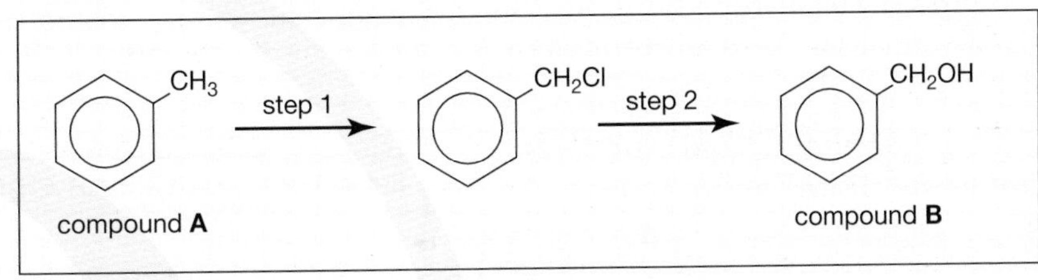

a) For step 1:
   (i) State the reagents and conditions used.
   (ii) Name the mechanism of the reaction.
   (iii) Write an equation for the reaction. (4)

b) For step 2:
   (i) State the reagents and conditions used.
   (ii) Name the mechanism.
   (iii) Write an equation for the reaction. (4)

1 Alcohols

At GCSE you will have studied the preparation of ethanol by the fermentation of sugars using yeast as a catalyst. Ethanol is a member of the alcohol homologous series. In this unit you will study the oxidation reactions of alcohols.

Naming and classifying alcohols

Alcohols form a homologous series and contain the functional group **—OH**. Alcohols are based on the alkanes with one of the H atoms replaced by the —OH group. The suffix **-ol** is used to indicate the alcohol group in the names of the alcohols. The general formula of the alcohols is $C_nH_{2n+1}OH$.

The names, structural formula and structures of some alcohols are shown in the Table 1.

Name	Structural formula	Structure
Methanol	CH_3OH	
Ethanol	CH_3CH_2OH	

Table 1 *Naming alcohols*

The higher alcohols have position isomers. Position isomers occur when the functional group can be on different carbon atoms (the carbon chain is the same). The position of the functional group is indicated by a number, as in Table 2.

Q

1 Name each of the following alcohols:
   a) $CH_3CH(OH)CH_2CH_2CH_3$
   b) $CH_3CH_2CH_2CH_2CH_2OH$
   c) $CH_3CH_2CH(OH)CH(CH_3)CH_3$

2 Explain in terms of intermolecular forces why alcohols have higher boiling points than the corresponding alkanes.

Name	Structural formula	Structure
Propan-1-ol	$CH_3CH_2CH_2OH$	
Propan-2-ol	$CH_3CH(OH)CH_3$	
Butan-1-ol	$CH_3CH_2CH_2CH_2OH$	
Butan-2-ol	$CH_3CH_2CH(OH)CH_3$	
Methyl-propan-1-ol	$CH_3CH(CH_3)CH_2OH$	
Methyl-propan-2-ol	$(CH_3)_3COH$	

Table 2 *Naming position isomers*

Alcohols are classified as **primary**, **secondary** and **tertiary**. The classification relates to the number of carbon atoms directly attached to the carbon atom carrying the —OH group.

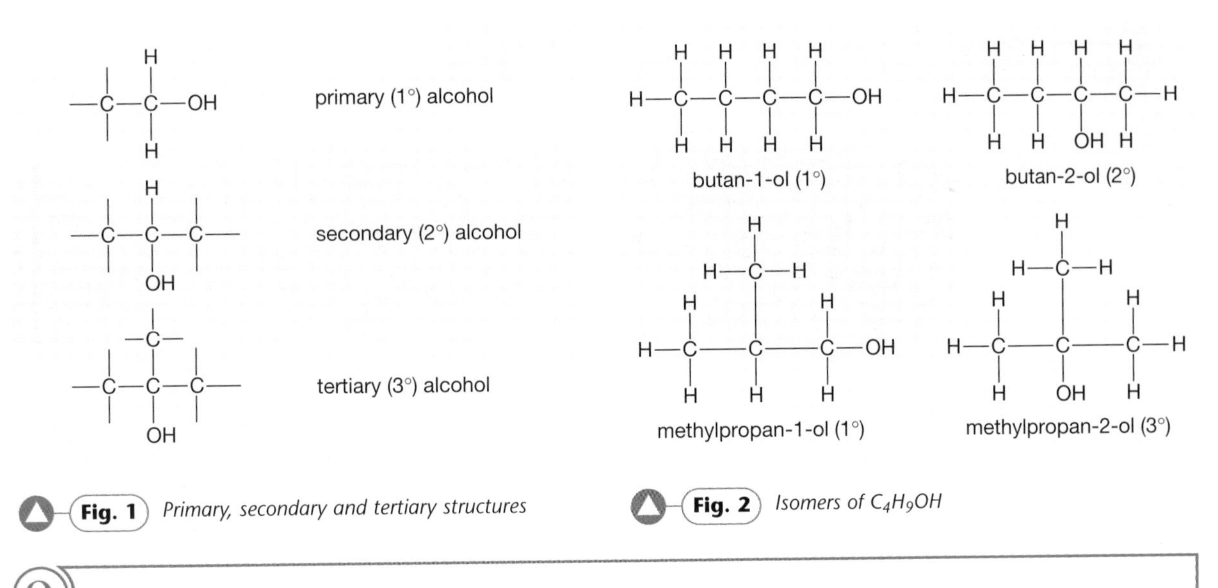

Fig. 1 *Primary, secondary and tertiary structures*

Fig. 2 *Isomers of C_4H_9OH*

> **Q**
>
> 3 Draw the structures of the alcohols containing a branched chain that share the molecular formula $C_5H_{11}OH$. Name each alcohol, and state whether the alcohol is primary, secondary or tertiary.

Industrial manufacture of ethanol

Ethanol, CH_3CH_2OH, can be made by two different industrial methods: direct hydration of ethene (see page 206) or fermentation. Each method has advantages and disadvantages (Table 3).

In the fermentation reaction a solution of a plant sugar together with yeast is kept at a temperature of approximately 37 °C. The yeast contains enzymes that catalyse the reaction. The reaction is slow and takes several days. Fermentation is a batch process and produces a mixture of compounds. In a batch process the reactants are mixed together: once the reaction is over, the product is removed and the process is restarted.

$$\underset{\text{sugar}}{C_6H_{12}O_6} \rightarrow \underset{\text{ethanol}}{2CH_3CH_2OH} + 2CO_2$$

	Direct hydration route: continuous process	Fermentation route: batch process
Advantages	fast reaction	uses renewable resources
	pure product	low energy use
	high yield	
Disadvantages	uses finite resources	slow reaction
	high energy use	impure product
		low yield

Table 3 *Industrial manufacture of ethanol*

The ethanol is separated by fractional distillation.

> **Q**
>
> 4 Explain the terms *renewable* and *finite resources.*

Key Ideas 218 – 219

- Alcohols contain the —OH functional group.
- Alcohols can be primary, secondary or tertiary.
- Ethanol can be prepared industrially by a continuous process involving the direct hydration of ethene, and by a batch process involving the fermentation of sugars.

2 Oxidation of alcohols

The reaction of alcohols with oxidising agents can be used to distinguish between primary, secondary and tertiary alcohols. Primary and secondary alcohols will react with acidified potassium or sodium dichromate(VI) to form different products. Tertiary alcohols do not react with acidified potassium or sodium dichromate(VI).

Oxidation reactions

Primary, secondary and tertiary alcohols can be distinguished by simple oxidation reactions using **acidified potassium dichromate(VI)** as an oxidising agent.

The full equations for the redox reactions between potassium dichromate and alcohols are fairly complex, and simplified equations using [O] to represent the oxidant are acceptable at this level.

Primary alcohols

Primary alcohols are oxidised to form an **aldehyde**. If an excess of the oxidising agent is used, then the aldehyde is oxidised to a **carboxylic acid**.

$$\underset{\text{primary alcohol}}{RCH_2OH} + [O] \rightarrow \underset{\text{aldehyde}}{RCHO} + H_2O$$

For example:

$$\underset{\text{propan-1-ol}}{CH_3CH_2CH_2OH} + [O] \rightarrow \underset{\text{propanal}}{CH_3CH_2CHO} + H_2O$$

Using an excess of the oxidising agent:

$$\underset{\text{propanal}}{CH_3CH_2CHO} + [O] \rightarrow \underset{\text{propanoic acid}}{CH_3CH_2COOH}$$

The overall reaction for the formation of a carboxylic acid is:

$$RCH_2OH + 2[O] \rightarrow RCOOH + H_2O$$

For example:

$$CH_3CH_2CH_2OH + 2[O] \rightarrow CH_3CH_2COOH + H_2O$$

Fig. 1 *Reflux apparatus*

When acidified potassium dichromate(VI) acts as an oxidising agent, the *orange* solution of potassium dichromate(VI) changes to a *green* solution containing chromium(III) ions.

The oxidation of primary alcohols can be controlled by altering the reaction conditions:

- If an alcohol is oxidised using **distillation** apparatus, then any aldehyde formed will distil off immediately from the mixture.
- If an alcohol is oxidised under **reflux** (see Figure 1), then the aldehyde cannot escape and will be oxidised to form a carboxylic acid. When reactants are heated under reflux, any volatile components will rise up the condenser. These volatile components will then condense and return to the reaction mixture.

Secondary alcohols

Secondary alcohols are oxidised by acidified potassium dichromate(VI) to form **ketones**.

$$\underset{\text{secondary alcohol}}{RCH(OH)R'} + [O] \rightarrow \underset{\text{ketone}}{RCOR'} + H_2O$$

For example:

$$\underset{\text{butan-2-ol}}{CH_3CH_2CH(OH)CH_3} + [O] \rightarrow \underset{\text{butanone}}{CH_3CH_2COCH_3} + H_2O$$

Tertiary alcohols

Tertiary alcohols are not easily oxidised, and there is no reaction between tertiary alcohols and acidified potassium dichromate(VI). An oxidation reaction with a tertiary alcohol involves the breaking of a carbon–carbon bond, and acidified potassium dichromate(VI) is not a strong enough oxidising agent.

Q

1 Give the structural formulae of the products formed, if any, when the following alcohols are reacted with acidified potassium dichromate(VI):

a) butan-1-ol
b) butan-2-ol
c) methylpropan-1-ol
d) methylpropan-2-ol

2 Propan-2-ol can be converted to propanal, CH_3CH_2CHO, and propanoic acid, CH_3CH_2COOH. Write equations and give the essential conditions used in each of these conversions.

Key Idea 220 – 221

- **Alcohols can be distinguished by the oxidation reactions with acidified potassium dichromate(VI). Primary alcohols form aldehydes and carboxylic acids, secondary alcohols form ketones, and tertiary alcohols do not react.**

3 Aldehydes and ketones

We have seen that primary and secondary alcohols can be oxidised by acidified potassium(VI) dichromate to form aldehydes and ketones respectively. Aldehydes and ketones both contain the carbonyl group.

Aldehydes and ketones

Aldehydes and ketones both contain the **carbonyl functional group**. The carbonyl group is planar, with bond angles of 120°.

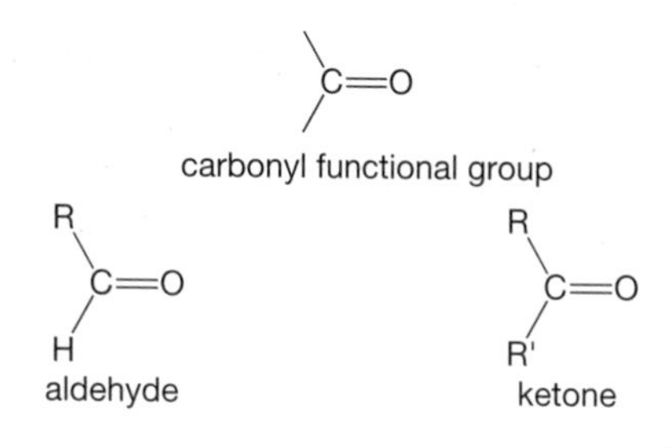

Fig. 1 *Carbonyl functional group*

The general formula of aldehydes and ketones is $C_nH_{2n}O$. Because they have a carbon—oxygen double bond, aldehydes and ketones are unsaturated. The *difference* between aldehydes and ketones is that aldehydes have a hydrogen atom attached to the carbonyl group.

- The aldehydes have this group at the *end* of the carbon chain. Their names end in **–al**.
- The ketones have this group in the *middle* of the carbon chain. Their names end in **–one**.

The higher ketones have position isomers, for example pentan-2-one, $CH_3CH_2CH_2COCH_3$, and pentan-3-one, $CH_3CH_2COCH_2CH_3$.

Notice that the aldehydes and the ketones have the same *general* formula, and therefore show **functional group isomerism**. For example, C_3H_6O could be propanal, CH_3CH_2CHO, or propanone, CH_3COCH_3.

Name	Structural formula	Structure
Methanal	$HCHO$	H—C(=O)—H
Ethanal	CH_3CHO	H—C(H)(H)—C(=O)—H
Propanal	CH_3CH_2CHO	H—C(H)(H)—C(H)(H)—C(=O)—H

Table 1 *Aldehydes*

Name	Structural formula	Structure
Propanone	CH_3COCH_3	CH_3 C=O CH_3
Butanone	$CH_3CH_2COCH_3$	CH_3CH_2 C=O CH_3

Table 2 *Ketones*

Q

1 Name the following aldehydes and ketones:
a) $CH_3CH_2CH_2CH_2CHO$ b) $CH_3CH_2COCH_2CH_3$ c) $CH_3CH(CH_3)CHO$

2 Write the structural formulae for the following compounds:
a) hexan-3-one b) 2-methylpentan-3-one c) 2,2-dimethylbutanal

Test for aldehydes

Aldehydes and ketones can be distinguished using a mild oxidising agent such as **Tollen's reagent** or **Fehling's solution**. The presence of the hydrogen atom attached to the carbonyl group allows aldehydes to be oxidised to carboxylic acids. This means that aldehydes are mild reducing agents.

Tollen's reagent contains silver nitrate dissolved in dilute aqueous ammonia. When Tollen's reagent is heated on a water bath with an aldehyde, a redox reaction occurs. The aldehyde is *oxidised* to a carboxylic acid, and the silver ions are *reduced* to silver metal. A positive result is the formation of a **silver mirror** in the test tube.

Fehling's solution is a blue solution containing copper(II) ions in an alkaline solution.

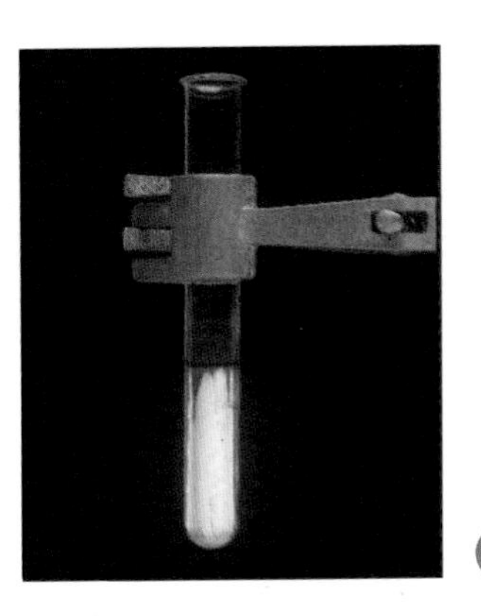

Fig. 2 *Silver mirror*

When an aldehyde is boiled with Fehling's solution, a redox reaction occurs. The aldehyde is oxidised to form a carboxylic acid, and the copper(II) ions are reduced to copper(I) oxide, Cu_2O. A positive test is when the blue solution turns brick-red due to the formation of Cu_2O.

Reduction of aldehydes and ketones

Aldehydes and ketones can be reduced to alcohols using $NaBH_4$ as a reducing agent. The equations are complex and the use of [H] to represent the reducing agent is adequate at this level.

Aldehydes are reduced to primary alcohols:

$$\underset{\text{aldehyde}}{RCHO} + 2[H] \rightarrow \underset{\text{primary alcohol}}{RCH_2OH}$$

For example:

$$\underset{\text{ethanal}}{CH_3CHO} + 2[H] \rightarrow \underset{\text{ethanol}}{CH_3CH_2OH}$$

Ketones are reduced to form secondary alcohols:

$$\underset{\text{ketone}}{RCOR'} + 2[H] \rightarrow \underset{\text{secondary alcohol}}{RCH(OH)R'}$$

For example:

$$\underset{\text{propanone}}{CH_3COCH_3} + 2[H] \rightarrow \underset{\text{propan-2-ol}}{CH_3CH(OH)CH_3}$$

Q

3 Give the structural formulae of the reduction products of the following aldehydes and ketones:

a) propanal
b) butanone
c) 3-methylbutanal
d) 3-methylpentan-2-one

Key Ideas 222 – 223

- **Aldehydes and ketones both contain the carbonyl group.**
- **Aldehydes and ketones are functional group isomers.**
- **Aldehydes and ketones can be reduced to alcohols by reaction with $NaBH_4$.**
- **Aldehydes and ketones can be distinguished using Tollen's reagent or by Fehling's solution.**

4 Losing water ...

In the alkene section we saw that alcohols can be formed from alkenes by a reaction with concentrated sulphuric acid followed by hydrolysis. This reaction can be reversed by eliminating water from an alcohol to form an alkene.

Elimination reactions

Alcohols undergo an **elimination reaction** to form alkenes when heated with concentrated sulphuric acid. Since water is eliminated, the reaction is also referred to as a **dehydration reaction**.

In the elimination reaction, the alcohol loses the —OH group and an H atom from an adjacent carbon atom:

$$CH_3CH_2CH_2OH \rightarrow CH_3CH{=}CH_2 + H_2O$$

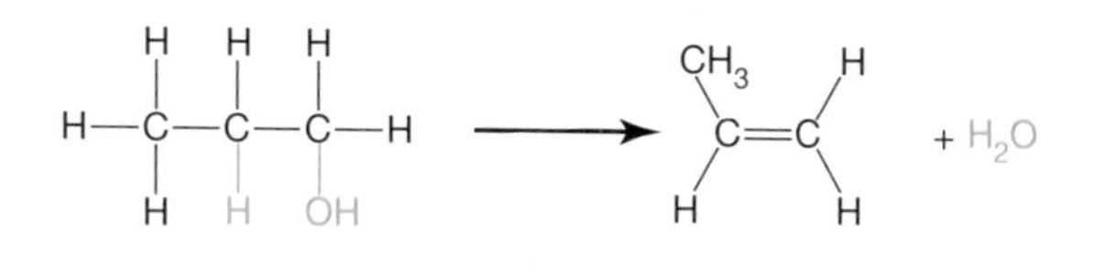

Fig. 1 *Loss of water from propan-1-ol*

The mechanism of the reaction involves three steps.

1 The concentrated sulphuric acid acts as a strong acid and protonates the —OH on the alcohol to form $—OH_2^+$. This is an important step as it changes the poor leaving group —OH into the good leaving group $—OH_2^+$.

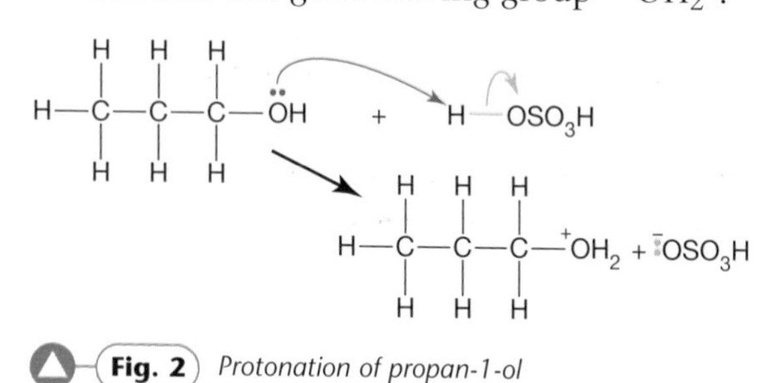

Fig. 2 *Protonation of propan-1-ol*

2 The $—OH_2^+$ is lost, leaving a carbocation intermediate.

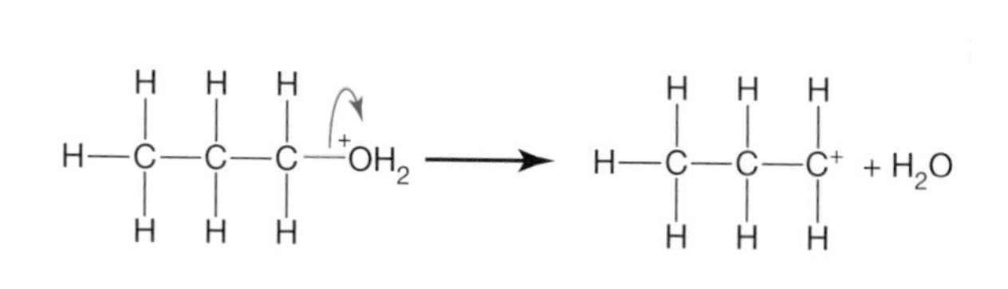

Fig. 3 *Loss of water from protonated alcohol*

3 The carbocation loses an H atom to form an alkene.

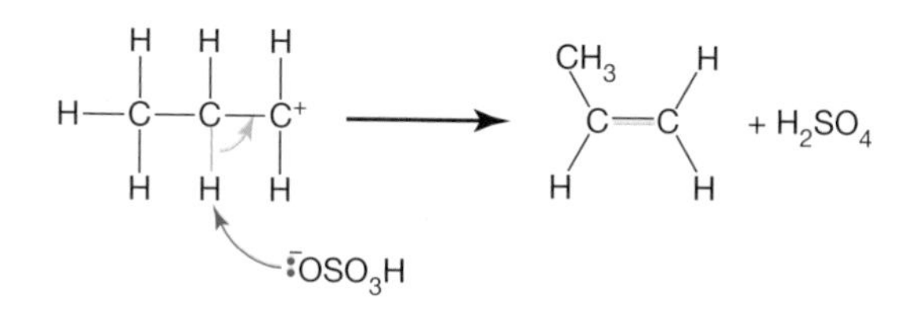

Fig. 4 *Loss of an H atom*

The full mechanism of the reaction is shown in Figure 5.

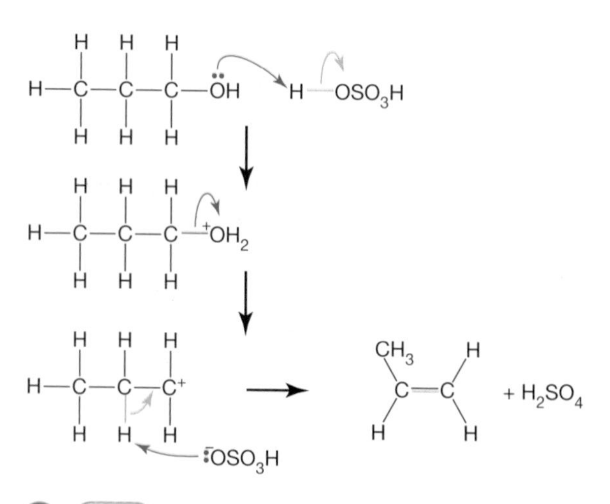

Fig. 5 *Full mechanism of the elimination reaction*

The sulphuric acid is re-formed at the end of the reaction, and therefore behaves as a catalyst.

In the elimination reaction of an alcohol, the alcohol loses both the —OH group and an H atom from an adjacent carbon atom. This can result in the formation of more than one alkene. For example, when concentrated sulphuric acid reacts with butan-2-ol there are two different alkenes formed, but-1-ene and but-2-ene.

To understand why two products are formed, we need to look at the mechanism of the reaction (Figure 6). In the carbocation formed by the loss of the —OH group, there are H atoms on both of the carbon atoms adjacent to the positive carbon atom. A hydrogen atom can be removed from either of these carbon atoms, resulting in the formation of two different alkenes.

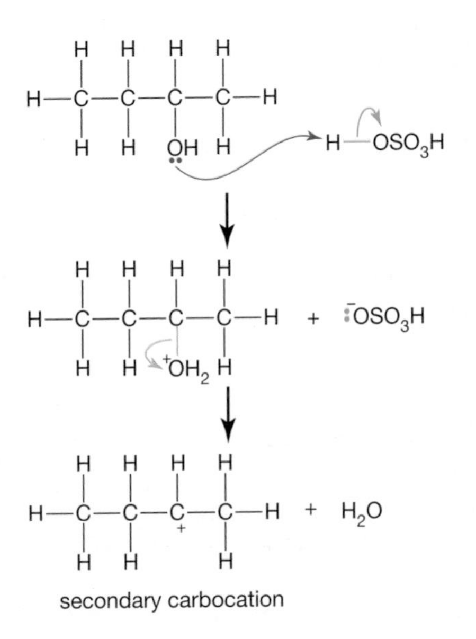

Fig. 6 *Formation of a carbocation from butan-1-ol*

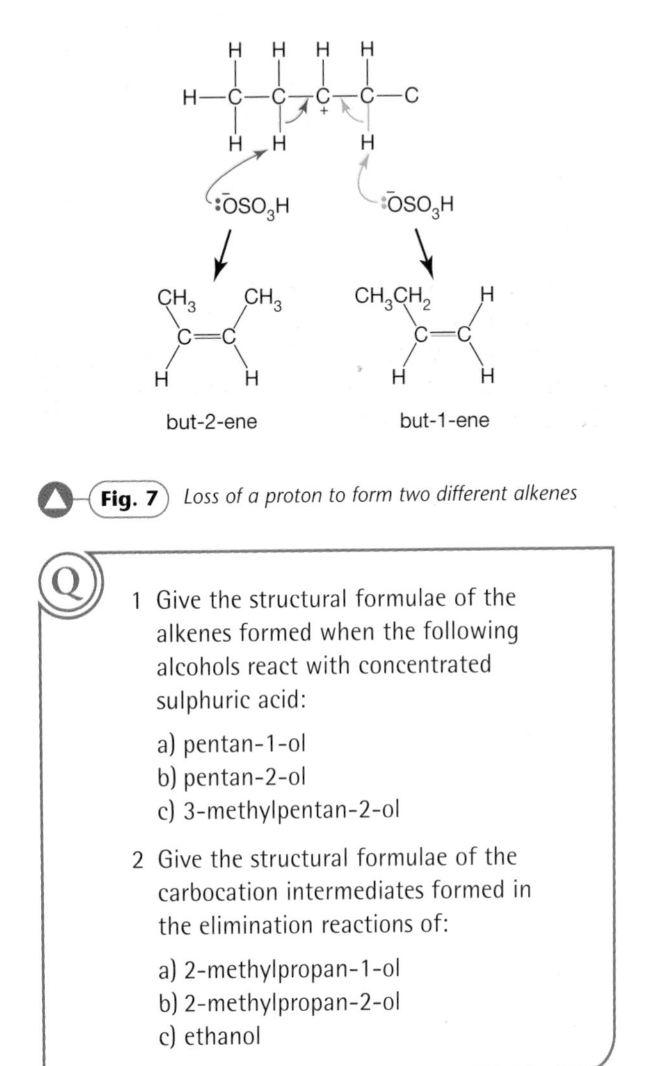

Fig. 7 *Loss of a proton to form two different alkenes*

Q

1 Give the structural formulae of the alkenes formed when the following alcohols react with concentrated sulphuric acid:

a) pentan-1-ol
b) pentan-2-ol
c) 3-methylpentan-2-ol

2 Give the structural formulae of the carbocation intermediates formed in the elimination reactions of:

a) 2-methylpropan-1-ol
b) 2-methylpropan-2-ol
c) ethanol

Key Ideas 224 – 225

- **Alcohols form alkenes in elimination reactions with concentrated sulphuric acid.**
- **The mechanism involves protonation of the alcohol, followed by the loss of a hydrogen atom from the carbocation intermediate.**

Unit 15 Questions

1. Name each of the following alcohols and state whether they are primary, secondary or tertiary alcohols.

   a) $CH_3CH_2CH_2CH_2CH_2OH$ (1)
   b) $CH_3CH_2CH(OH)CH_2CH_3$ (1)
   c) $CH_3CH_2CH_2CH_2OH$ (1)
   d) $(CH_3)_2CHCH_2OH$ (1)
   e) $(CH_3)_3COH$ (1)

2. Alcohols can be prepared by the reaction of haloalkanes with aqueous KOH. Write equations for the conversion of the following haloalkanes into alcohols. In each case name the haloalkane and the alcohol formed.

   a) $CH_3CH_2CH_2Br$ (3)
   b) $CH_3CH_2CH_2CHClCH_3$ (3)
   c) $CH_3CH_2CH(CH_3)CH_2I$ (3)

3. **a)** Alcohols undergo oxidation reactions when warmed with acidified potassium dichromate(VI). Write equations for the oxidation of the following alcohols. In each case name the organic products formed.
   (i) $CH_3CH_2CH_2CH_2OH$
   (ii) $CH_3CH_2CH(OH)CH_3$
   (iii) $CH_3CH_2CH_2OH$ (6)

   b) Explain why $(CH_3)_3COH$ is not oxidised by warming with acidified potassium dichromate(VI). (2)

4. When ethanol reacts with acidified potassium dichromate(VI), two different products can be formed depending upon the conditions used. Name the two different products and for each product state the conditions used in its formation. (4)

5. Alcohols react with concentrated sulphuric acid to form alkenes.

   a) Name the type of reaction. (1)
   b) Outline the mechanism of the reaction of butan-1-ol with concentrated sulphuric acid. (4)
   c) When butan-2-ol reacts with concentrated sulphuric acid, three different alkenes are formed.
   (i) Name, and draw the structures of, the three alkenes formed.
   (ii) Explain in terms of the mechanism of the reaction why the reaction of butan-2-ol forms three products. (9)

(6) The structure of cyclohexanol is shown in Figure 1.

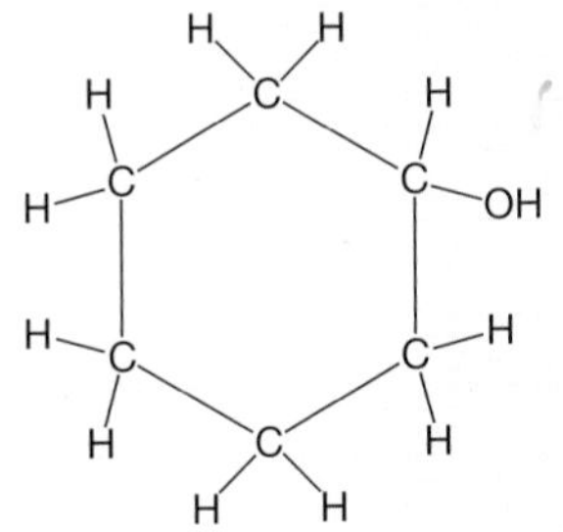

Fig. 1

a) State whether cyclohexanol is a primary, secondary or tertiary alcohol. (1)
b) Write an equation for the reaction of cyclohexanol with concentrated sulphuric acid, and name the product formed. (2)
c) Outline the mechanism for the reaction. (4)

(7) Ethanol can be prepared by fermentation and by direct hydration of ethene.

a) Give the essential conditions in the fermentation reaction. (3)
b) Give the essential conditions for the direct hydration reaction. (3)
c) Give *two* advantages and *two* disadvantages for each of the two different methods of preparing ethanol. (4)

(8) Name the following compounds:

a) $CH_3CH(CH_3)CH_2CHO$ (1)
b) $CH_3CH_2CH_2COCH_2CH_3$ (1)
c) $CH_3CH(CH_3)CH(CH_3)CHO$ (1)

(9) For each of the following aldehydes and ketones, write an equation for the reaction with $NaBH_4$ and name the product of the reaction.

a) propanal (2)
b) pentan-2-one (2)
c) 3-methylbutanal (2)

(10) For each of the following pairs of compounds, describe a simple chemical test to differentiate between the pair of compounds. In each case give the observations when each of the compounds is tested.

a) CH_3COCH_3 and CH_3CH_2CHO (2)
b) CH_3CH_2OH and CH_3COCH_3 (2)

Module 3 Questions

1 a) Name, and write the structures for, the *four* isomeric alcohols of molecular formula $C_4H_{10}O$. Classify these alcohols as primary, secondary or tertiary alcohols. (12)

b) Describe how each of these alcohols reacts when warmed with acidified sodium dichromate(VI) solution by giving the name and structure of the organic product in each case. How does the type of product relate to the classification of the starting alcohol? (10)

2 a) Explain the meaning of the term *stereoisomerism*. Discuss the stereoisomerism shown by but-2-ene and explain how it arises. (6)

b) Outline the mechanism for the reaction between but-2-ene and HBr. Explain why there is only one product in this reaction. (8)

3 Organic reactions may involve substitution, addition or elimination.

a) Name *two* types of mechanisms involving substitution, and show the mechanisms for these reactions using methane reacting with chlorine and bromoethane reacting with potassium cyanide as examples. (10)

b) If concentrated sulphuric acid is added to but-1-ene and the products are hydrolysed, a mixture of two alcohols results. Name and draw the structures for these alcohols, and explain why the mixture does not contain equal proportions of each alcohol. (9)

c) Write an equation for the reaction between 1-bromopropane and hot, ethanolic potassium hydroxide, showing clearly the structure of the product. Discuss the role of the hydroxide ions in this reaction and explain how this role would change if the solution were aqueous rather than ethanolic. (6)

4 Haloalkanes undergo nucleophilic substitution reactions.

a) Explain the term *nucleophilic substitution* and why haloalkanes undergo this type of reaction. (6)

b) Outline the mechanism for the reaction between bromoethane and ammonia to form ethylamine. Why is excess ammonia required? (7)

c) Identify **X** by name and formula in the reaction sequence shown below:

$CH_3Cl \rightarrow \mathbf{X} \rightarrow CH_3COOH$ (2)

5 In some countries, such as Brazil, car fuel is a mixture of petrol and ethanol. Petrol is obtained from the fractional distillation of petroleum.

a) Give a reason why the use of ethanol as a fuel should be recommended. (2)

b) Write an equation, and give the conditions necessary, for the industrial production of ethanol from ethene. Discuss the advantages and disadvantages of this process compared to the manufacture of ethanol by fermentation. (6)

c) Write an equation to show the complete combustion of a molecule of ethanol in oxygen. Use this equation and the data below to calculate the standard enthalpy of combustion of ethanol. (7)

Bond	Mean bond enthalpy/kJ mol^{-1}
C—H	413
C—O	358
C—C	347
C=O	805
O=O	498
O—H	464

d) Impurities present in petroleum fractions may contain sulphur. What environmental problems may arise from the burning of petrol obtained from petroleum? (2)

6 Ketones cannot be oxidised to carboxylic acids, so an indirect synthetic route is required to make an acid from a ketone. One example is shown in Figure 1. Compound **A** is an aromatic ketone and compound **E** is an aromatic carboxylic acid.

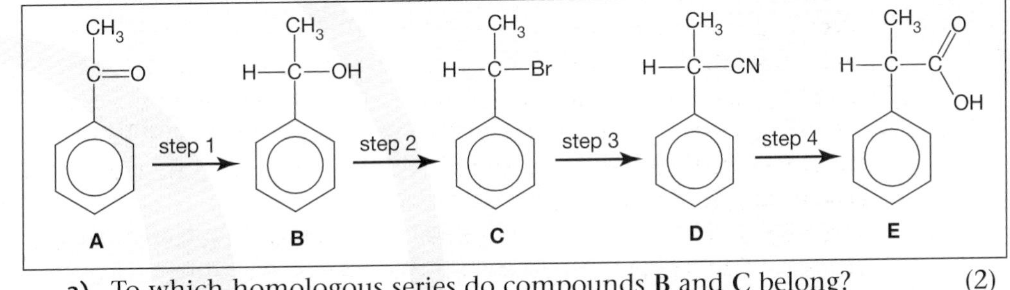

Fig. 1

a) To which homologous series do compounds **B** and **C** belong? (2)

b) Give reagents and essential conditions, and name the reaction type, for each of the steps 1, 3 and 4. (9)

c) Describe a test you could use to distinguish a ketone and an aldehyde. (4)

7 Using the two pairs of examples given below, explain how altering the reaction conditions can alter the organic product of a reaction. In each case state the conditions used and the type of reaction that occurs. Identify the organic products by name and formula.

a) 2-bromopropane and potassium hydroxide (8)

b) propan-1-ol and acidified potassium dichromate(VI) (7)

8 The dehydration of butan-2-ol by concentrated sulphuric acid results in the formation of *three* isomers of butene.

a) Name and draw the structures of the three alkenes. (6)

b) Outline the mechanism for the formation of one of these products. (7)

2

9 a) Explain, using an example, the meaning of the terms *homolytic fission* and *free radical*. (4)
b) Thermal cracking is a free-radical process. Discuss the industrial importance of this process. (2)
c) One of the products of thermal cracking is ethene. Discuss, using equations, how ethene is used in the manufacture of ethanol, antifreeze and plastics. (6)

10 Show, by giving equations and stating the reagents, how the following conversions may be carried out:

propene → 2-bromopropane → propan-2-ol → propanone

In each case name the type of reaction that occurs. Outline the mechanism for the first step, and explain why the reaction mixture may contain some 1-bromopropane. (15)

11 Epoxyethane and ethane-1,2-diol are two commercially important chemicals produced industrially from ethene.

a) Outline, using equations and stating essential reaction conditions, how epoxyethane and ethane-1,2-diol are produced. (4)
b) Give a use for ethane-1,2-diol. (1)
c) Write an equation for the reaction between epoxyethane and ethanol. Give two commercial uses of the product of this reaction. (4)

12 a) What is a *carbocation*? (1)
b) The conversions of but-1-ene into 2-bromobutane, and butan-2-ol into but-1-ene, both involve carbocations. Outline the mechanisms of these reactions. (14)

13 a) Explain the meaning of the terms *empirical formula* and *molecular formula*. (2)
b) The percentage composition by mass of a hydrocarbon **X** is C = 85.7% and H = 14.3%. When vaporised, 0.21 g of **X** occupies a volume of 85 cm^3 at standard temperature and pressure. Use these data to calculate the molecular formula of **X**. (5)
c) Hydrocarbon **X** has three structural isomers. Give the names of and structure formulae for these three isomers. (6)

14 a) Explain the meaning of the term *homologous series*. (2)
b) Table 1 shows the trend in boiling points of the first six straight-chain alkanes.

Alkane	Boiling point /°C
Methane	–164
Ethane	–89
Propane	–42
Butane	–1
Pentane	36
Hexane	69

Table 1

(i) Explain this trend in boiling points.
(ii) Predict, with reasons, whether $CH_3CH_2C(CH_3)_3$ will have a higher or lower boiling point than hexane.
(iii) Predict, with reasons, whether ethanol will have a higher or lower boiling point than propane. (10)

15 **a)** (i) What feature of haloalkane molecules enables them to undergo nucleophilic substitution reactions?
(ii) Explain why iodoethane undergoes these reactions more readily than choroethane. (4)

b) Name the reagent needed and write an equation for the reaction that occurs when 1-bromobutane is converted to:
(i) butan-1-ol (ii) pentanenitrile (iii) but-1-ene. (6)

16 Describe the chemical tests you could carry out to distinguish between the following pairs of organic compounds. In each case, state the reagent you would use and the observations you would make for each compound. Write the structural formula for the organic compound that gives a positive result for the test described.

a) Hexane and hex-2-ene. (4)
b) Butanal and butanone. (5)
c) Methylpropan-1-ol and methylpropan-2-ol. (5)

17 Discuss the different types of isomerism that occur in carbon compounds. Illustrate your answer by reference to isomers of each of the following molecules:

C_5H_{12} C_3H_7Cl C_4H_8 C_4H_8O

In each case name the type of isomerism, and give the names and structures of the different isomers. (25)

18 Compound **A** (C_4H_9Br) reacts with warm, aqueous potassium hydroxide solution to produce compound **B** ($C_4H_{10}O$). When heated with acidified sodium dichromate(VI) solution, compound **B** is converted into compound **C** (C_4H_8O) which does *not* react with Tollen's solution.

If the potassium hydroxide solution is ethanolic, compound **A** undergoes a different reaction to form compound **D** (C_4H_8). When heated with concentrated sulphuric acid, followed by hydrolosis, compound **D** is converted into compound **B**.

Identify compounds **A** to **D** by name and formula, and name the types of reaction that occur at each stage. (12)

Appendix A

Periodic Table

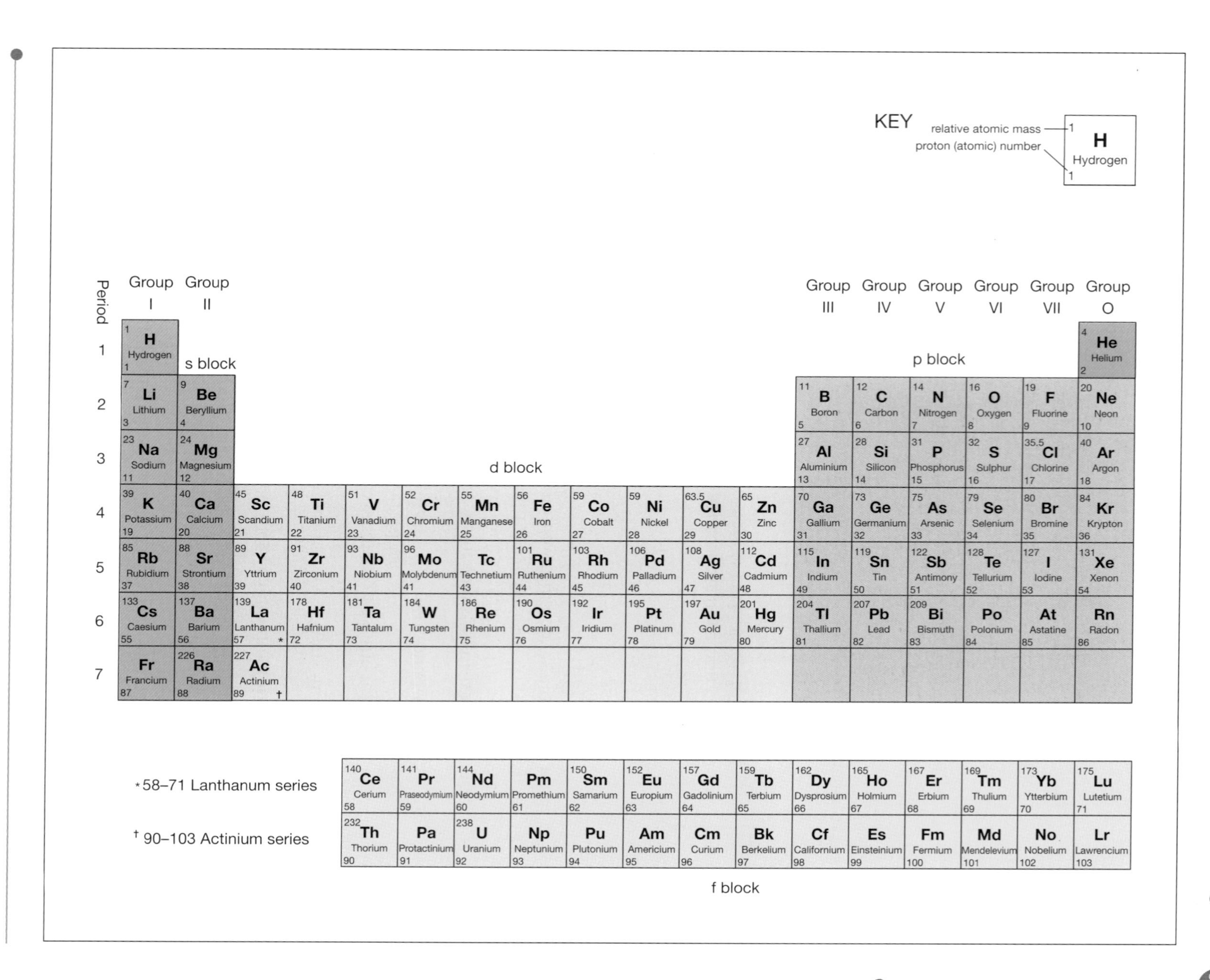

Appendix B

A table of common ions

Cations		Anions	
Name	Formula	Name	Formula
aluminium	Al^{3+}	bromide	Br^-
ammonium	NH_4^+	carbonate	CO_3^{2-}
barium	Ba^{2+}	chlorate(I)	ClO^-
calcium	Ca^{2+}	chlorate(V)	ClO_3^-
chromium(III)	Cr^{3+}	chloride	Cl^-
cobalt(II)	Co^{2+}	fluoride	F^-
copper(II)	Cu^{2+}	hydrogencarbonate	HCO_3^-
hydrogen	H^+	hydroxide	OH^-
iron(II)	Fe^{2+}	iodide	I^-
iron(III)	Fe^{3+}	nitrate	NO_3^-
lithium	Li^+	nitrite	NO_2^-
magnesium	Mg^{2+}	oxide	O^{2-}
potassium	K^+	sulphate	SO_4^{2-}
silver	Ag^+	sulphide	S^{2-}
sodium	Na^+	sulphite	SO_3^{2-}
zinc	Zn^{2+}	thiosulphate	$S_2O_3^{2-}$

Appendix C

Types of calculation involving moles

1 Calculating relative molecular mass, M_r (formula mass)

For these calculations you must be able to write formulae. Atomic masses, are quoted in the Periodic Table.

e.g. Calculate the M_r for ammonium sulphate.

Formula = $(NH_4)_2SO_4 = (2 \times 14) + (8 \times 1) + 32 + (4 \times 16) = 132$

2 Converting moles into grams (mass) and vice versa

No. of moles = mass/M_r Mass = no. of moles × M_r

The ability to use moles confidently is required for most of the subsequent calculations.

One mole = the molecular mass in grams

e.g. Calculate the number of moles in 13.2 g of ammonium sulphate.

no. of moles = 13.2/132 = 0.1

e.g. Calculate the number of grams in 0.25 moles of ammonium sulphate.

no. of grams = 0.25 × 132 = 33 g

3 Reacting masses (GCSE)

For these calculations you must be able to write the balanced equation for the reaction.

e.g. Calculate the mass of calcium oxide produced when 10 g of calcium carbonate is heated.

$CaCO_3 \rightarrow CaO + CO_2$ (thermal decomposition)

From the equation,

1 mole of calcium carbonate → 1 mole of calcium oxide

i.e. 100 g → 56 g

so 10 g → 5.6 g i.e. 56 × 10/100.

4 Calculating molarities

(Molarity = moles per decimetre cubed = mol dm^{-3})

For an individual solution

You can use a formula or work from first principles.

M = (mass × 1000)/(M_r × V) where V is in cm^3

e.g. Calculate the molarity of a solution of ammonium sulphate if 13.2 g is dissolved in 250 cm^3.

13.2/132 = 0.1 moles in 250 cm^3 ∴ 0.4 mol dm^{-3}

From a titration

Here you will need the balanced equation. Most reactions will be a 1:1 ratio of reactants.

e.g. What volume of 0.2 M NaOH is required to neutralise 25 cm^3 of 0.1 M HCl?

$NaOH + HCl \rightarrow NaCl + H_2O$
$0.2 \times V = 0.1 \times 25 \qquad V = 0.1 \times 25 / 0.2 = 12.5\,cm^3$

If the reaction is a different ratio of moles, this must be taken into account.

e.g. What volume of 0.2 M NaOH is required to neutralise 25 cm^3 of 0.1 M H_2SO_4?

$2NaOH + H_2SO_4 \rightarrow Na_2SO_4 + 2H_2O$

No. of moles of H_2SO_4 **=** $\boldsymbol{V \times M/1000}$ $= 25 \times 0.1 / 1000 = 0.0025$
No. of moles of NaOH required $= 2 \times 0.0025 = 0.005$
So $0.005 = V \times 0.2 / 1000$
$V = 0.005 \times 1000 / 0.2 = 25\,cm^3$

5 Gas calculations

Using $\boldsymbol{pV = nRT}$

e.g. What volume is occupied by 0.1 mol of a gas at 27 °C and 100 kPa pressure?

$V = nRT/p$ where n = number of moles
R = gas constant
T = temperature in degrees kelvin
$= 0.1 \times 8.31 \times (273 + 27) / 100 = 2.5\,dm^3 = 2500\,cm^3$

6 Empirical and molecular formulae

When M_r is given

e.g. Compound X contains 1.2 g of carbon, 0.2 g of hydrogen and 1.6 g of oxygen.
If the $M_r = 60$, calculate the molecular formula.

$C:H:O = 1.2/12 : 0.2/1 : 1.6/16$
$= 0.1 : 0.2 : 0.1$
$= 1 : 2 : 1$ Empirical formula $= CH_2O$
Empirical mass $= 12 + 2 + 16 = 30$
So molecular formula $= 2 \times$ empirical formula $= (CH_2O)_2 = C_2H_4O_2$

When M_r has to be calculated

(This combines calculations in 5 and 6.)

e.g. The % composition by mass of X is C = 85.7%, H = 14.3%. Calculate the empirical formula. 0.21 g of X occupies a volume of 84 cm^3 at stp. Calculate the M_r for X, and hence calculate the molecular formula.

$C:H = 85.7/12 : 14.3/1$
$= 7.14 : 14.3$
$= 1 : 2$ Empirical formula $= CH_2$ Empirical mass $= 14$

$\boldsymbol{M_r = (m \times R \times T)/(V \times p)}$ where stp = standard temperature and pressure
$= 0.21 \times 8.31 \times 273 \times 1000/84 \times 100$
$= 56.7$ (this works out at 56 if $p = 101.3$ kPa)
$56/14 = 4$
So molecular formula $= 4 \times$ empirical formula $= (CH_2)_4 = C_4H_8$

Glossary

acid A species that can donate a hydrogen ion (a proton donor).

activation energy, E_a The minimum energy required for a collision between particles to result in a reaction.

addition reaction A reaction in which a molecule is added across a double bond, resulting in a saturated product.

aliphatic An organic molecule in which the carbon atoms are arranged in a straight or branched chain.

alkali A soluble base.

amphoteric Describes a compound that can react as an acid and a base.

aromatic Describes an organic molecule that contains at least one benzene ring.

atomic number The number of protons in the nucleus of an atom.

base A species that can accept a hydrogen ion (a proton acceptor).

batch process A process in which the products are removed at the end of the reaction. The reaction is then started again with a new supply of reactants.

bond angle The angle formed between two bonds attached to the same atom.

bond dissociation energy The energy needed to break a particular covalent bond.

calorimeter The apparatus in which heat energy changes are measured.

carbocation A very reactive species with a positive charge on a carbon atom.

carbonium ion An alternative name for a **carbocation**.

catalyst A substance that alters the rate of a chemical reaction, without itself being changed by the reaction.

chain isomers Isomers in which the carbon atoms in the chain are arranged in a different pattern.

compound A substance made up of more than one sort of atom bonded together.

continuous process A reaction in which more reactants are added as the products are removed.

co-ordinate bond A covalent bond in which both members of the pair of electrons are provided by one of the atoms. Also called a **dative bond**.

covalent bond A bond formed by the sharing of a pair of electrons.

cracking A process by which large molecules are broken down into smaller molecules.

dative bond An alternative name for a **co-ordinate bond.**

dehydration A reaction that involves the loss of a water molecule.

delocalised electrons Electrons that are free to move and are not confined to a particular atom or bond.

diatomic molecule A molecule containing two atoms.

electronegativity A measure of the ability of an atom to attract the pair of electrons in a covalent bond.

electronic structure The arrangement of the electrons in the energy levels of an atom or ion.

electrophile An electron-deficient species that can accept a lone pair of electrons.

element A substance made up of one sort of atom.

elimination reaction A reaction in which a double bond is formed as a small molecule is lost from a saturated compound.

empirical formula The simplest whole-number ratio of the atoms present in a molecule.

endothermic reaction A reaction that absorbs heat energy (from the surroundings).

enthalpy change, ΔH The heat energy change in a reaction occurring at constant pressure.

equilibrium A reaction in which the rates of the forward and reverse reactions are equal.

exothermic reaction A reaction that releases heat energy (to the surroundings).

fractional distillation A process that separates the hydrocarbons present in petroleum according to their boiling points.

free radical A very reactive species with an unpaired electron.

functional group The atom or group of atoms responsible for the chemical properties of an organic molecule.

functional group isomerism Isomers that contain different functional groups.

geometrical isomers Isomers in which the presence of a double bond results in the different orientation of some atoms or groups of atoms within the molecules.

halogen Any element from Group 7 of the Periodic Table.

Hess's Law The enthalpy change in a reaction is independent of the route of the reaction.

homologous series A group of compounds that have the same functional group and similar chemical properties and that show a gradation in physical properties.

hydrocarbon A compound containing the elements carbon and hydrogen only.

hydrogenation A reaction in which hydrogen is added to a compound.

hydrolysis A reaction with water.

intermediate A species that is formed and used up during a reaction.

intermolecular forces The weak forces of attraction between molecules.

ion A species that has a charge.

ionic bond The force of attraction between oppositely charged ions.

ionisation energy The energy required to remove an electron from a gaseous atom.

isomers Compounds with the same molecular formula but different properties.

isotopes Atoms of the same element that have different numbers of neutrons.

Le Chatelier's Principle When one or more of the factors affecting the position of a chemical equilibrium change, then the equilibrium moves so as to oppose the change.

lone pair A pair of electrons in the outer occupied energy level which are not involved in bonding.

kelvin A unit of temperature.

mass number The sum of the protons and neutrons in an atom.

mean bond enthalpy The average bond energy for a particular bond in different compounds.

mechanism The individual steps by which a reaction occurs.

metallic bonding The force of attraction between the delocalised outer electrons and the resulting positive centres of the metal atoms.

molarity The concentration of a solution in mol/dm^3.

mole The working quantitative unit used in calculations. One mole of a substance is the amount of that substance which contains as many particles as there are atoms in exactly 12 g of carbon-12.

molecular formula A formula which shows the actual numbers of atoms present in a molecule.

molecular ion A molecule that has lost or gained electrons.

molecule A compound formed by covalent bonding.

neutralisation A reaction in which an acid and a base react together to form a salt and water.

noble gas Any of the elements in Group 0 of the Periodic Table.

nuclear charge The charge of the nucleus due to the number of protons present.

nucleophile A species that can donate a lone pair of electrons to an electron-deficient carbon atom.

oxidation A reaction in which a species loses electrons.

oxidation state (number) The number of electrons of an atom used in forming bonds.

oxidising agent An electron acceptor.

pascal A unit of pressure.

Periodic trend A repeating pattern of properties shown by the elements in a Period of the Periodic Table.

polar bond A covalent bond in which the pair of electrons are not shared equally.

polymer A large molecule made by joining together many thousands of smaller molecules.

position isomers Isomers in which the functional group is attached to different carbon atoms.

redox reaction A reaction in which oxidation and reduction occur.

reducing agent An electron donor.

reduction A reaction in which a species gains electrons.

reflux A method of prolonged heating without loss by evaporation.

relative atomic mass of an element The average mass of an atom of an element (taking into account the natural occurrence of its isotopes) compared to the mass of one atom of carbon-12.

relative molecular mass The average mass of a molecule compared to the mass of one atom of carbon-12.

saturated compound A compound in which the covalent bonds are all single bonds.

shielding The effect of the inner electrons screening the outer electrons from the nuclear charge.

species A general term used to denote an atom, an ion or a molecule.

standard solution A solution whose concentration is known exactly.

stereoisomers Molecules with the same structural formula which differ in the orientation of some atoms or groups of atoms.

structural formula A formula which shows how the atoms are arranged in the molecule.

structural isomers Molecules with the same molecular formula but with different arrangements of atoms within the molecule.

substitution A reaction in which an atom or a group of atoms is replaced by a different atom or group of atoms.

transition metal An element with partially filled d sub-levels in its atoms.

unsaturated compound A compound containing one or more multiple covalent bonds.

Answers: Module 1

1 Atomic structure

Pages 4 – 5

1. 9 protons, 10 neutrons and 9 electrons.
2.

Element	Atomic number	Mass number	Number of protons	Number of neutrons	Number of electrons
Potassium	19	40	**19**	**21**	**19**
Oxygen	**8**	**16**	8	8	**8**
Chlorine	**17**	37	**17**	20	**17**
Phosphorus	**15**	31	**15**	**16**	15
Aluminium	13	**27**	**13**	14	**13**

Pages 6 – 7

1. **a)** Carbon-12: 6 protons, 6 neutrons and 6 electrons.
   b) Carbon-14: 6 protons, 8 neutrons and 6 electrons.
2. The different isotopes have different masses and hence diffuse at different speeds. The isotope with the smaller mass diffuses more quickly.
3. Isotopes are atoms of the same element with different numbers of neutrons *or* with the same atomic number but different mass numbers.
4. *Two of*: different melting points, boiling points, densities, and rates of diffusion.

Pages 8 – 9

1. So that the atoms can be accelerated and deflected.
2. Accelerated by an electrical field; deflected by a magnetic field.

Pages 10 – 11

1. $$\frac{(64 \times 48.9) + (66 \times 27.8) + (67 \times 4.1) + (68 \times 18.6) + (70 \times 0.62)}{(48.9 + 27.8 + 4.1 + 18.6 + 0.62)} = 65.46$$
2. m/z values: 32, 34 and 36.
3. Both have the same m/z value of 16.

Pages 12 – 13

1. Sub-levels: 3s, 3p and 3d.
2. **a)** s block **b)** d block **c)** p block **d)** p block **e)** s block **f)** p block.

Pages 14 – 15

1. a) $1s^2 2s^2 2p^6 3s^2 3p^5$. b) $1s^2 2s^2 2p^6 3s^2 3p^6 4s^2$. c) $1s^2 2s^2 2p^4$.
2. a) Carbon. b) Titanium. c) Zinc.
3. a) $1s^2 2s^2 2p^5$ b) $1s^2 2s^2 2p^6 3s^2 3p^4$
   c) $1s^2 2s^2 2p^6 3s^2 3p^6 3d^3 4s^2$

Pages 16 – 17

1. a) b) c)

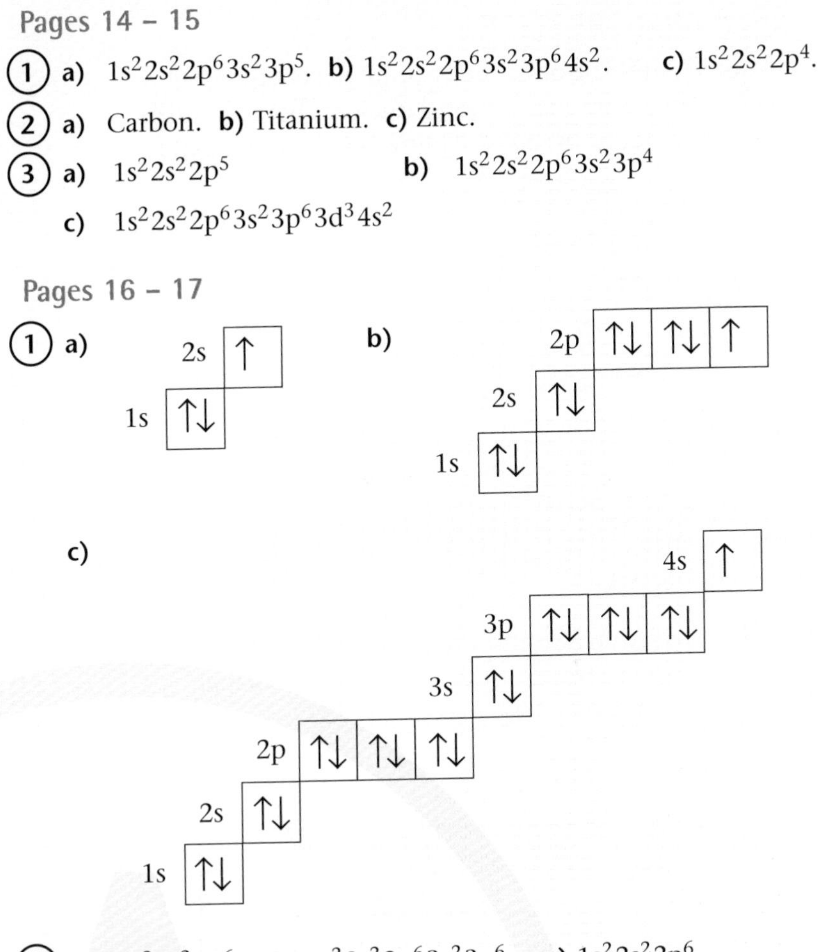

2. a) $1s^2 2s^2 2p^6$. b) $1s^2 2s^2 2p^6 3s^2 3p^6$. c) $1s^2 2s^2 2p^6$.
3. a) $1s^2 2s^2 2p^6 3s^2 3p^6$. b) $1s^2 2s^2 2p^6 3s^2 3p^6 3d^{10} 4s^2 4p^6$. c) $1s^2 2s^2 2p^6 3s^2 3p^6$.
4.

	Atomic number	Mass number	Number of protons	Number of neutrons	Number of electrons	Electronic structure
Mg	**12**	**24**	**12**	12	**12**	$1s^2 2s^2 2p^6 3s^2$
Al^{3+}	**13**	27	**13**	**14**	10	$\mathbf{1s^2 2s^2 2p^6}$
S^{2-}	**16**	**32**	16	16	**18**	$\mathbf{1s^2 2s^2 2p^6 3s^2 3p^6}$
Sc^{3+}	21	45	**21**	**24**	**18**	$\mathbf{1s^2 2s^2 2p^6 3s^2 3p^6}$
Ni^{2+}	**28**	**58**	**28**	30	26	$\mathbf{1s^2 2s^2 2p^6 3s^2 3p^6 3d^8}$

5. Fe^{2+}: $1s^2 2s^2 2p^6 3s^2 3p^6 3d^6$; Fe^{3+}: $1s^2 2s^2 2p^6 3s^2 3p^6 3d^5$. Fe^{3+} has greater stability because it has a half-filled 3d sub-level.

Pages 18 – 19

1. $Al(g) \rightarrow Al^+(g) + e^-$
2. The electron removed from boron is in a higher sub-level (2p) than the electron removed from beryllium (2s).

Pages 20 – 21

(1) The electron removed from sodium is from a lower principal energy level and is closer to the nucleus. Potassium has a greater nuclear charge, but the effect of the nuclear charge is reduced by the shielding action of the extra inner electrons.

(2)

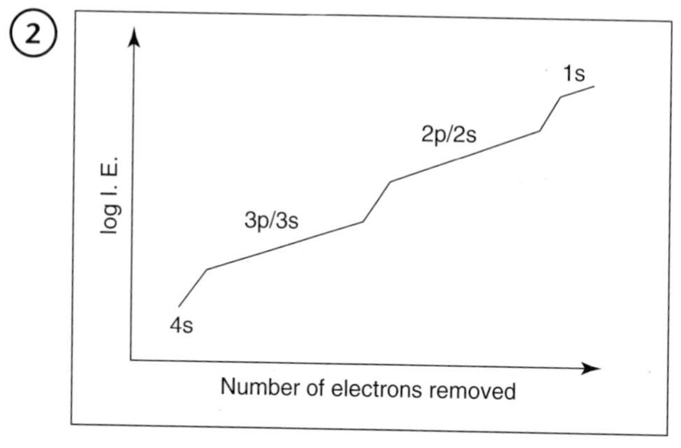

(3) The first three electrons removed from aluminium are from the third principal energy level. The fourth electron is from the second principal energy level, which is nearer the nucleus.

Pages 22 – 23: Unit 1 Questions

(1)

	Relative mass	Relative charge
Proton	1	+1
Neutron	1	0
Electron	1/836	–1

(2) *Atomic number*: The number of protons in the nucleus of an atom (1). *Mass number*: The sum of protons and neutrons in the nucleus of an atom (1).

(3) **a)** Atoms of the same element (1) that have different numbers of neutrons (1).
b) Same number of electrons and same electronic structure (1); chemical properties of an element depend upon electronic structure (1). Different mass numbers and hence different masses (1); physical properties such as melting point, boiling point, density and rate of diffusion depend upon mass (1).

(4) **a)** $1s^2 2p^6 3s^2 3p^5$. **b)** $1s^2 2s^2 2p^6 3s^2$. **c)** $1s^2 2s^2 2p^6 3s^2 3p^6 4s^1$.
d) $1s^2 2s^2 2p^6 3s^2 3p^6 3d^{10} 4s^1$. **e)** $1s^2 2s^2 2p^6 3s^2 3p^6$. **f)** $1s^2$.
g) $1s^2 2s^2 2p^6$. **h)** $1s^2 2s^2 2p^6 3s^2 3p^6 3d^3$.

(5) **a)** (i) Ionised by an electron gun (1), which 'knocks' an electron from the atom, forming a positive ion (1).
(ii) Accelerated by an electric field (1); ions are focused into a beam (1).
(iii) Deflected by a magnetic field (1); for ions with the same charge moving at the same velocity, deflection depends upon mass (*or* deflection depends upon m/z value) (1).
(iv) Detected by ion-current detector (1); current proportional to number of ions striking the detector (1).

b) (i) $A_r = \frac{\text{mean (average) mass of an atom}}{\text{mass of one atom of carbon-12}} \times 12$

(ii) $A_r = \frac{(5.8 \times 54) + (91.6 \times 56) + (2.2 \times 57) + (0.33 \times 58)\ (1)}{(5.8 + 91.6 + 2.2 + 0.33)\ (1)} = 55.91\ (1)$

6 a) Energy needed to remove an electron (1) from a gaseous atom (1) (or in mole quantities).

b) $Na(g) \rightarrow Na^+(g) + e^-$ (equation, 1; state symbols, 1)

c) The electron removed from sodium is from a higher (1) principal energy level (1) (3s) than the electron removed from neon (2p).

d) The electron removed is in each case a 3s electron (1), but the nuclear charge of magnesium is greater than that of sodium (1).

e) The electron removed from aluminium is from a higher (1) sub-level (1) (3p) than the electron removed from magnesium (3s).

f) The value is lower. In both cases the electron removed is from the 3p sub-level (1), but in the case of sulphur the electron removed is a paired electron (1).

7 a) There is a large range in values of the ionisation energies (1).

b) There is always an increase in the value of successive ionisation energies (1), since after an electron has been removed the remaining electrons are pulled closer to the unchanged nucleus (1), hence more energy is needed to remove the next electron (1).

c) The first four ionisation energies show a steady increase (1) in value, indicating that the electrons are in the same energy level (1). The fifth ionisation energy is much larger in value, indicating that the electron is in a lower energy level (1). This suggests that carbon has two electrons in the first energy level and four electrons in the second energy level (1).

2 Bonding

Pages 24 – 25

1 a)

b)

c)

d)

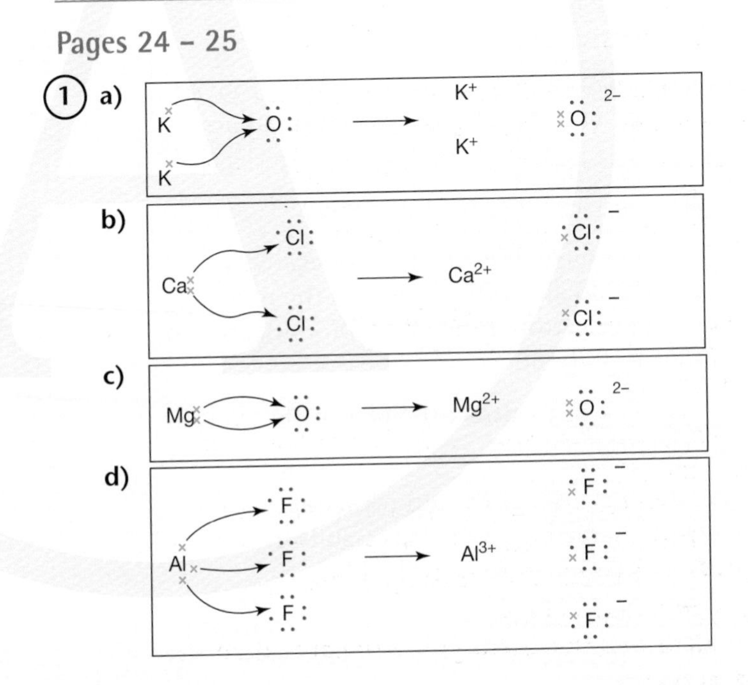

(2) Melting point depends upon the strength of the ionic bonds. Ionic bonds are the electrostatic force of attraction between oppositely charged ions. The melting point of NaCl is greater than that of KCl because the Na^+ ion is smaller than the K^+ ion and this leads to stronger ionic bonding. MgO has a higher melting point than NaCl and KCl since both the ions in MgO carry twice the charge of the ions in NaCl and KCl, and hence the ionic bonds are much stronger.

Pages 26 – 27

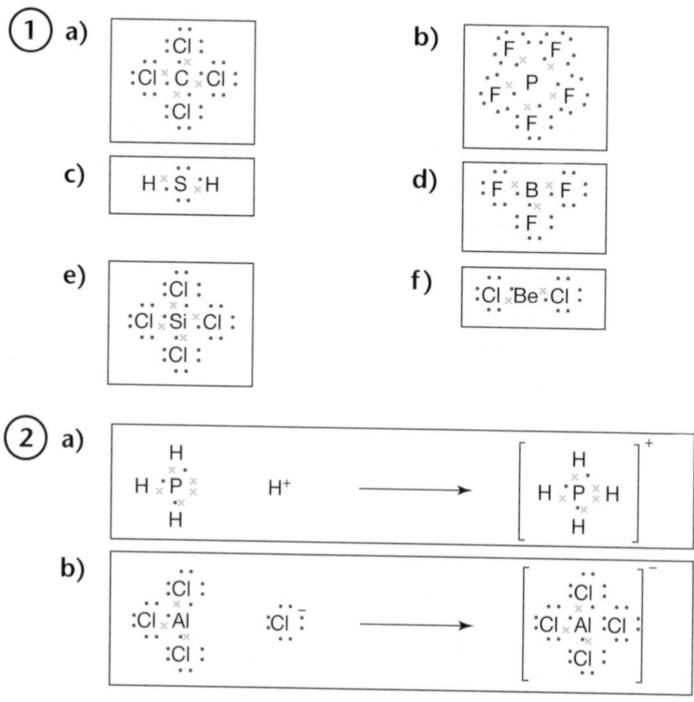

Pages 28 – 29

(1) Chlorine has a smaller atomic radius than bromine. The pair of electrons in the covalent bond are closer to the nucleus of the chlorine atom, and are more strongly attracted.

(2) Oxygen has a smaller atomic radius, and a greater nuclear charge.

(3) Polar bonds are formed when the atoms have different electronegativity values. The electrons in the covalent bond are 'pulled' towards the more electronegative atom. Chlorine is more electronegative than bromine, so the H—Cl bond is more polar than the H—Br bond.

(4) A measure of the relative ability of an atom to attract the pair of electrons in a covalent bond (*or* to withdraw electron density).

(5) Across a Period, the electronegativity increases as the atomic radius decreases and the size of the nuclear charge increases. This results in increasing attraction for the pair of electrons in the covalent bond.

(6) When the atoms have different electronegativities, the pair of electrons in a covalent bond are unequally shared. They are drawn towards the more electronegative atom, making the bond polar.

Pages 30 – 31

(1) The sodium atom has three occupied principal energy levels: $1s^2 2s^2 2p^6 3s^1$. The electron removed from sodium to form the sodium ion is the 3s electron, which leaves the ion with only two occupied principal energy levels.

(2) Both ions have the same electronic structure, $1s^2 2s^2 2p^6$, but the magnesium ion has a greater nuclear charge, so the outer electrons are pulled inwards: this makes the Mg^{2+} ion smaller than the Na^+ ion.

(3) The oxide ion has the same nuclear charge as the oxygen atom but has two extra electrons, so the effect of the nuclear charge is less.

(4) The lithium ion, Li^+, is the smallest of the Group I metal ions and has the greatest size-to-charge ratio. This makes the lithium ion very polarising, and the large iodide ion is polarised by the lithium ion. This results in lithium iodide having some covalent character. Covalent compounds dissolve in non-polar solvents, but are generally insoluble in water.

Pages 32 – 33

(1)

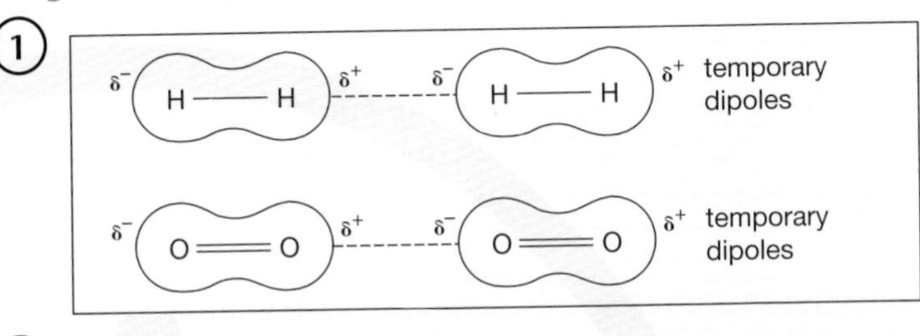

(2) ICl has the higher boiling point. Bromine, Br_2, is non-polar whereas ICl is polar (Cl and I have different electronegativities). There are van der Waals forces between the bromine molecules: there are permanent dipole–dipole forces between the molecules of iodine monochloride. The relative molecular masses are very similar, but ICl has stronger intermolecular forces.

Pages 34 – 35

(1) **b)** CH_3OH and **c)** NH_3

(2) Ammonia has hydrogen bonding between the molecules. This is the strongest form of intermolecular force, so ammonia has a higher boiling point than the larger PH_3 compound, which has the weaker permanent dipole–dipole forces.

Pages 36 – 37

(1) In a solid the particles vibrate about a fixed position. As the temperature increases, the vibrations become more violent and some of the forces between the particles are loosened or broken, allowing the particles to move around. The solid changes to a liquid (melts). As the temperature increases further, the forces between the liquid particles are broken and the particles escape to form a gas (boil). In a gas the particles move with rapid, random motion.

Pages 38 – 39

1. In magnesium there are two mobile electrons per atom, whereas in sodium there is only one mobile electron per atom. There are more electrons available in magnesium to conduct electricity.

2. They are both metals. Metallic bonding is the force of attraction between the delocalised electrons and the 'positive' centres. Aluminium has three delocalised electrons per atom, compared to the one delocalised electron per atom in sodium. Hence metallic bonding in aluminium is stronger.

3. The force of attraction between oppositely charged ions.

4. Cs$^{\times}$ → $\cdot$Cl: ⟶ Cs^{+} [:Cl:]$^{-}$

5. Metals conduct electricity by the movement of the delocalised electrons, in both the solid and liquid states. Ionic compounds can only conduct electricity when the ions are free to move: the ions are free to move when the ionic compound is dissolved in water or is molten.

Pages 40 – 41

1. Carbon (as diamond or graphite) has a giant atomic structure and a very large number of covalent bonds must be broken to break down the giant structure. Iodine exists as simple I_2 molecules with weak van der Waals forces between the molecules.

2. Diamond has a giant rigid tetrahedral atomic structure, with each carbon atom forming four covalent bonds. Diamond is very hard and is a non-conductor of electricity. Graphite exists as plates with each carbon forming three covalent bonds. The plates can slide over each other, making graphite a soft substance. The 'spare' electrons make graphite a good conductor of electricity.

Pages 42 – 43

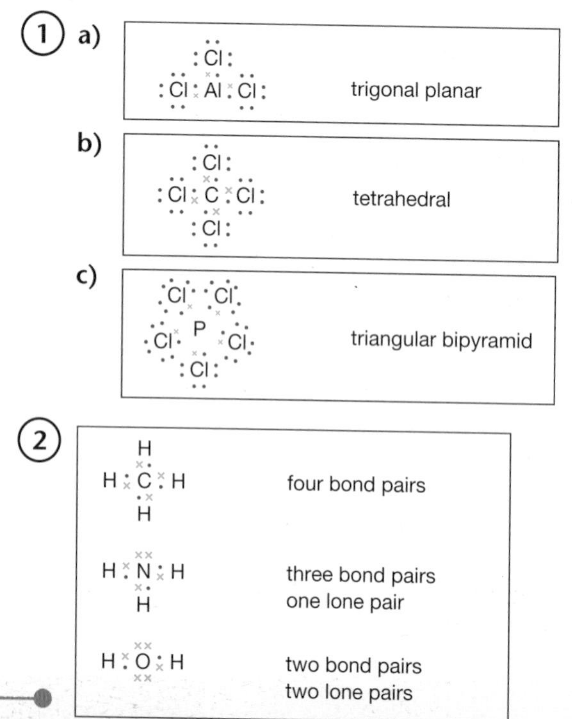

Pages 44 – 45

a) Trigonal planar. **b)** Tetrahedral. **c)** Tetrahedral. **d)** Distorted tetrahedral (pyramidal). **e)** Distorted tetrahedral (V-shaped). **f)** Distorted tetrahedral (pyramidal). **g)** Distorted tetrahedral (V-shaped). **h)** Distorted triangular bipyramid. **i)** Distorted octahedral. **j)** Distorted triangular bipyramid.

Pages 46 – 47: Unit 2 Questions

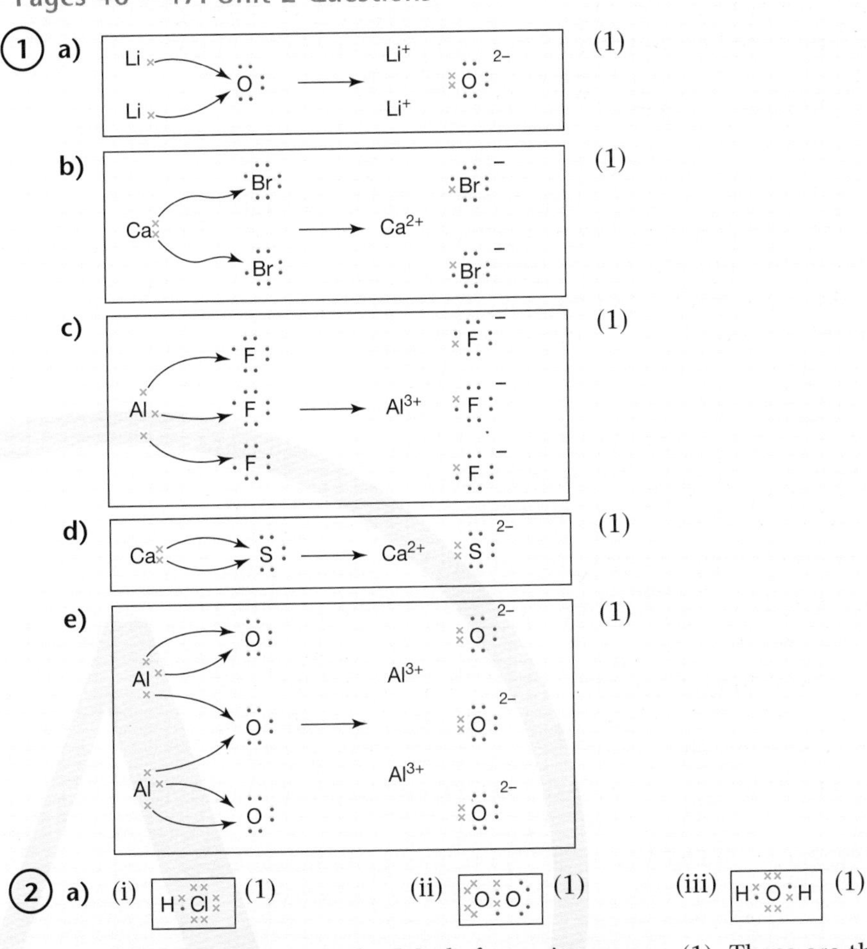

b) There are only van der Waals forces in oxygen (1). These are the weakest intermolecular forces (1). HCl is a polar molecule (1), and has the stronger permanent dipole–dipole forces (1).

c) There is hydrogen bonding in water (1). This is the strongest form of intermolecular forces (1). In HCl there are polar bonds (1), and hence permanent dipole–dipole forces (1), but these are weaker than hydrogen bonds.

3 **a)** A measure of the ability of an atom to attract the pair of electrons in a covalent bond (*or* to withdraw electron density) (1).

b) Electronegativity increases (1), since the atomic radius decreases (1) and the nuclear charge increases (1).

c) A covalent bond (1) in which the pair of electrons are not equally shared (1).

d) In Br_2 the atoms have the same electronegativity (1), but in HBr the atoms have different electronegativity values (1).

4 a) In diamond the carbon atoms are covalently bonded in a giant atomic structure (1). Each carbon atom forms four (1) covalent bonds (1), resulting in a giant tetrahedral structure. In graphite each carbon atom forms three covalent bonds (1), resulting in plates (1) of hexagonally arranged carbon atoms. The bonds between the plates are longer than the bonds between the carbon atoms in the plates (1) (*or* the distance between the plates is longer than the distance between the carbon atoms in the plates).

b) An atom of graphite only uses three of its outer electrons in forming covalent bonds (1); the 'spare' electron allows graphite to conduct electricity (1). In diamond each carbon atom uses its four outer electrons in covalent bonding.

5 a) Giant ionic structure (1), with the sodium and chloride ions forming a face-centred cubic structure (1).

b) (i) In solid NaCl the ions vibrate (1) about fixed positions (1).
(ii) In liquid NaCl the ions are free to move (1) randomly (1).

c) Iodine consists of covalently bonded I_2 molecules (1). The molecules in the solid are held together by van der Waals forces (1).

d) When solid iodine is heated, the weak van der Waals forces are broken (1), and a purple vapour containing separate I_2 molecules is formed (1). The ionic bonds in NaCl are strong forces of attraction (1), and are not weakened or broken when gently heated (1).

6 a) A co-ordinate bond is formed between a species with a lone pair of electrons (1) and a species with an empty (vacant) atomic orbital (1).

b) (2)

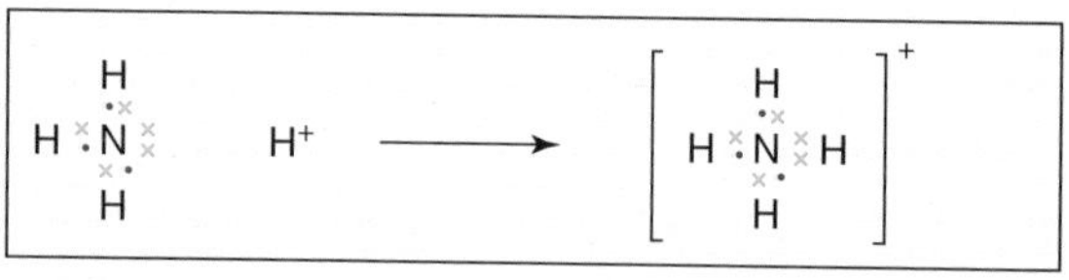

H
H ˣ• N ˣˣ   H⁺ ⟶ [H ˣ• N ˣ• H]⁺
H

7 a) van der Waals (1), permanent dipole–dipole (1), and hydrogen bonding (1).

b) In H_2 the temporary induced δ^+ charge (1) on one H_2 molecule is attracted to the temporary induced δ^- charge (1) on an adjacent molecule.
The C—Cl bond is a polar bond. In CH_3Cl the permanent δ^+ charge (1) on the carbon atom on one molecule is attracted to the permanent δ^- charge (1) on a chlorine atom in an adjacent molecule.
In NH_3 the lone pair (1) on the nitrogen atom is attracted to the δ^+ charge (1) on a hydrogen of an adjacent molecule.

8

	Shape	Bond angle(s)	
BF_3	trigonal planar	120°	(2)
CCl_4	tetrahedral	109.5°	(2)
PF_5	triangular bipyramid	90° and 120°	(2)
SF_6	octahedral	90°	(2)
NCl_3	pyramidal	107°	(2)
PCl_4^+	tetrahedral	109.5°	(2)
ICl_4^+	distorted triangular bipyramid	(approx.) 90° and 120°	(2)

9 a)

b)

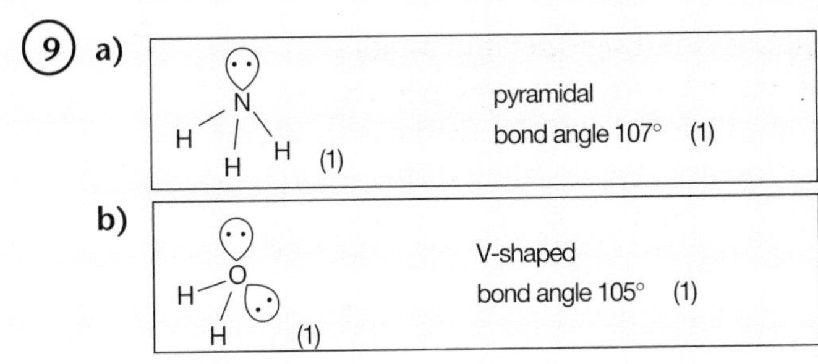

c) The greater mutual repulsion (1) caused by the extra (1) lone pair of electrons in water results in a reduced bond angle.

10 a) (i) $AlCl_3$: trigonal planar (1), bond angle 120° (1).
(ii) PH_3: pyramidal (distorted tetrahedral) (1), bond angle 107° (1).

b) (i) There is a lone pair of electrons on the phosphorus atom (1), and a vacant orbital on the aluminium atom (1).

(ii)

```
   :Cl: H
:Cl: Al :P: H
   :Cl: H
```

(iii) Both are 109.5° (1) since the aluminium atom and the phosphorus atom now have four bond pairs (1).

11 a) MgO: both compounds are ionic (1). The magnesium ion is smaller than the calcium ion (1), hence the ionic bond is stronger (1).

b) NaCl: the bonding in NaCl is ionic (1). The bonding in HCl is covalent, with permanent dipole forces between the molecules (1): these intermolecular forces are weaker than ionic bonds (1).

c) SO_2: both are covalent molecules with polar bonds (1). CO_2 is linear, has no overall dipole and has van der Waals forces (1); SO_2 has permanent dipole forces between the molecules (1), which are stronger than van der Waals forces.

3 Periodicity

Pages 48 – 49

1 In both cases the outer electrons are in the third principal energy level. The greater nuclear charge of the sulphur nucleus draws the electrons closer.

2 The electron removed is from the same principal energy level, but the nuclear charge steadily increases across the Period, making it increasingly difficult to remove an electron.

3 Across the Period the atomic radius decreases and the nuclear charge increases. This results in a greater attraction for the electrons in a covalent bond.

Pages 50 – 51

1 The Group II metals lose two electrons to form dipositive ions when they react with water. In barium the two outer electrons are further from the nucleus and are more easily lost.

(2) The bonding in the elements is metallic bonding. Metallic bonding is the force of attraction between delocalised electrons and 'positive' centres. Each element has two delocalised electrons. Going down the Group, the atomic radius increases and the strength of the metallic bonding decreases.

(3) $Ba(s) + 2H_2O(l) \rightarrow Ba(OH)_2(aq) + H_2(g)$ Barium hydroxide and hydrogen.

Pages 52 – 53

(1) Both are tetrahedral.

(2) An amphoteric compound can react both as an acid and as a base.
As an acid: $Ba(OH)_2(aq) + 2NaOH(aq) \rightarrow Ba(OH)_4^{2-}(aq) + 2Na^+(aq)$
As a base: $Ba(OH)_2(aq) + 2HCl(aq) \rightarrow BaCl_2(aq) + 2H_2O(l)$

(3) Barium hydroxide is more soluble in water, and hence there is a greater concentration of hydroxide ions.

(4) $Ba(OH)_2(aq) + 2HCl(aq) \rightarrow BaCl_2(aq) + 2H_2O(l)$
$Ca(OH)_2(s) + 2HCl(aq) \rightarrow CaCl_2(aq) + 2H_2O(l)$

Pages 54 – 55: Unit 3 Questions

(1) **a)** Both have metallic bonding, but the bonds in aluminium are stronger than those in magnesium (1). Aluminium has three delocalised electrons per atom, compared to two delocalised electrons per atom in magnesium (1).
b) Both are simple molecular compounds with van der Waals forces (1), but S_8 has a greater mass than P_4 (1), and therefore has a higher melting point.
c) Silicon exists as a giant atomic structure (1). A large number of covalent bonds have to be overcome to break down the giant structure (1).
d) Argon exists as separate atoms (1) with weak van der Waals forces (1).

(2) **a)** Electronegativity increases (1) across a Period, since the atomic radius decreases (1) and the nuclear charge increases (1).
b) Electronegativity decreases (1) down the Group, since the atomic radius increases (1) and the inner electrons shield the covalent pair from the effect of the nucleus (1).

(3) **a)** Atomic radius decreases (1) across the Period, as the outer electrons are in the same principal energy level (1) but the nuclear charge increases (1).
b) Atomic radius increases (1) down the Group, since more principal energy levels are occupied (1). The inner electrons shield the outer electrons from the effect of the increased nuclear charge (1).

(4) Dissolve the sulphate in water (1), and add hydrochloric acid (1) followed by a few drops of barium chloride solution (1). A white precipitate (1) confirms the presence of a sulphate.

(5) **a)** The reaction with water becomes more vigorous (1) down the group since the outer electrons are further from the nucleus (1) and are more easily lost (1).
b) (i) $Ca(s) + 2H_2O(l) \rightarrow Ca(OH)_2(s) + H_2(g)$ (1).
(ii) Low solubility of $Ca(OH)_2$ (1).
(iii) Squeaky 'pop' with lighted splint (1).

(6) **a)** Solubility increases down the Group (1).
b) Strength as a base increases down the Group (1).
c) As solubility increases, the concentration of hydroxide ions in solution increases (1); it is the hydroxide ions that make the hydroxides act as bases (1).

(7) **a)** Able to act as an acid and as a base (1).
b) As a base: $Ba(OH)_2(aq) + H_2SO_4(aq) \rightarrow BaSO_4(s) + 2H_2O(l)$ (1)
As an acid: $Ba(OH)_2(aq) + 2KOH(aq) \rightarrow Ba(OH)_4^{2-}(aq) + 2K^+(aq)$ (1)

(8) The beryllium ion, Be^{2+}, is the smallest Group (II) metal ion and has the largest size-to-charge ratio (1). The highly polarising (1) beryllium ion polarises the large chloride ion, and this results in sharing of the electrons from the chloride ion (1).

(9) **a)** Mg: $1s^22s^22p^63s^2$ (1); Ca: $1s^22s^22p^63s^23p^64s^2$ (1).
b) The outermost electrons are in an s sub-level (1).
c) $Ca(g) \rightarrow Ca^+(g) + e^-$ (1)
$Ca^+(g) \rightarrow Ca^{2+}(g) + e^-$ (1)
$Ca^{2+}(g) \rightarrow Ca^{3+}(g) + e^-$ (1)
d) The first two electrons removed are from the fourth principal energy level (1), but the third electron is from the third principal energy level (1). The electron removed is closer to the nuceus, and is more tightly held (1).

4 Amount of substance

Pages 56 – 57

(1) **a)** 0.5 kg. **b)** 0.025 kg. **c)** 0.005 kg.

(2) **a)** $2\,dm^3$. **b)** $50\,dm^3$. **c)** $2 \times 10^{-3}\,dm^3$.

(3) **a)** 16. **b)** 256.

Pages 58 – 59

(1) 1 mole of carbon atoms = $1.9925 \times 10^{-23} \times 6.023 \times 10^{23} = 12.0008\,g$

(2) **a)** A_r of Li = 7
1 mol = 7 g
2 mol = $7 \times 2 = 14$ g

b) A_r of Fe = 56
1 mol = 56 g
0.5 mol = $56 \times 0.5 = 28$ g

(3) **a)** A_r of K = 39
39 g = 1 mol
39/39 = 1 mol

b) A_r of Mg = 24
24 g = 1 mol
2.4/24 = 0.1 mol

Pages 60 – 61

1

	Formula	Calculation	M_r
a)	$Ca(OH)_2$	$40 + (16 \times 2) + (1 \times 2)$	74
b)	NaCl	$23 + 35.5$	58.5
c)	$ZnCO_3$	$65 + 12 + (16 \times 3)$	125
d)	CaO	$40 + 16$	56
e)	$MgSO_4$	$24 + 32 + (16 \times 4)$	120
f)	$AlCl_3$	$27 + (35.5 \times 3)$	133.5

(2) a) 1 mol of $Ca(OH)_2$ = 74 g
2 mol = 74 × 2 = 148 g
b) 1 mol of NaCl = 58.5 g
0.25 mol = 58.5 × 0.25 = 14.625 g
c) 1 mol of $ZnCO_3$ = 125g
0.01 mol = 125 × 0.01 = 1.25g

(3) a) 1 mol of CaO = 56 g
112/56 = 2 mol
b) 1 mol of $MgSO_4$ = 120 g
12/120 = 0.1 mol
c) 1 mol of $AlCl_3$ = 133.5 g
33.375/133.5 = 0.25 mol

(4)

Name	Formula	M_r	Name	Formula	M_r
Oxygen	O_2	**32**	Ammonia	NH_3	**17**
Nitrogen	N_2	**28**	Carbon dioxide	CO_2	**44**
Chlorine	Cl_2	**71**	Methane	CH_4	**16**
Fluorine	F_2	**38**	Sulphur dioxide	SO_2	**64**
Hydrogen	H_2	**2**	Nitrogen dioxide	NO_2	**46**

(5) a) 1 mol of Cl_2 = 71 g
7.1/71 = 0.1 mol
b) 1 mol of H_2 = 2 g
20/2 = 10 mol
c) 1 mol of SO_2 = 64 g
3.2/64 = 0.05 mol

(6) a) 1 mol of F_2 = 38 g
2 mol = 38 × 2 = 76 g
b) 1 mol of N_2 = 28 g
0.4 mol = 28 × 0.4 = 11.2 g
c) 1 mol of CO_2 = 44 g
0.05 mol = 44 × 0.05 = 2.2 g

Pages 62 – 63

(1) Mass of lead = 16.41 – 10.20 = 6.21 g
Mass of oxygen = 17.37 – 16.41 = 0.96 g
So 6.21 g of lead combines with 0.96 g of oxygen
$\frac{6.21}{207}$ mol of lead combines with $\frac{0.96}{16}$ mol of oxygen
= 0.03 = 0.06
Ratio of Pb:O = 0.03:0.06 = 1:2
Formula = PbO_2

Pages 64 – 65

(1) 2.31 g of S combines with (5 – 2.31) g of Fe
2.31 g of S combines with 2.69 g of Fe
$\frac{2.31}{32}$ mol of S combines with $\frac{2.69}{56}$ mol of Fe
0.072 mol of S combines with 0.048 mol of Fe
i.e. 3 mol of S combines with 2 mol of Fe
Molar ratio Fe:S = 2:3
Formula = Fe_2S_3

(2) 2.52 g of Mg combines with (4.20 – 2.52) g of O

2.52 g of Mg combines with 1.68 g of O

$\frac{2.52}{24}$ mol of Mg combines with $\frac{1.68}{16}$ mol of O

0.105 mol of Mg combines with 0.105 mol of O

i.e. 1 mol of Mg combines with 1 mol of O

Molar ratio of Mg : O = 1 : 1

Formula = MgO

(3) Molar ratio of Na : C : O = $\frac{4.6}{23} : \frac{1.2}{12} : \frac{4.8}{16}$

= 0.2 : 0.1 : 0.3

= 2 : 1 : 3

Formula = Na_2CO_3; sodium carbonate

(4) The mass of oxygen present in compound **A** = (6.8 – 3.6) g = 3.2 g

Molar ratio of Ca : S : O = $\frac{2.0}{40} : \frac{1.6}{32} : \frac{3.2}{16}$ = 0.05 : 0.05 : 0.20

Simplest ratio of Ca : S : O = 1 : 1 : 4

A = $CaSO_4$

(5) 6.95 g of $FeSO_4.xH_2O$ contains 3.8 g of $FeSO_4$

So the mass of H_2O present = (6.95 – 3.8) g = 3.15 g

M_r of H_2O = 18, M_r of $FeSO_4$ = 152

Molar ratio of $FeSO_4 : H_2O$ = $\frac{3.8}{152} : \frac{3.15}{18}$ = 0.025 : 0.175

Simplest ratio = 1 : 7 i.e. 1 mol of $FeSO_4$ combines with 7 mol of H_2O

Formula = $FeSO_4.7H_2O$, so $x = 7$

Pages 66 – 67

(1) The molar ratio of C : H : O in **X** = $\frac{1.2}{12} : \frac{0.2}{1} : \frac{1.6}{16}$

= 0.1 : 0.2 : 0.1

= 1 : 2 : 1

Empirical formula of **X** = CH_2O

Empirical formula mass = 12 + (1 × 2) + 16 = 30

M_r = 60 so molecular formula = (empirical formula)$_2$ = $(CH_2O)_2$ = $C_2H_4O_2$

(2) If 20% of compound by mass is calcium, then 80% equals bromine

Molar ratio of Ca : Br = $\frac{20}{40} : \frac{80}{80}$

= 0.5 : 1

= 1 : 2

Formula = $CaBr_2$

(3) If the hydrocarbon contains 85.7% by mass of carbon, then 14.3% equals hydrogen.

Molar ratio of C:H $= \frac{85.7}{12} : \frac{14.3}{1}$

$= 7.14 : 14.3$

$= 1 : 2$

Empirical formula = CH_2
Empirical formula mass = 14
$M_r = 56$
56/14 = 4 so molecular formula = (empirical formula)$_4$ = $(CH_2)_4$ = C_4H_8

Pages 68 – 69

(1) **a)** $Ba(NO_3)_2(aq) + Na_2SO_4(aq) \rightarrow BaSO_4(s) + 2NaNO_3(aq)$
b) $Ba^{2+}(aq) + SO_4^{2-}(aq) \rightarrow BaSO_4(s)$

(2) **a)** $Mg(s) + \mathbf{2}HCl(aq) \rightarrow MgCl_2(aq) + H_2(g)$
b) $CH_4(g) + \mathbf{2}O_2(g) \rightarrow CO_2(g) + \mathbf{2}H_2O(l)$
c) $Fe_2O_3(s) + \mathbf{3}CO(g) \rightarrow \mathbf{2}Fe(s) + \mathbf{3}CO_2(g)$
d) $Mg^{2+}(aq) + \mathbf{2}OH^-(aq) \rightarrow Mg(OH)_2(s)$
e) $Cl_2(g) + \mathbf{2}e^- \rightarrow \mathbf{2}Cl^-(aq)$

(3) **a)** $2Mg(s) + O_2(g) \rightarrow 2MgO(s)$
b) $2NaOH(aq) + H_2SO_4(aq) \rightarrow Na_2SO_4(aq) + 2H_2O(l)$
or $OH^-(aq) + H^+(aq) \rightarrow H_2O(l)$
c) $Ca(s) \rightarrow Ca^{2+}(aq) + 2e^-$
d) $2AgNO_3(aq) + CuCl_2(aq) \rightarrow 2AgCl(s) + Cu(NO_3)_2(aq)$
or $Ag^+(aq) + Cl^-(aq) \rightarrow AgCl(s)$

Pages 70 – 71

(1) $NH_3(g) + 2O_2(g) \rightarrow HNO_3(aq) + H_2O(l)$
1 mol — 1 mol
17 g — 63 g
17 tonnes — 63 tonnes
$17 \times 2 = 34$ tonnes $\rightarrow 63 \times 2 = 126$ tonnes

(2) $CaCO_3(s) \rightarrow CaO(s) + CO_2(g)$
1 mol — 1 mol
100 g — 56 g
100 kg — 56 kg
$\frac{16.8}{56} = 0.3$ mol of calcium oxide
0.3 mol of $CaCO_3$ = (100×0.3) kg = 30 kg

Pages 72 – 73

(1) $C_3H_8(g) + 5O_2(g) \rightarrow 3CO_2(g) + 4H_2O(g)$
1 mol — 5 mol
One mole of any gas at stp occupies 22.4 dm^3
25 dm^3 of propane = 25/22.4 mol = 1.116 mol
This reacts with 5×1.116 mol of oxygen = 5.58 mol
$5.58 \times 22.4 = 125$ dm^3 of oxygen

or: If molar ratio = 1 : 5, then molar volume ratio = 25 : 25 × 5 = 125 dm^3

(2) $CaCO_3(s) + 2HCl(aq) \rightarrow CaCl_2(aq) + H_2O(l) + CO_2(g)$
1 mol   1 mol
100 g   22.4 dm^3
112 cm^3 = 0.112 dm^3 = 0.112/22.4 mol = 0.005 mol
so 0.005 mol of limestone is needed = (0.005 × 100) g = 0.5 g

Pages 74 – 75

Using $pV = nRT$

(1) $$V = \frac{nRT}{p} = \frac{0.1 \times 8.31 \times (273 + 35)}{150 \times 1000}$$
$= 1.7 \times 10^{-3}\,m^3$ or 1.7 dm^3

(2) $$p = \frac{nRT}{V} = \frac{71 \times 8.31 \times 298 \times 1000}{71 \times 50}$$
$= 4.953 \times 10^4$ Pa or 49.53 kPa

(3) $$T = \frac{pV}{nR} = \frac{100 \times 1000 \times 20 \times 44}{4.4 \times 8.31 \times 1000} = 2407\,K$$

Pages 76 – 77

Using $M_r = \frac{mRT}{pV}$

(1) $$M_r = \frac{12.02 \times 8.31 \times (273 + 37)}{172 \times 1000 \times 0.003} = 60$$

(2) $$M_r = \frac{0.636 \times 8.31 \times 310 \times 1000 \times 1000}{150 \times 1000 \times 154} = 71$$

(3) $$M_r = \frac{1.6 \times 8.31 \times 301 \times 1000}{100 \times 1 \times 1000} = 40$$

(4) $$M_r = \frac{0.24 \times 8.31 \times (273 + 100) \times 1000 \times 1000}{134 \times 99 \times 1000} = 56.1 \text{ (take as 56)}$$
The empirical formula of **Z** is CH_2 so the empirical mass = 14
56/14 = 4 so the molecular formula = (empirical formula)$_4$ = $(CH_2)_4$ = C_4H_8

Pages 78 – 79

(1) **a)** 63. **b)** 40. **c)** 111.

(2) Using $n = \frac{VM}{1000}$

a) $n = \frac{1000 \times 1}{1000}$ = 1 mol = 63 g of HNO_3

b) $n = \frac{1000 \times 2}{1000}$ = 2 mol = (2 × 40) g = 80 g of NaOH

c) $n = \frac{500 \times 2}{1000}$ = 1 mol = 111 g of $CaCl_2$

d) $n = \frac{100 \times 5}{1000}$ = 0.5 mol = (0.5 × 40) g = 20 g of NaOH

Page 80 – 81

1. Using $m = \frac{VM \times M_r}{1000}$

Compound	M	V (cm^3)	M_r	m (g)
a) KOH	1	1000	56	56
b) NaCl	0.1	500	58.5	2.925
c) $AgNO_3$	0.5	250	170	21.25
d) $NaNO_3$	0.2	100	85	1.7

2. The M_r of sulphamic acid (NH_2SO_3H) = 97

a) Moles (n) = mass/M_r

$= 5.210/97 = 0.0537\,\text{mol}$

b) Using $M = n \times 1000/V$

$$M = \frac{0.0537 \times 1000}{250} = 0.215$$

3. The M_r of washing soda ($Na_2CO_3.10H_2O$) = 286

n = mass/M_r

$= 2.86/286 = 0.01\,\text{mol}$

Using $M = n \times 1000/V$

$$= \frac{0.01 \times 1000}{100} = 0.1\,\text{mol}\,\text{dm}^{-3}$$

Alternatively, using $M = \frac{m \times 1000}{V \times M_r}$ the calculation can be performed in one step:

$$M = \frac{2.86 \times 1000}{100 \times 286} = 0.1\,\text{mol}\,\text{dm}^{-3}$$

4. The M_r of copper(II) sulphate ($CuSO_4$) = 159.5

n = mass/M_r

$$= \frac{159.5 \times 1000}{159.5} = 10\,\text{mol}$$

$$V = \frac{n \times 1000}{M}$$

$$V = \frac{10 \times 1000}{1} = 10000\,\text{cm}^3$$

If each bottle holds 500 cm^3 then the number of bottles of 1 M copper(II) sulphate which can be made up is 10000/500 = 20 bottles.

Pages 82 – 83

1. First write the equation with the information given underneath:

$NaOH + HCl \rightarrow NaCl + H_2O$

1 mol 1 mol

25 cm^3 20 cm^3

0.2 M M = ?

Using $n = VM/1000$

Moles of NaOH = $25 \times 0.2/1000 = 0.005$

Since 1 mol of NaOH reacts with 1 mol of HCl then the moles of acid = 0.005

For HCl: $0.005 = \frac{20 \times M}{1000}$

Rearranging: $M = \frac{0.005 \times 1000}{20} = 0.25$

So the concentration of the acid is 0.25 mol dm^{-3}

(2) $KOH + HNO_3 \rightarrow KNO_3 + H_2O$
1 mol   1 mol
0.15 M 0.2 M
$V = ?$ $V = 32 \text{ cm}^3$
Moles of alkali = moles of acid = $32 \times 0.2/1000 = 0.0064$

For KOH: $0.0064 = \dfrac{V \times 0.15}{1000}$

Rearranging: $V = \dfrac{0.0064 \times 1000}{0.15} = 42.67 \text{ cm}^3$

(3) $NaHCO_3(aq) + HCl(aq) \rightarrow NaCl(aq) + CO_2(g) + H_2O(l)$
1 mol   1 mol
25 cm^3 28.5 cm^3
0.1 M $M = ?$
Moles of acid = moles of alkali = $25 \times 0.1/1000 = 0.0025$

For HCl: $0.0025 = \dfrac{28.5 \times M}{1000}$

Rearranging: $M = \dfrac{0.0025 \times 1000}{28.5} = 0.088 \text{ mol dm}^{-3}$

Pages 84 – 85

(1) **a)** $H_2SO_4(aq) + 2NaOH(aq) \rightarrow Na_2SO_4(aq) + 2H_2O(l)$
1 mol   2 mol
30 cm^3 25 cm^3
$M = ?$ $M = 0.2$
Moles of NaOH = $25 \times 0.2/1000 = 0.005$
but 2 mol of NaOH reacts with 1 mol of H_2SO_4 so moles of acid = $0.005/2 = 0.0025$

For H_2SO_4: $0.0025 = \dfrac{30 \times M}{1000}$

Rearranging: $M = \dfrac{0.0025 \times 1000}{30} = 0.083 \text{ mol dm}^{-3}$

b) $H_2SO_4(aq) + 2NaOH(aq) \rightarrow Na_2SO_4(aq) + 2H_2O(l)$
1 mol   2 mol
25 cm^3 26.5 cm^3
0.025 M $M = ?$

Moles of $H_2SO_4 = \dfrac{25 \times 0.025}{1000} = 0.000625 = 6.25 \times 10^{-4}$

Moles of NaOH which react with this amount of acid
$= 6.25 \times 10^{-4} \times 2 = 1.25 \times 10^{-3}$

For NaOH: $1.25 \times 10^{-3} = \dfrac{26.5 \times M}{1000}$

Rearranging: $M = \dfrac{1.25 \times 10^{-3} \times 1000}{26.5} = 0.047 \text{ mol dm}^{-3}$

(2) $CaCO_3(s) + 2HCl(aq) \rightarrow CaCl_2(aq) + CO_2(g) + H_2O(l)$
1 mol 2 mol
5.0 g 2 M
$V = ?$
Moles of $CaCO_3$ = mass/M_r = 5.0/100 = 0.05
0.05 mol of $CaCO_3$ reacts with (2 × 0.05) mol of HCl
Moles of HCl = 0.1

For HCl: $= 0.1 = \frac{V \times 2}{1000}$

Rearranging: $V = \frac{0.1 \times 1000}{2} = 50\,cm^3$

(3) **a)** The M_r of sodium carbonate (Na_2CO_3) = 106
Moles of Na_2CO_3 = 2.601/106 = 0.0245

$0.0245 = \frac{250 \times M}{1000}$

Rearranging: $M = \frac{0.0245 \times 1000}{250} = 0.098\ mol\ dm^{-3}$

b) $Na_2CO_3(aq) + 2HCl(aq) \rightarrow 2NaCl(aq) + CO_2(g) + H_2O(l)$
1 mol 2 mol
25 cm^3 26.35 cm^3
0.098 M $M = ?$

Moles of $Na_2CO_3 = \frac{25 \times 0.098}{1000} = 0.00245$

0.00245 mol of Na_2CO_3 reacts with (2 × 0.00245) mol of HCl
Moles of HCl = 0.0049

$0.0049 = \frac{26.35 \times M}{1000}$

Rearranging: $M = \frac{0.0049 \times 1000}{26.35} = 0.186\ mol\ dm^{-3}$

(4) $Na_2CO_3(aq) + 2HCl(aq) \rightarrow 2NaCl(aq) + CO_2(g) + H_2O(l)$
1 mol 2 mol
25 cm^3 30 cm^3
$M = ?$ $M = 0.1$
Moles of HCl = 30 × 0.1/1000 = 0.003
0.003 mol of HCl reacts with 0.003/2 mol of Na_2CO_3 = 0.0015

For Na_2CO_3: $0.0015 = \frac{25 \times M}{1000}$

Rearranging: $M = \frac{0.0015 \times 1000}{25} = 0.06$

i.e. the concentration of Na_2CO_3 in the solution is $0.06\ mol\ dm^{-3}$
$= 0.06$ mol per 1000 cm^3
In 250 cm^3 there are 0.06/4 mol = 0.015
Converting this amount to grams:
mass = $n \times M_r$
= 0.015 × 106
= 1.59 g
In 4.00 g of washing soda, 1.59 g is Na_2CO_3

$\% = \frac{1.59 \times 100}{4.00} = 39.75\%$

Pages 86 – 87: Unit 4 Questions

1 **a)** relative atomic mass = $\frac{\text{the average mass of an atom} \times 12}{\text{the mass of one atom of carbon-12}}$

or relative atomic mass = average mass per atom of an element (1) relative to carbon-12 where C = 12.000 (1).

b) ^{12}C is the standard reference (1).

c) One mole = $1.9925 \times 10^{-23} \times L$

$12.000 = 1.9925 \times 10^{-23} \times L$ (1)

$L = \frac{12.000}{1.9925 \times 10^{-23}} = \frac{12.000}{1.9925} \times 10^{23} = 6.023 \times 10^{23}$ (1)

2

Chemical name	Formula	State at room temperature and pressure	Relative molecular mass
Water	**H_2O**	**liquid**	18
Ammonia	NH_3	gas	**17**
Methane	**CH_4**	**gas**	16
Sulphuric acid	**H_2SO_4**	liquid	**98**
Hydrated copper(II) sulphate	$CuSO_4.5H_2O$	solid	**249.5**
Calcium carbonate	**$CaCO_3$**	**solid**	**100**
Butane	C_4H_{10}	**gas**	**58**

($\frac{1}{2}$ each)

3 **a)** M_r of glucose = 180

Moles = 1.8/180 = 0.01 mol (1)

b) M_r of ethanol = 46

Moles = 0.92/46 = 0.02 mol (1)

1 mol of any gas at stp occupies 22.4 dm^3

448 cm^3 = 0.448 dm^3

Moles = 0.448/22.4 (1)

= 0.02 mol of carbon dioxide (1)

c) $C_6H_{12}O_6 \rightarrow C_2H_6O + CO_2$

0.01 mol 0.02 mol 0.02 mol

1 2 2

Balanced equation: $C_6H_{12}O_6 \rightarrow 2C_2H_6O + 2CO_2$ (1)

4 Using: $pV = nRT$

$n = pV/RT$ (1)

$= \frac{100 \times 1000 \times 120}{1000 \times 8.31 \times (273 + 18)}$ (1)

= 4.9623 (1)

mass = $n \times A_r$

= 4.9623 × 4 = 19.85 g (1)

(5) **a)** Molar ratio of C:H:O in compound **X** = $\frac{1.12}{12}:\frac{0.187}{1}:\frac{1.49}{16}$ (1)
$= 0.093:0.187:0.093$
$= 1 : 2 : 1$ (1)

Empirical formula of **X** = CH_2O (1)

b) $M_r = \frac{mRT}{pV}$ (1) $= \frac{0.126 \times 8.31 \times 373 \times 1000}{172 \times 37.8}$ (1)
$= 60$ (1)

c) Empirical formula mass = 12 + 2 +16 = 30 (1)
Molecular formula = empirical formula × 2 = $C_2H_4O_2$ (1)

(6) First calculate the empirical formula.

$\frac{62.04}{12}$ mol of carbon: = 5.17 mol

$\frac{10.41}{1}$ mol of hydrogen: = 10.41 mol

$\frac{27.55}{16}$ mol of oxygen: = 1.72 mol (1)

The molar ratio of C:H:O = 5.17:10.41:1.72

The simplified ratio of C:H:O = $\frac{5.17}{1.72} : \frac{10.41}{1.72} : \frac{1.72}{1.72}$
$= 3 : 6 : 1$ (1)

So the empirical formula = C_3H_6O (1).

Now calculate the molecular formula using the M_r value given in the question.
Empirical mass = (12 × 3) + (1 × 6) + 16 = 58 (1).

But M_r = 58.
So the molecular formula = empirical formula = C_3H_6O (1).

(7) **a)** Molar ratio of C:H:O in compound **Z** = $\frac{52.2}{12}:\frac{13.0}{1}:\frac{34.8}{16}$ (1)
$= 4.35 : 13.0 : 2.175$
$= 2 : 6 : 1$ (1)

Empirical formula of **Z** is C_2H_6O (1)
1 mol of any gas at stp occupies 22.4 dm^3
11.2 cm^3 = 11.2×10^{-3} dm^3

Moles of **Z** = $\frac{11.2 \times 10^{-3}}{22.4} = 5.0 \times 10^{-4}$ (1)

If 5.0×10^{-4} mol = 0.023 g

then 1 mol (M_r) = $\frac{0.023}{5.0 \times 10^{-4}} = 46$ (1)

Empirical formula mass = (2 × 12) + (6 × 1) + 16 = 46
so molecular formula = empirical formula = C_2H_6O (1)

b) $C_2H_6O + 3O_2 \rightarrow 2CO_2 + 3H_2O$ (1)

c) 1 mol of **Z** burns in 3 mol of oxygen
2 mol of **Z** burns in 6 mol of oxygen (1)

$V = \frac{nRT}{p} = \frac{6 \times 8.31 \times 398}{120 \times 1000}$ (1)
$= 0.165\ m^3$ (1)

(8) **a)** $Mg(s) + H_2SO_4(aq) \rightarrow MgSO_4(aq) + H_2(g)$ (1)

b) From the equation, 1 mol of Mg reacts with 1 mol of acid
Moles of acid = 100 × 1/1000 = 0.1 (1)
1 mol of Mg = 24 g
0.1 mol of Mg = 24 × 0.1 = 2.4 g (1)

c) Moles of Mg = 4.8/24 = 0.2
From the equation, 1 mol of Mg produces 1 mol of H_2
so 0.2 mol of H_2 are produced (1)
1 mol of any gas at stp occupies 22.4 dm^3
0.2 mol = 22.4 × 0.2 = 4.48 dm^3 = 4480 cm^3 (1)

(9) a) The number of hydrogens is not the same on both sides of the equation (1). The charges on the ions are not the same (1).

b) M_r of sodium carbonate = 106.
Moles of sodium carbonate = 5.202/106 = 0.049 (1)
$n = VM/100$

$$0.049 = \frac{500 \times M}{1000}$$

Rearranging: $M = \frac{0.049 \times 1000}{500}$ (1)

$= 0.098\ mol\ dm^{-3}$ (1)

c) $Na_2CO_3(aq) + 2HCl(aq) \rightarrow 2NaCl(aq) + CO_2(g) + H_2O(l)$

1 mol 2 mol
25 cm^3 24.5 cm^3
$M = 0.098$ $M = ?$

Moles of Na_2CO_3 = 25 × 0.098/1000 = 0.00245 (1)
0.00245 mol of Na_2CO_3 reacts with (0.00245 × 2) mol of HCl = 0.0049 (1)

For HCl: $0.0049 = \frac{24.5 \times M}{1000}$

Rearranging: $M = \frac{0.0049 \times 1000}{24.5}$

$= 0.200$ (1)

d) $Na_2CO_3(aq) + 2HCl(aq) \rightarrow 2NaCl(aq) + CO_2(g) + H_2O(l)$

2 mol 2 mol

Moles of HCl = 100 × 0.200/1000 = 0.02
so moles of NaCl formed = 0.02 (1)
M_r of NaCl = 58.5
mass = 0.02 × 58.5 = 1.17 g (1)

Answers: Module 2

5 Energetics

Pages 94 – 95

1. 'Bond dissociation energy' refers to a particular bond. 'Mean bond enthalpy' refers to an average value for a particular bond in a range of compounds. The H—H bond exists only in the H_2 molecule.

2. Bonds broken: 436 + 242 = 678
Bonds formed: 2 × 431 = 862
Exothermic reaction: $-184\,kJ\,mol^{-1}$

Pages 96 – 97

1. **a)** $2Na(s) + C(s) + 1\frac{1}{2}O_2(g) \rightarrow Na_2CO_3(s)$
b) $2C(s) + 3H_2(g) + \frac{1}{2}O_2(g) \rightarrow CH_3CH_2OH(l)$
c) $N_2(g) + 2H_2(g) + 1\frac{1}{2}O_2(g) \rightarrow NH_4NO_3(s)$

2. **a)** $C_3H_8(g) + 5O_2(g) \rightarrow 3CO_2(g) + 4H_2O(l)$
b) $CH_3OH(l) + 1\frac{1}{2}O_2(g) \rightarrow CO_2(g) + 2H_2O(l)$
c) $C_3H_6(g) + 4\frac{1}{2}O_2(g) \rightarrow 3CO_2(g) + 3H_2O(l)$

Pages 98 – 99

1.
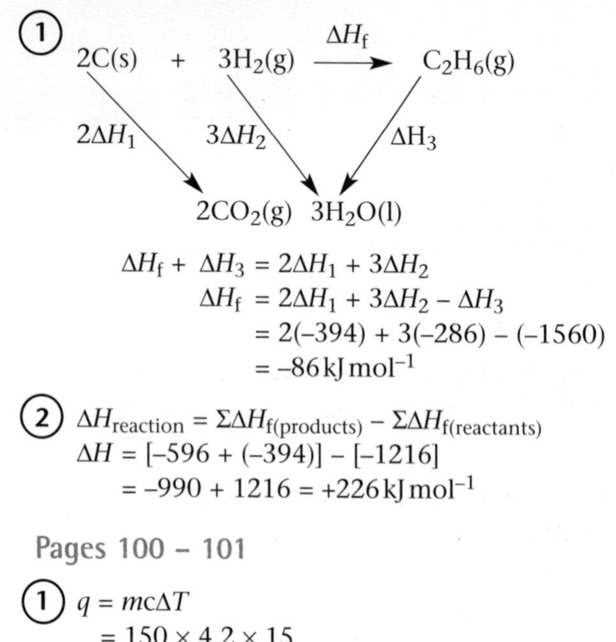

$$\Delta H_f + \Delta H_3 = 2\Delta H_1 + 3\Delta H_2$$
$$\Delta H_f = 2\Delta H_1 + 3\Delta H_2 - \Delta H_3$$
$$= 2(-394) + 3(-286) - (-1560)$$
$$= -86\,kJ\,mol^{-1}$$

2. $\Delta H_{reaction} = \Sigma\Delta H_{f(products)} - \Sigma\Delta H_{f(reactants)}$
$\Delta H = [-596 + (-394)] - [-1216]$
$= -990 + 1216 = +226\,kJ\,mol^{-1}$

Pages 100 – 101

1. $q = mc\Delta T$
$= 150 \times 4.2 \times 15$
$= 9450\,J\ (9.45\,kJ)$

② $q = mc\Delta T = 75 \times 4.2 \times 12.5 = 3937.5\,\text{J}$

This is for 0.15 g of ethanol ($M_r = 46$)

Therefore for one mole of ethanol, $q = \dfrac{3937.5 \times 46}{0.15} = 1\,207\,500\,\text{J} = 1207.5\,\text{kJ}$

Enthalpy of combustion is $-1207.5\,\text{kJ mol}^{-1}$

Pages 102 – 103

① $q = mc\Delta T = 70 \times 4.2 \times 5 = 1470\,\text{J}$
Moles of acid ($MV/1000$) = $(1 \times 30)/1000 = 0.03$ moles

Therefore for one mole of acid, $q = \dfrac{1470}{0.03} = 49\,000\,\text{J} = 49\,\text{kJ}$

Enthalpy of neutralisation is $-49\,\text{kJ mol}^{-1}$

② $q = mc\Delta T = 40 \times 4.2 \times 7.3 = 1226.4\,\text{J}$
This is for 3.41 g of Na_2CO_3 ($M_r = 106$)

Therefore for one mole, $q = \dfrac{1226.4 \times 106}{3.41} = 38\,122\,\text{J} = 38.1\,\text{kJ}$

Enthalpy change is $-38.1\,\text{kJ mol}^{-1}$

Pages 104 – 105: Unit 5 Questions

① The total energy in a closed system is constant (1).

② **a)** Heat energy change (1) at constant pressure (1)
b) Temperature of 298 K (1) and pressure of 100 kPa (1).

③ **a)** Average energy needed to break a particular bond (1) in a range of different situations (1).
b) Bonds broken: 436 + 193 = 629 (1)
Bonds formed: = 2 × 366 = 732 (1)
Exothermic reaction: $\Delta H = -103\,\text{kJ mol}^{-1}$ (1)
c) The Br—Br bond (1); lower bond energy (1).
d) The H—H bond only exists in H_2 the Br—Br bond only exists in Br_2 (1).

④

Bonds broken:			Bonds formed:	
	1 C—C	348	4 C=O	2972
	5 C—H	2060	6 O—H	2778
	1 C—O	360		
	1 O—H	463		
	3 O=O	1488		
	Total	4719 (2)	Total	5750 (2)

Exothermic; (4719 – 5750) = $-1031\,\text{kJ mol}^{-1}$ (1)

⑤ The enthalpy change (1) in a reaction is independent of the route (1).
This law allows the calculation of enthalpy changes that cannot be determined directly by experiment (1).

⑥ **a)** Enthalpy change when one mole (1) of a substance undergoes complete combustion (1) under standard conditions.
b) $4C(s) + 5H_2(g) \rightarrow C_4H_{10}(g)$ (2)

c)

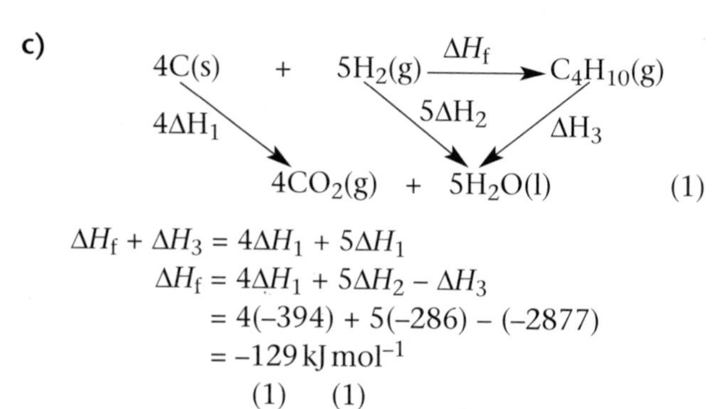

(1)

$$\Delta H_f + \Delta H_3 = 4\Delta H_1 + 5\Delta H_1$$
$$\Delta H_f = 4\Delta H_1 + 5\Delta H_2 - \Delta H_3$$
$$= 4(-394) + 5(-286) - (-2877)$$
$$= -129\,\text{kJ mol}^{-1}$$
(1) (1)

7 **a)** Enthalpy change in the formation of one mole (1) of a compound from its elements (1), in their standard states and under standard conditions.

b) ΔH reaction = $\Sigma\Delta H_f$ (products) – $\Sigma\Delta H_f$ (reactants) (1)

ΔH = [–1146 + (–394) + (–286)] – [2(–959)] (1)
= [–1826] – [–1918]
= +92 kJ mol^{-1} (1)

8 **a)** To ensure that all the copper sulphate reacted (1).

b) $q = mc\Delta T$ (1) = 25 × 4.2 × 21 = 2205 J (1).

c) Moles = $MV/1000$ (1) = (0.5 × 25)/1000 = 0.0125 moles (1)

d) Molar enthalpy = 2205/0.0125 (1) = 176 400 J = –176 kJ mol^{-1} (1)

e) Heat loss; use a lid on the calorimeter (1); lag the calorimeter (1).

9 **a)** In the first reaction gaseous water is formed, and in the second reaction liquid water is formed (1). Energy is required to change water from a liquid to a gas.

b) The difference in enthalpy change is 44 kJ mol^{-1} (1). Therefore 44 kJ mol^{-1} of energy are needed to change two moles of water from liquid into a gas (1). Hence the enthalpy change for one mole is +22 kJ mol^{-1} (1).

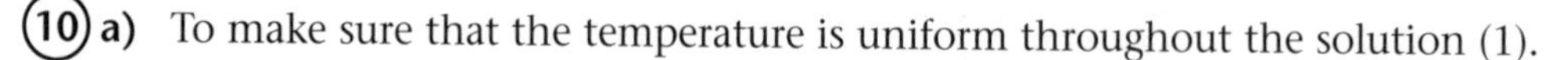

10 **a)** To make sure that the temperature is uniform throughout the solution (1).

b)

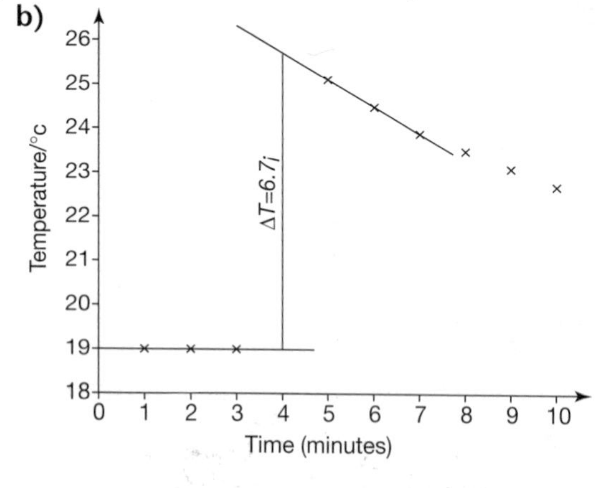

Sensible curves on axes (1); correct plotting of points (1); correct extrapolations (1).

c) From the cooling curve, ΔT = 6.7°C (1).

d) Heat energy released, $q = mc\Delta T$ (1).
q = 100 × 4.2 × 6.7 (1) = 2814 J (1).

e) Moles of acid = $MV/1000$ (1).
moles = (1 × 50)/1000 = 0.05 mol (1).

f) Molar enthalpy change = 2814/0.05 (1) = 56280J = –56.3 kJ mol^{-1} (1).

6 Kinetics

Pages 106 – 107

1. There is an energy barrier to a reaction (activation energy): only collisions between particles with energy equal to or greater than the activation energy will result in a reaction.

2. If the concentration is increased, then there are more particles per unit volume. This will increase the frequency of collisions.

Pages 108 – 109

1. The total number of particles (molecules) present in the gas.

2. *Three of:* the peak will move to a lower energy; the peak is narrower and higher; the peak is of a lower energy; there is a disproportionate decrease in the number of molecules with higher energies.

Pages 110 – 111: Unit 6 Questions

1. **a)** There is an energy barrier to a reaction (1). Only those particles with enough energy to overcome the energy barrier can react (1).
   b) *Three of:* increasing the concentration; increasing the surface area of a solid; increasing the pressure of a gas; increasing the temperature (1 each).
   c) The minimum energy (1) that a particle must have for a collision to result in a reaction (1).

2. **a)** A substance that alters the rate (1) of a chemical reaction without itself being changed (1).
   b) A catalyst can speed up a chemical reaction by providing an alternative reaction pathway (1) with a lower activation energy (1).
   c)

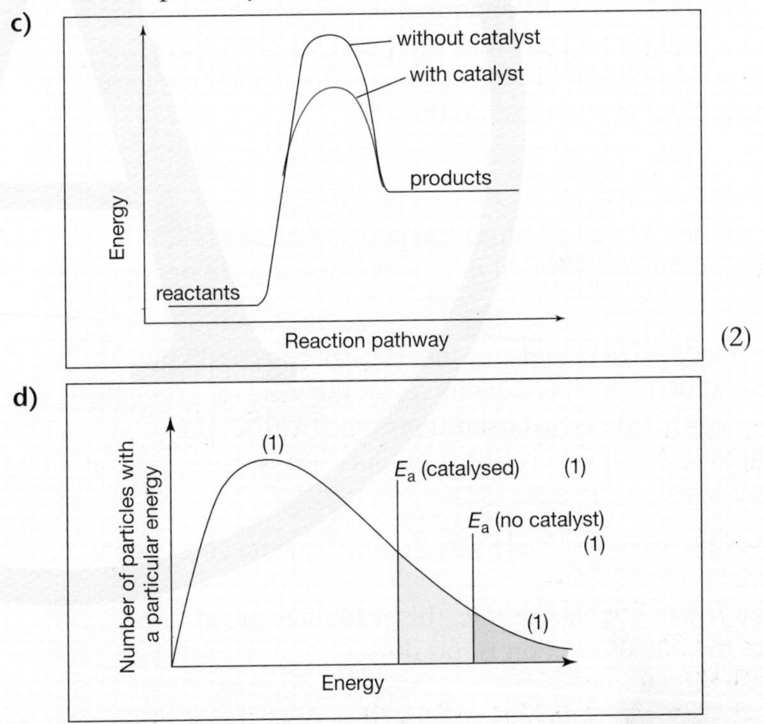

(2)

d)

3. a) T_1 (1)
   b) The total number (1) of particles present in the gas (1).
   c) There are some particles (1) with very high energies (1).
   d) At higher temperatures there are more particles (1) with energy greater than the activation energy (1).

4. a) To make sure that all the magnesium ribbon reacted (1).
   b) 15 seconds (1).
   c) (i) Steeper slope (1); same final volume (1).
   (ii) Steeper slope (1); same final volume (1).
   (iii) Same initial slope (1); final volume is 50 cm^3 (1).

5. a) **B** (1). b) **A** (1). c) **C** (1). d) **D** (1).

7 Equilibria

Pages 112 – 113

1. *Dynamic*: Reactions continue to occur. *Equilibrium*: No change in the overall concentrations of the reactants.

2. a) Homogeneous. b) Heterogeneous. c) Homogeneous. d) Heterogeneous.

Pages 114 – 115

1. Decreasing the temperature of the reaction will favour the reverse reaction – the one with the negative enthalpy change (exothermic). The equilibrium shifts to oppose the decrease in temperature.

2. a) No change. b) Pushes the equilibrium to the right. c) Pushes the equilibrium to the right.

3. Increasing the temperature favours the endothermic reaction which removes added heat energy: the equilibrium will move to the right. Increasing the pressure moves the equilibrium to the side of the reaction which has fewer gaseous moles: the equilibrium will move to the left.

Pages 116 – 117

1. When one or more of the factors that affect the position of a chemical equilibrium are changed, then the equilibrium will move so as to oppose the change.

2. Higher pressures increase the yield, but this incurs extra costs and safety considerations; decreasing the temperature increases the yield, but the rate of the reaction slows down. The actual conditions used (440 °C and 200 atmos) are compromise conditions which produce a reasonable yield at a reasonable rate and a reasonable cost.

3. a) It is not needed as the yield is already very good. The extra costs would not be recovered.
   b) Since the forward reaction is exothermic, the yield increases at lower temperatures but the rate of reaction is too slow.

Pages 118 – 119: Unit 7 Questions

1. The products (1) can re-form the original reactants (1).

2. Both the forward and reverse reactions continue to occur (dynamic) (1), but the overall concentrations remain the same (equilibrium) (1).

3. *Homogeneous*: All the reactants are in the same state (phase) (1).
Heterogeneous: The reactants are in different states (1).

4. **a)** When one or more of the factors that affect the position of a chemical equilibrium change (1), then the equilibrium moves so as to oppose the change (1).
b) (i) $X(g) + Y(g) \rightarrow Z(g)$ ΔH +ve
Increasing the pressure: equilibrium moves to the right (1)
Increasing the temperature: equilibrium moves to the right (1)
(ii) $A(g) + B(g) \rightarrow C(g) + D(g)$ ΔH –ve
Increasing the pressure: no change (1).
Increasing the temperature: equilibrium moves to the left (1).
(iii) $S(g) \rightarrow R(g) + T(g)$ ΔH +ve
Increasing the pressure: equilibrium moves to the left (1).
Increasing the temperature: equilibrium moves to the right (1).

5. **a)** Reaction involves an increase in the number of gaseous moles (1), and the equilibrium moves to the left (fewer gaseous moles) when the pressure increases (1).
b) Forward reaction is endothermic and is favoured by an increase in temperature because it removes the added heat energy (1): the equilibrium moves to the right (1).

6. **a)** All the reactants are in the same state (1).
b) (i) As pressure increases, the reacting particles are pushed closer together (it increases the concentration) (1); this increases the rate of collisions, and hence the rate at which equilibrium is reached (1).
(ii) The increase in pressure pushes the equilibrium to the side with fewer gaseous moles (1): equilibrium moves to the right (1).
c) (i) Increasing the temperature increases the percentage of particles with energy equal to or greater than the activation energy (1); this increases the rate at which equilibrium is reached (1).
(ii) Increase in temperature favours the endothermic reaction (1): equilibrium moves to the left (1).
d) (i) The catalyst provides an alternative reaction pathway with a lower activation energy (1): this increases the rate at which equilibrium is reached (1).
(ii) The catalyst speeds up the forward and reverse reactions equally (1): there is no change in the position of the equilibrium (1).
e) If more hydrogen is added then the equilibrium shifts to remove the added hydrogen (1): the equilibrium moves to the right (1).

7. The reaction is exothermic (1) and lower temperatures increase the yield (1), but at lower temperatures the rate of reaction is too slow (1). The reaction involves a reduction in gaseous moles (1), and higher pressures increase the yield (1). Higher pressures are more expensive and can cause safety

problems (1). Compromise conditions are chosen to produce a good yield at a reasonable rate and at a reasonable cost (1).

8 **a)** (i) Decrease in moles (1).
(ii) Yield increases as pressure increases (1). Increase in pressure favours reaction with a decrease in gaseous moles (1).
b) (i) Exothermic (1).
(ii) Yield increases as temperature decreases (1). Exothermic reactions are favoured by a decrease in temperature (1).

9 **a)** Reaction 1 (1). **b)** Reaction 3 (1). **c)** Reaction 4 (1). **d)** Reaction 2 (1).

8 Redox reactions

Pages 120 – 121

1 **a)** Reduced. **b)** Reduced. **c)** Oxidised. **d)** Oxidised.

2 **a)** $Cu(s) + 2AgNO_3(aq) \rightarrow Cu(NO_3)_2(aq) + 2Ag(s)$
or $Cu(s) + 2Ag^+(aq) \rightarrow Cu^{2+}(aq) + 2Ag(s)$
b) $Cu(s) \rightarrow Cu^{2+}(aq) + 2e^-$
The copper atom has lost electrons so it has been oxidised.

Pages 122 – 123

1 **a)** Na = +1, Cl = –1. **b)** Li = +1, O = –2. **c)** H = +1, S = –2. **d)** Ca = 0. **e)** O_2 = 0. **f)** P = +5, F = –1.

Pages 124 – 125

1 **a)** S = +4. **b)** Mn = +7. **c)** Cl = +5. **d)** N = +3.

2 **a)** SO_3^{2-}: S = +4 SO_4^{2-}: S = +6 HSO_3^-: S = +4 S^{2-}: S = –2
b) The maximum oxidation state of sulphur is +6 (this equals its Group number in the Periodic Table).

Pages 126 – 127

1 **a)** $FeSO_4$ **b)** $FeCl_3$ **c)** $CuCO_3$ **d)** P_2O_3

2

Name	Formula	Variable oxidation state
Lead(II) oxide	**PbO**	**+2**
Lead(IV) oxide	PbO_2	+4
Sodium chlorate(I)	NaClO	**+1**
Iron(III) hydroxide	**$Fe(OH)_3$**	+3

3 **a)** $Fe(NO_3)_2$. **b)** $CrCl_3$. **c)** $CuSO_3$.

4 **a)** Iron(III) sulphate. **b)** Cobalt(II) chloride. **c)** Sodium nitrite or sodium nitrate(III).

Pages 128 – 129

1 **a)** $2I^- + Cl_2 \rightarrow I_2 + 2Cl^-$
–1 0 0 –1

The iodide ion is oxidised to iodine; the chlorine is reduced to the chloride ion.

b) $8HI + H_2SO_4 \rightarrow H_2S + 4H_2O + I_2$

$-1 \quad +6 \quad -2 \quad 0$

The iodide is oxidised to iodine; the sulphur in sulphuric acid is reduced to sulphide.

c) $SO_2 + 2H_2O + 2Cu^{2+} + 2Cl^- \rightarrow H_2SO_4 + 2H^+ + 2CuCl$

$+4 \quad +2 \quad +6 \quad +1$

The sulphur in sulphur dioxide is oxidised; the copper(II) ions are reduced to copper(I).

d) $Cr_2O_7^{2-} + 8H^+ + 3SO_3^{2-} \rightarrow 2Cr^{3+} + 4H_2O + 3SO_4^{2-}$

$+6 \quad +4 \quad +3 \quad +6$

The chromium in the dichromate ion is reduced to the chromium(III) ion; the sulphur in the sulphite ion is oxidised to sulphate.

e) $5Fe^{2+} + MnO_4^- + 8H^+ \rightarrow 5Fe^{3+} + Mn^{2+} + 4H_2O$

$+2 \quad +7 \quad +3 \quad +2$

The iron(II) ion is oxidised to iron(III); the manganese in the manganate(VII) ion is reduced to manganese(II).

(2) **a)** $\mathbf{Fe_2O_3} + 3CO \rightarrow 2Fe + 3CO_2$

oxidising agent

b) $2\mathbf{Mg} + O_2 \rightarrow 2MgO$

reducing agent

c) $Zn + \mathbf{Fe^{2+}} \rightarrow Zn^{2+} + Fe$

oxidising agent

d) $\mathbf{SO_2} + 2H_2O + 2Cu^{2+} + 2Cl^- \rightarrow H_2SO_4 + 2H^+ + 2CuCl$

reducing agent

e) $Cr_2O_7^{2-} + 8H^+ + 3\mathbf{SO_3^{2-}} \rightarrow 2Cr^{3+} + 4H_2O + 3SO_4^{2-}$

reducing agent

Pages 130 – 131

(1) **a)** $Ca \rightarrow Ca^{2+} + 2e^-$

b) $Fe^{2+} \rightarrow Fe^{3+} + e^-$

(2) **a)** $Cu^{2+} + 2e^- \rightarrow Cu$

b) $Al^{3+} + 3e^- \rightarrow Al$

(3) $SO_2 + 2H_2O \rightarrow H_2SO_4 + 2e^- + 2H^+$

(4) $ClO_3^- + 6e^- + 6H^+ \rightarrow Cl^- + 3H_2O$

Pages 132 – 133

(1) $H_2SO_4(aq) + 6HI(g) \rightarrow S(s) + 4H_2O(l) + 3I_2(s)$

$+6 \quad -1 \quad 0 \quad 0$

The sulphur is reduced and the iodide is oxidised, so this is a redox reaction.

(2) **a)** $MnO_2 + 4H^+ + 2Cl^- \rightarrow Mn^{2+} + 2H_2O + Cl_2$

Oxidising agent: MnO_2. Reducing agent: $2Cl^-$.

b) $SO_2 + 2H_2O + 2Fe^{3+} \rightarrow SO_4^{2-} + 4H^+ + 2Fe^{2+}$

Oxidising agent: SO_2. Reducing agent: $2Fe^{3+}$.

c) $IO_3^- + 6H^+ + 5I^- \rightarrow 3I_2 + 3H_2O$

Oxidising agent: IO_3^-. Reducing agent: $5I^-$.

Pages 134 – 135: Unit 8 Questions

1. a) Oxidation is the loss of electrons (1).
   b) $2Br^- \rightarrow Br_2 + 2e^-$ (1)
   c) $H_2SO_4 + 2H^+ + 2e^- \rightarrow SO_2 + 2H_2O$ (2)
   d) $H_2SO_4 + 2H^+ + 2Br^- \rightarrow Br_2 + SO_2 + 2H_2O$ (2)

2. a) (i) +5 (1). (ii) –1 (1).
   b) (i) +5 (1). (ii) +3 (1).
   c) Reduction is electron gain (1).
   d) An oxidising agent is an electron acceptor (1).
   e) (i) Potassium bromate(V) (1). (ii) Potassium nitrate(III) (1).

3. a) Sulphur dioxide (1).
   b) Copper(II) ions (1).
   c) $SO_2 + 2H_2O \rightarrow SO_4^{2-} + 4H^+ + 2e^-$ (2)

4. a) $2Ce^{4+}(aq) + C_2O_4^{2-}(aq) \rightarrow 2Ce^{3+}(aq) + 2CO_2(g)$ (2)
   b) 0.0025 mol of ethanedioate ions will react with (2×0.0025) mol of cerium (IV) (1).
   Using $n = VM/1000$

   $$2 \times 0.0025 = \frac{V \times 0.100}{1000} \quad (1)$$

   $$\text{Rearranging: } V = \frac{2 \times 0.0025 \times 1000}{0.100} = 50\,\text{cm}^3 \ (1)$$

5. a) The calcium metal sinks and dissolves in the water (1). The production of hydrogen gas causes fizzing (1) (some of the calcium hydroxide may precipitate if enough calcium is added).
   b) $Ca(s) \rightarrow Ca^{2+}(aq) + 2e^-$ (1)
   c) Oxidising agent (1).

6. a) (i) +4 (1). (ii) +6 (1). (iii) +6 (1).
   b) $IO_3^- + 6H^+ + 5e^- \rightarrow \frac{1}{2}I_2 + 3H_2O$ (2)
   c) Reducing agent (1).
   d) (i) +1 (1). (ii) Iodine(I) chloride (1). (iii) $ICl + KI \rightarrow KCl + I_2$ (1).

7. a) $Cl_2 + 2e^- \rightarrow 2Cl^-$ (1)
   b) $SO_3^{2-} + H_2O \rightarrow SO_4^{2-} + 2e^- + 2H^+$ (2)
   c) $SO_3^{2-} + H_2O + Cl_2 \rightarrow SO_4^{2-} + 2Cl^- + 2H^+$ (1)
   d) The sulphate(IV) ions are reduced to sulphur (1), and the sulphide is oxidised to S (1).

8. a) SO_2, S and H_2S (3).
   b) I_2 (1).
   c) $NaHSO_4$, Na_2SO_4, HI and H_2O (4):
   d) $H_2SO_4 + 8e^- + 8H^+ \rightarrow H_2S + 4H_2O$ (2).

9 Group VII, the halogens

Pages 136 – 137

1. The iodine molecule is bigger than the fluorine molecule, so the van der Waals forces holding the molecules together are stronger. This means that iodine has a higher melting point, because more energy is needed to break the forces.

(2) Fluorine is more electronegative than iodine.

(3) Fluorine is the smallest halogen atom. Lack of shielding, and a smaller distance over which the nucleus acts, mean that the nucleus can easily attract electrons. Electronegativity is the power of an atom to attract a pair of electrons in a covalent bond.

(4) An oxidising agent needs to be able to attract electrons. The bromine atom is smaller than the iodine atom, so the effect of the nuclear charge is greater because there is less shielding and a smaller distance over which the nucleus acts. Bromine can therefore attract electrons more readily than iodine.

Pages 138 – 139

(1) Chlorine is a better oxidising agent than iodine because it can attract electrons more readily. This is because the effect of the nuclear charge is greater, as there is less shielding and a smaller distance over which the nucleus acts. The relative strength of chlorine and iodine as oxidising agents can be illustrated using displacement reactions: chlorine will displace iodine from a solution of iodide ions, but iodine cannot displace chlorine.

(2) Solution **X** contains bromide ions, so could for example be potassium bromide.

$Cl_2(aq) + 2Br^-(aq) \rightarrow 2Cl^-(aq) + Br_2(aq)$

(3) $2Br^- \rightarrow Br_2 + 2e^-$

This equation shows that bromide ions have been oxidised, because they have lost electrons.

Pages 140 – 141

(1) A reducing agent is an electron donor.

(2) $2HBr(aq) + H_2SO_4(aq) \rightarrow Br_2(l) + SO_2(g) + 2H_2O(l)$

+6 (S in H_2SO_4) +4 (S in SO_2)

(3) **a)** $H_2SO_4 + 2H^+ + 2e^- \rightarrow SO_2 + 2H_2O$
b) $H_2SO_4 + 6H^+ + 6e^- \rightarrow S + 4H_2O$
c) $H_2SO_4 + 8H^+ + 8e^- \rightarrow H_2S + 4H_2O$

(4) Bromide ions can only reduce H_2SO_4 to SO_2, a change in oxidation state of 2; but iodide ions can reduce H_2SO_4 to H_2S, a change in oxidation state of 8.

Pages 142 – 143

(1) X^- is Br^-.

(2) A white precipitate is formed, which dissolves in aqueous ammonia, leaving a colourless solution.

(3) Silver nitrate solution is added, followed by dilute aqueous ammonia. With HCl a white precipitate is observed, which readily dissolves in dilute ammonia. With HBr a cream precipitate is observed, which is insoluble in dilute ammonia solution.

Pages 144 – 145

1. Salt contains chloride ions:
$Cl_2(g) + H_2O(l) \rightleftharpoons 2H^+(aq) + Cl^-(aq) + ClO^-(aq)$
The chloride ions would shift the equilibrium to the left, releasing harmful chlorine gas.

2. **a)** $Cl_2(aq) + 2e^- \rightarrow 2Cl^-(aq)$
b) $Cl_2(aq) + 2H_2O(l) \rightarrow 2ClO^-(aq) + 2e^- + 4H^+(aq)$

3. | $NaClO$ | $NaClO_3$ |
| --- | --- |
| +1 | +5 |
| sodium chlorate(I) | sodium chlorate(V) |

Pages 146 – 147

1. Cleaners may contain acid (H^+ ions):
$2H^+(aq) + Cl^-(aq) + ClO^-(aq) \rightleftharpoons Cl_2(g) + H_2O(l)$
These ions would shift the equilibrium to the right, causing the release of harmful chlorine gas.

2. **a)** $Cl_2(aq) + 2I^-(aq) \rightarrow 2Cl^-(aq) + I_2(s)$
b) $I_2(aq) + 2S_2O_3^{2-}(aq) \rightarrow 2I^-(aq) + S_4O_6^{2-}(aq)$

3. Take a known volume of each bleach, e.g. 10 cm^3. The volume must be exactly the same in each case to ensure a fair test, so measure carefully using a pipette. Dilute the bleach with water, e.g. to 250 cm^3 in a standard flask. Titrate 25 cm^3 samples, to which excess KI has been added, against standard sodium thiosulphate solution, using starch as the indicator. The sample of bleach that requires the most thiosulphate solution contains the most chlorine.

Pages 148 – 149: Unit 9 Questions

1. **a)** Add aqueous chlorine dropwise to the solution of sodium bromide in a fume cupboard. Bromine will be released (1). (It could be separated from the aqueous solution by solvent extraction using a non-polar solvent.)

b) The colourless solution of sodium bromide would turn yellow/orange (1).
c) $Cl_2(aq) + 2Br^-(aq) \rightarrow 2Cl^-(aq) + Br_2(aq)$ (1)
0   −1   −1   0
The chlorine is reduced to chloride ions, as shown by the decrease in oxidation states (1); and the bromide ions are oxidised, as shown by the increase in oxidation states (1).

2. Down Group VII, the boiling points increase (1). This is because as the atoms get bigger (1), the strength of the van der Waals forces between the molecules increases (1).

3. **a)** Electronegativity is the power of an atom to attract electrons (1) from a covalent bond (1).
b) Down Group VII, the electronegativity decreases (1). This is because as the atoms get bigger (1), the effect of the nuclear charge decreases (1) (due to increased shielding), so the atom attracts electrons less readily.

4. a) +6 (1)
   b) +4 (1)
   c) $H_2SO_4 + 2H^+ + 2e^- \rightarrow SO_2 + 2H_2O$ (2)
   d) $2Br^- \rightarrow Br_2 + 2e^-$ (1)
   e) Oxidising agent (1)
   f) $H_2SO_4 + 2H^+ + 2Br^- \rightarrow Br_2 + SO_2 + 2H_2O$ (2)

5. a) Cl^- (1) b) Br^- (1) c) I^- (1) d) Cl^- (1) e) I^- (1)

6. AgCl will dissolve in both dilute and concentrated ammonia solution (1).
   AgBr is insoluble in dilute ammonia (1), but will dissolve in concentrated ammonia solution (1).
   AgI is insoluble in both dilute and concentrated ammonia solution (1).

7. a) To test for chlorine, use damp blue litmus paper (1). The paper will turn red and then white (1).
   b) To test for chloride ions, use acidified silver nitrate solution (1), followed by aqueous ammonia (1). A white precipitate will form (1), which dissolves in aqueous ammonia (1).

8. a) (i) $Cl_2(g) + H_2O(l) \rightleftharpoons 2H^+(aq) + Cl^-(aq) + ClO^-(aq)$ (1)
   (ii) $Cl_2(g) + 2NaOH(aq) \rightarrow H_2O(l) + NaCl(aq) + NaClO(aq)$ (1)
   b) (i) Sodium chlorate(I) (1). (ii) Bleach (1).

9. a) (i) The blue colour of the solution (1) would turn brown (1). A black precipitate of iodine (1) would form.
   (ii) Copper(I) iodide (1).
   (iii) Reducing agent (1).

   b) (i) Number of moles = $VM/1000 = \frac{50 \times 0.02}{1000}$ (1)
   $= 0.001\,\text{mol}$ (1)

   (ii) From the equation 2 mol of copper ions → 1 mol of iodine so 0.001 mol of copper ions → 0.0005 mol of iodine(1) = 5.0×10^{-4}
   (iii) $I_2(aq) + 2S_2O_3^{2-}(aq) \rightarrow 2I^-(aq) + S_4O_6^{2-}(aq)$
   Moles of thiosulphate = $2 \times 5.0 \times 10^{-4} = \frac{V \times 0.100}{1000}$

   Rearranging: $V = \frac{2 \times 5 \times 10^{-4} \times 1000}{0.100} = 10\,\text{cm}^3$ (1)

   (iv) Starch solution (1).

10. a) (i) $Cl_2(g) + H_2O(l) \rightleftharpoons 2H^+(aq) + Cl^-(aq) + ClO^-(aq)$ (2)
   (ii) Add a piece of blue litmus paper (1). It would turn red and then white (1), as it is bleached by chlorine.
   b) (i) The pale green colour would fade (1) (the solution becomes colourless).
   (ii) The addition of alkali will shift this equilibrium to the right (1) as the OH^- ions react with the H^+ ions to form water (1). Less chlorine is now present, so its colour fades.
   or If chlorine is reacted with an excess of cold, dilute sodium hydroxide solution, the equilibrium shifts so far to the right that the reaction goes to completion.

(iii) $H^+(aq) + OH^-(aq) \rightarrow H_2O(l)$
or $Cl_2(g) + 2OH^-(aq) \rightarrow H_2O(l) + Cl^-(aq) + ClO^-(aq)$

c) (i) The green colour of the solution would intensify (1).
(ii) Addition of H^+ ions will shift the equilibrium to the left (1), so more chlorine will be present (1), and the solution will be greener.
(iii) $2H^+(aq) + Cl^-(aq) + ClO^-(aq) \rightarrow Cl_2(g) + H_2O(l)$ (1)

10 Extraction of metals

Pages 150 – 151

1. Iron and aluminium.
2. Carbon monoxide, carbon dioxide, oxides of sulphur.

Page 152 – 153

1. Iron ore, coke, limestone and air.
2. $Fe_3O_4 + 4C \rightarrow 3Fe + 4CO$
   $Fe_3O_4 + 4CO \rightarrow 3Fe + 4CO_2$
3. Limestone is added to remove acidic impurities, mainly silicon dioxide. The limestone decomposes, forming the basic oxide, CaO, which reacts with SiO_2 forming slag:
   $CaCO_3 \rightarrow CaO + CO_2$
   $CaO + SiO_2 \rightarrow CaSiO_3$
4. Molten iron is mixed with scrap iron. Some magnesium is added to remove any sulphur. Pure oxygen is then blown into the molten iron. The non-metals are oxidised by the oxygen. Volatile oxides escape as fumes; lime, CaO, is then added to remove any non-volatile oxides.
5. They can be used to heat the air blasted into the furnace.

Pages 154 – 155

1. Cryolite lowers the melting point of the aluminium oxide, and lowers the working temperature of the cell. This saves money.
2. The cost of the large quantities of electricity used, and the replacement of the positive electrodes which burn away.
3. Negative electrode: $Al^{3+} + 3e^- \rightarrow Al$ Positive electrode: $2O^{2-} \rightarrow O_2 + 4e^-$
4. $TiO_2 + 2Cl_2 + 2C \rightarrow TiCl_4 + 2CO$
   $TiCl_4 + 4Na \rightarrow Ti + 4NaCl$
   or $TiCl_4 + 2Mg \rightarrow Ti + 2MgCl_2$
5. It is a batch process; and sodium, chlorine and argon are expensive to produce.

Pages 156 – 157: Unit 10 Questions

1. **a)** Bauxite (1).

b) Aluminium oxide (1).
c) Negative (1).
d) (i) $2O^{2-} \rightarrow O_2 + 4e^-$ (2) (ii) $Al^{3+} + 3e^- \rightarrow Al$ (2)
e) The carbon burns away (1) in the oxygen (1) produced at the electrode.
f) Large quantities of electricity are needed (1); the positive electrodes have to be replaced regularly (1).

2 a) Iron ore (1); limestone (1); coke (1); air (1).
b) $Fe_2O_3 + 3C \rightarrow 2Fe + 3CO$ (1)
$Fe_2O_3 + 3CO \rightarrow 2Fe + 3CO_2$ (1)
c) $CaCO_3 \rightarrow CaO + CO_2$ (1)
$CaO + SiO_2 \rightarrow CaSiO_3$ (1)
d) Carbon (1).

3 The molten iron is mixed with scrap iron (1), and magnesium is added to remove any sulphur (1). Pure oxygen gas (1) is blown into the mixture to oxidise the non-metal impurities (1). Volatile oxides escape from the mixture (1). Lime is added to remove any non-volatile non-metal oxides (1). Calculated quantities of carbon and metals are then added to make the desired steel (1).

4 a) Rutile (1).
b) Titanium(IV) oxide (1).
c) $TiO_2 + 2Cl_2 + 2C \rightarrow TiCl_4 + 2CO$ (2)
d) (i) $TiCl_4 + 4Na \rightarrow Ti + 4NaCl$ (2) (ii) Heat (1); a 'blanket' of argon (1).
e) Strong (1); resistant to corrosion (1).
f) *Three of*: Batch process (1); high cost of sodium (1), chlorine (1) and argon (1).
g) The titanium reacts with carbon (1) to form a carbide (1), which makes the metal brittle.

5 *Batch*: Reactants are mixed, and at the end of the reaction the product is removed (1). The titanium(IV) chloride reacts with the sodium metal (1). When the reaction is finished, the titanium metal is removed and the reaction is started again with a fresh supply of titanium(IV) chloride and sodium (1). *Continuous*: Reactants are added as the product is removed (1): more aluminium oxide is added to the cell (1) as the aluminium metal is removed (1).

6 a) Carbon monoxide (1), toxic gas (1); carbon dioxide (1), greenhouse gas (1); oxides of sulphur (1), acid rain (1).
b) Carbon monoxide/carbon dioxide (1); toxic/greenhouse gas (1).
c) Carbon dioxide (1), greenhouse gas (1); chlorine (1), toxic gas (1).

7 a) Carbon reduction (1); haematite (1); continuous (1); impure (carbon) (1).
b) Electrolysis (1); bauxite (1); continuous (1); pure (1).
c) Reduction by a more reactive metal (1); rutile (1); batch (1); pure (1).

8 a) Iron: low cost (1); named desirable property, e.g. strength (1). Aluminium: e.g. low density (1); good electrical/thermal properties (1); corrosion resistance (1). (*Max. 2 each.*)
b) Reduces energy consumption (1); reduces environmental problems associated with ore extraction (1); reduces demand on finite resources (1). (*Max. 2.*)

Answers: Module 3

11 Nomenclature and isomerism

Pages 164 – 165

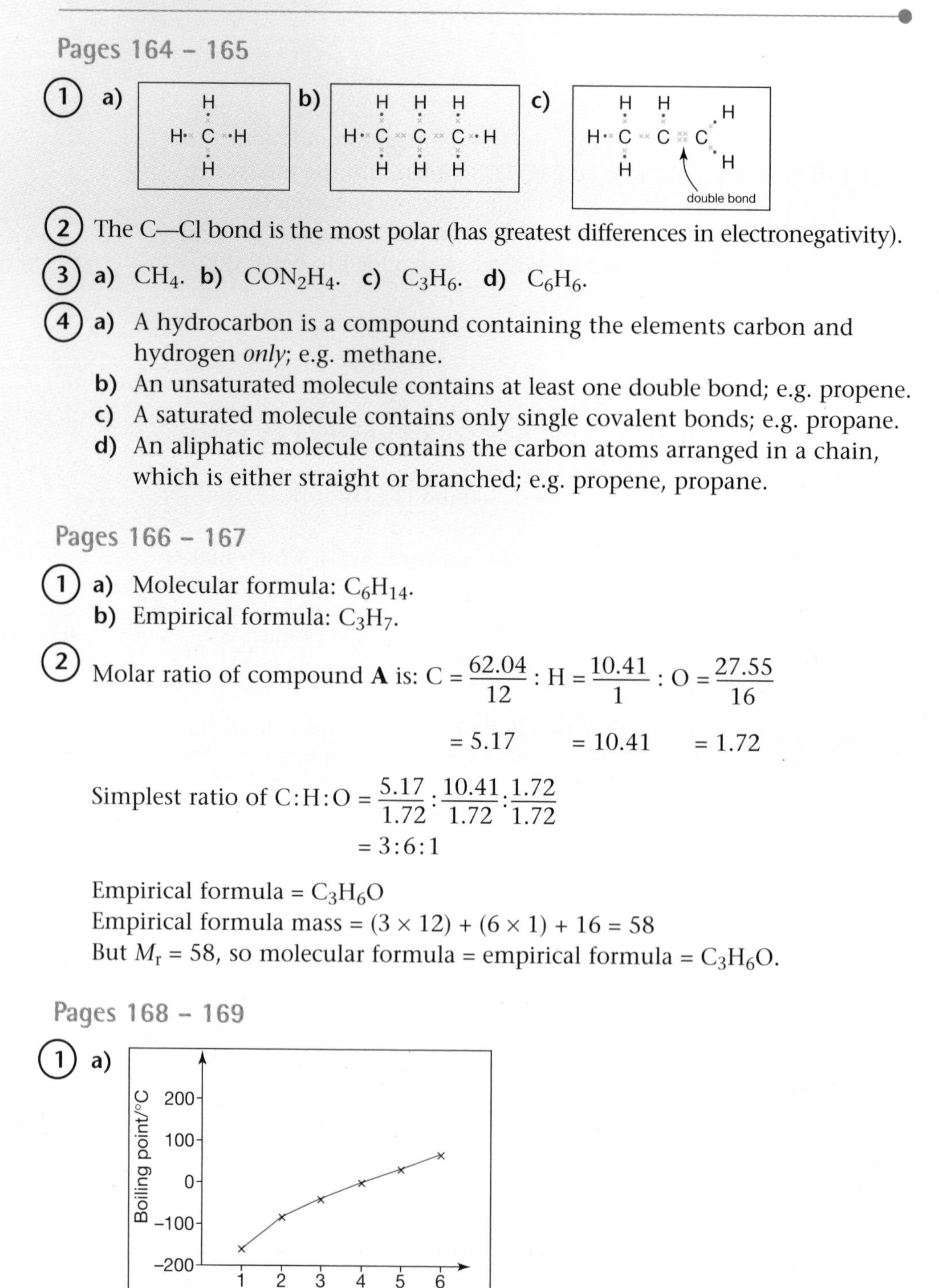

(1) a) 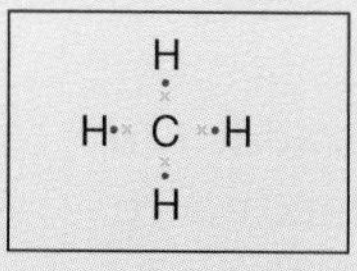

b) 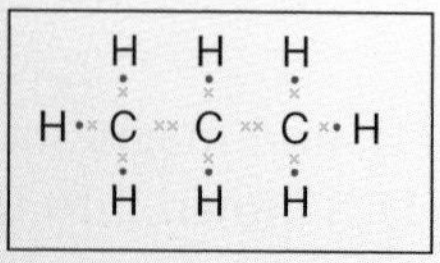

c) 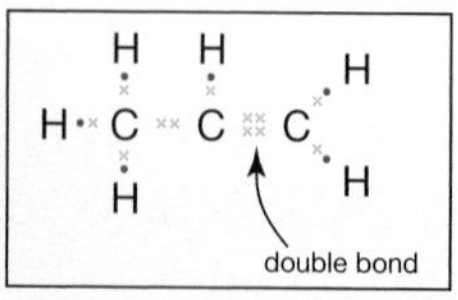

(2) The C—Cl bond is the most polar (has greatest differences in electronegativity).

(3) **a)** CH_4. **b)** CON_2H_4. **c)** C_3H_6. **d)** C_6H_6.

(4) **a)** A hydrocarbon is a compound containing the elements carbon and hydrogen *only*; e.g. methane.
b) An unsaturated molecule contains at least one double bond; e.g. propene.
c) A saturated molecule contains only single covalent bonds; e.g. propane.
d) An aliphatic molecule contains the carbon atoms arranged in a chain, which is either straight or branched; e.g. propene, propane.

Pages 166 – 167

(1) **a)** Molecular formula: C_6H_{14}.
b) Empirical formula: C_3H_7.

(2) Molar ratio of compound **A** is: $C = \frac{62.04}{12} : H = \frac{10.41}{1} : O = \frac{27.55}{16}$

$$= 5.17 \qquad = 10.41 \qquad = 1.72$$

Simplest ratio of $C:H:O = \frac{5.17}{1.72} : \frac{10.41}{1.72} : \frac{1.72}{1.72}$
$= 3:6:1$

Empirical formula = C_3H_6O
Empirical formula mass = $(3 \times 12) + (6 \times 1) + 16 = 58$
But $M_r = 58$, so molecular formula = empirical formula = C_3H_6O.

Pages 168 – 169

(1) a) 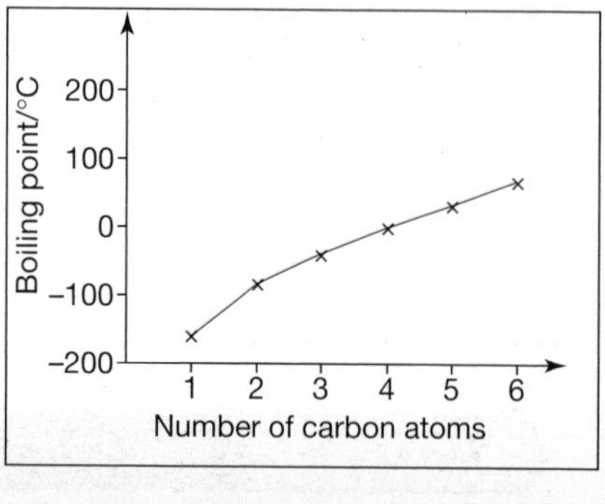

(2) **a)** (i) Alkene. (ii) Alkane. (iii) Haloalkane. (iv) Alcohol. (v) Carboxylic acid.

b) (i) (ii) (iii) (iv) (v)

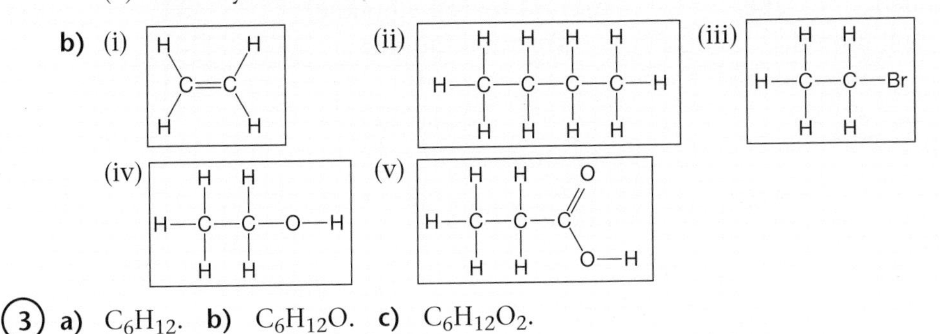

(3) **a)** C_6H_{12}. **b)** $C_6H_{12}O$. **c)** $C_6H_{12}O_2$.

Pages 170 – 171

(1)

Name	Formula	Homologous series
Ethane	**C_2H_6**	**alkanes**
Propene	C_3H_6	**alkenes**
Ethanoic acid	CH_3COOH	**carboxylic acids**
Propanal	CH_3CH_2CHO	aldehydes
Butanol	**C_4H_9OH**	**alcohols**

(2) **a)** The molecular formula gives the actual number of the different types of atoms present in one molecule; e.g. the molecular formula of ethene is C_2H_4 – each molecule contains two carbon atoms and four hydrogen atoms.

b) The structural formula shows how the atoms are arranged in a molecule; e.g. the structural formula of ethene is

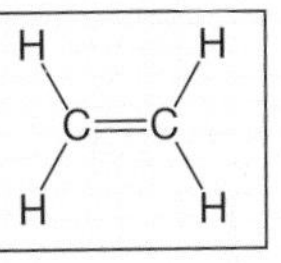

c) A homologous series is a group of organic molecules which can all be represented by the same general formula, which show similar chemical properties, and which show a trend in physical properties; e.g. the alkenes are a homologous series.

d) The functional group of a molecule is an atom or group of atoms which determines the chemical properties of the molecule; e.g. the functional group of the alkenes is C=C.

Pages 172 – 173

(1) Heptane has 9 isomers.

(2) The three structural isomers of pentane are as shown on the right.

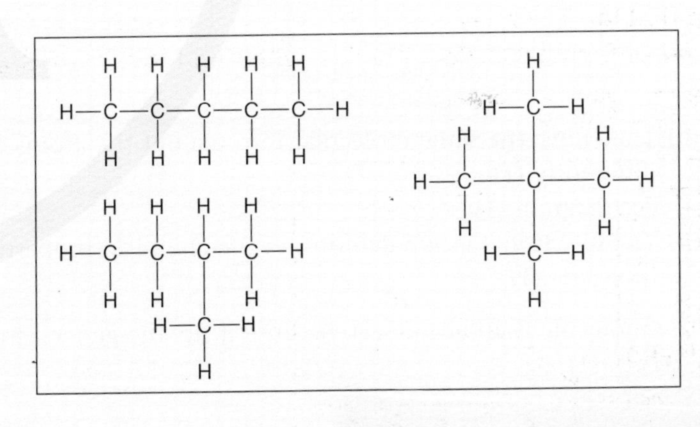

(3) a) is a branched isomer.

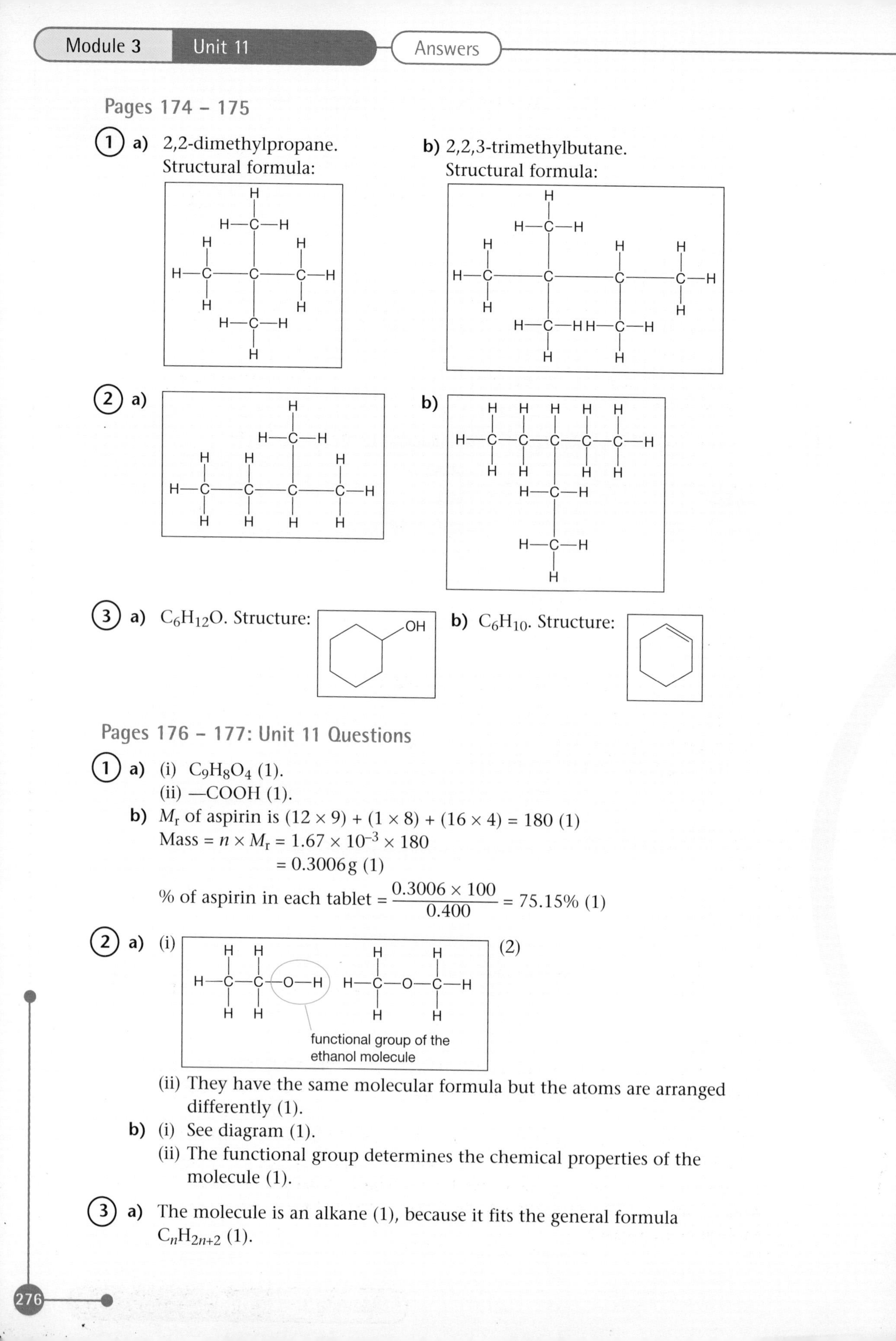

Pages 174 – 175

1 a) 2,2-dimethylpropane.
Structural formula:

b) 2,2,3-trimethylbutane.
Structural formula:

2 a)

b)

3 a) $C_6H_{12}O$. Structure:

b) C_6H_{10}. Structure:

Pages 176 – 177: Unit 11 Questions

1 a) (i) $C_9H_8O_4$ (1).
(ii) —COOH (1).

b) M_r of aspirin is $(12 \times 9) + (1 \times 8) + (16 \times 4) = 180$ (1)
Mass $= n \times M_r = 1.67 \times 10^{-3} \times 180$
$= 0.3006\,g$ (1)

% of aspirin in each tablet $= \dfrac{0.3006 \times 100}{0.400} = 75.15\%$ (1)

2 a) (i) (2)

(ii) They have the same molecular formula but the atoms are arranged differently (1).

b) (i) See diagram (1).
(ii) The functional group determines the chemical properties of the molecule (1).

3 a) The molecule is an alkane (1), because it fits the general formula C_nH_{2n+2} (1).

b) (i)

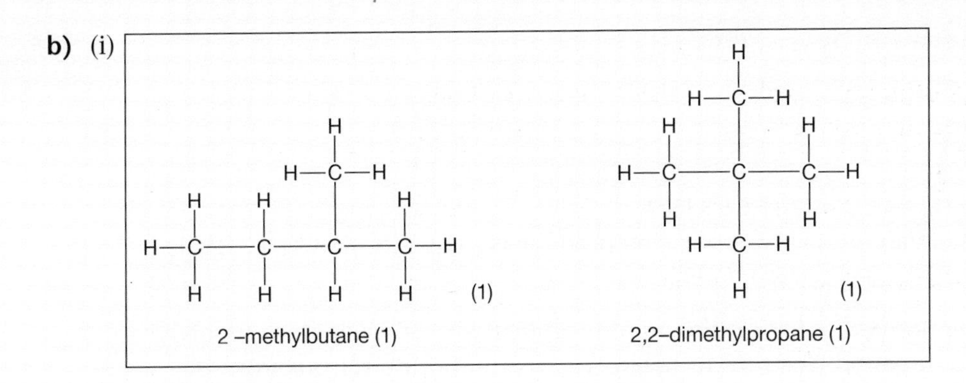

(ii) 109.5° (1).

c) The bonds can rotate (1), so the groups which appear to be branches are actually part of the straight chain of pentane (1).

4 **a)** In **X** the mass of oxygen present = 9.75 – (3.90 + 0.65) = 5.20 g (1)

The molar ratio of C:H:O $= \frac{3.90}{12} : \frac{0.65}{1} : \frac{5.20}{16}$ (1)

$= 0.325 : 0.65 : 0.325$

$= 1 : 2 : 1$

Empirical formula = CH_2O (1)

b) Empirical formula mass = 12 + 2 + 16 = 30 (1)
$M_r = 60$
so the molecular formula = empirical formula × 2 = $C_2H_4O_2$ (1)

c)

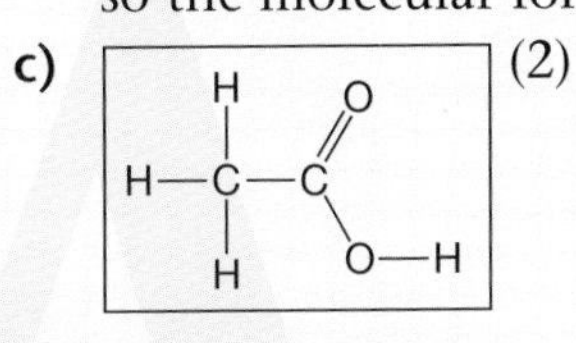

(2)

5 **a)** c (1).
b) e (1).
c) d (1).
d) g or b (1).
e) b (1).

12 Petroleum and alkanes

Pages 178 –179

1 **a)** Hexane is a larger molecule than ethane, so the van der Waals forces between the molecules are stronger and more energy is needed to break them.

b) Methylpropane and butane are isomers: they contain the same number of atoms, but methylpropane is a branched molecule. This means it has a more spherical shape, so the contact with neighbouring molecules is reduced and consequently the van der Waals forces between the molecules are weaker.

2) a) $C_4H_{10} + 6\frac{1}{2}O_2 \rightarrow 4CO_2 + 5H_2O$
   b) $C_4H_{10} + 4\frac{1}{2}O_2 \rightarrow 4CO + 5H_2O$

Pages 180 – 181

1) a) $C_2H_6 + Cl_2 \rightarrow C_2H_5Cl + HCl$
   b) There are probably about 7 by-products: $C_2H_4Cl_2$, $C_2H_3Cl_3$, $C_2H_2Cl_4$, C_2HCl_5, C_2Cl_6, HCl, C_4H_{10}.

2) During homolytic fission the bond breaks equally, so each atom involved has an unpaired electron. In heterolytic fission both electrons from the bond are transferred to one of the atoms.

3) A carbocation is a species which contains a carbon atom that carries a positive charge.

Pages 182 – 183

1) The two electrons from the covalent bond are shared equally between the two chlorine atoms.

2) Once a free radical is formed, free radicals continue to be used and formed.

3) a) $Cl_2 \rightarrow 2Cl\cdot$
   b) $C_2H_6 + Cl\cdot \rightarrow HCl + C_2H_5\cdot$
   c) $C_2H_5\cdot + Cl\cdot \rightarrow C_2H_5Cl$

Pages 184 – 185

1) An energy source which once used up cannot be replaced.

2) Methane

3) Their different boiling points.

4) $C_4H_{10} + 6\frac{1}{2}O_2 \rightarrow 4CO_2 + 5H_2O$

Pages 186 – 187

1) Kerosene.

2) 220–375 °C.

3) $C_6H_4 \rightarrow C_{14}H_{30}$

4) The components may decompose before they boil.

Pages 188 – 189

1) They are surplus to demand.

2) $C_{12}H_{26} \rightarrow C_2H_4 + C_3H_6 + \underset{\text{heptane}}{C_7H_{16}}$

3) The molecules are more likely to be branched or aromatic. These help the fuel to burn more efficiently (they prevent auto-ignition).

Pages 190 – 191

1. H_2SO_3: sulphurous acid (or sulphuric(IV) acid).
   H_2SO_4: sulphuric acid (or sulphuric(VI) acid).

2. The spark from the engine provides enough energy to overcome the activation energy for the reaction between the nitrogen and oxygen from the air to form nitrogen monoxide:
   $N_2(g) + O_2(g) \rightarrow 2NO(g)$
   Nitrogen monoxide readily combines with oxygen in the air or engine to form nitrogen dioxide:
   $2NO(g) + O_2(g) \rightarrow 2NO_2(g)$

3. Free radicals.

Pages 192 – 193

1. The d block (the transition metals).

2. It does not rust, so is less likely to need replacing.

3. From +2 to 0 ($NO \rightarrow N_2$).

4. **a)** Reduction. **b)** Oxidation.

5. Lead would reduce the efficiency of the catalyst (the catalyst would be 'poisoned').

6. Carbon dioxide; it is a greenhouse gas.

Pages 194 – 195: Unit 12 Questions

1. **a)** (i) Free-radical substitution (1).
   (ii) Step 1 is initiation (1); steps 2 and 3 are propagation (2); step 4 is termination (1).
   b) The reaction will continue (1) until all the free radicals have been used up (1).
   c) There are too many by-products (1) for this to be a viable method of preparation for chloroethane. The removal of these would be time-consuming and costly (1).
   An alternative method of preparation is from ethene:
   $C_2H_4 + HCl \rightarrow C_2H_5Cl$ (1)
   d) The six positions on the molecule where the chlorine atom could reside are all equivalent (2).

2. **a)** (i) A saturated molecule contains only single covalent bonds (1). A hydrocarbon is a molecule containing only the elements carbon and hydrogen (1).
   (ii) The alkanes (1).
   b) (i) Diesel (1).
   (ii) It would burn with a smoky flame (1) and it would be difficult to ignite (1).
   (iii) Refinery gas (1).
   c) Ships/power stations (1).
   d) Bitumen/tar (1).

3 a) (i) Free-radical mechanism (1). (ii) $C_{12}H_{26} \rightarrow \mathbf{C_8H_{18}} + 2C_2H_4$ (1).
b) (i) Carbocation mechanism (1). (ii) Zeolite (1).
(iii) Reaction can occur at a lower temperature, so energy is conserved.
c) (i) Compound **X** is 2,4-dimethylpentane (1).
(ii) Compound **Y** is an alkene (1).

4 a) (i) (2)

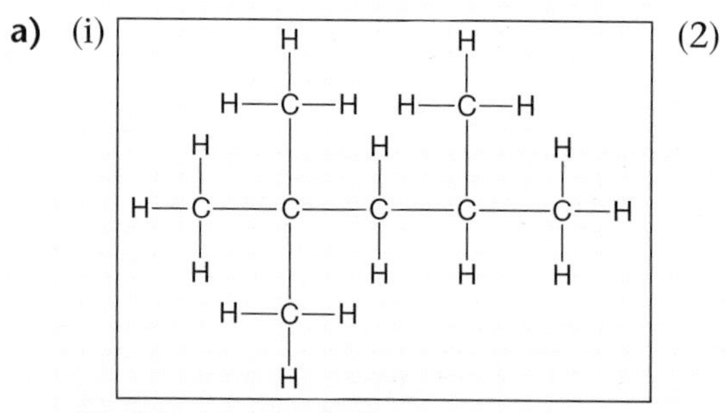

(ii) 2,2,4-trimethylpentane (1).
(iii) Because it is a branched-chain molecule (1), it helps the petrol to burn more efficiently (prevents auto-ignition) (1).
b) (i) $C_8H_{18}(l) + 12\frac{1}{2}O_2(g) \rightarrow 8CO_2(g) + 9H_2O(l)$ (2)
(ii) $12\frac{1}{2}$ dm^3 (1)
c) (i) $C_8H_{18}(l) + 8\frac{1}{2}O_2(g) \rightarrow 8CO\ (g) + 9H_2O(l)$ (2)
(ii) Carbon monoxide is toxic (1): it reduces the oxygen-carrying capacity of the blood (1).
(iii) Inside the catalytic converter, carbon monoxide is oxidised to carbon dioxide (1):
$2CO + O_2 \rightarrow 2CO_2$
or $2CO + 2NO \rightarrow 2CO_2 + N_2$ (1)

13 Alkenes

Pages 196 – 197

1 a) b) c)

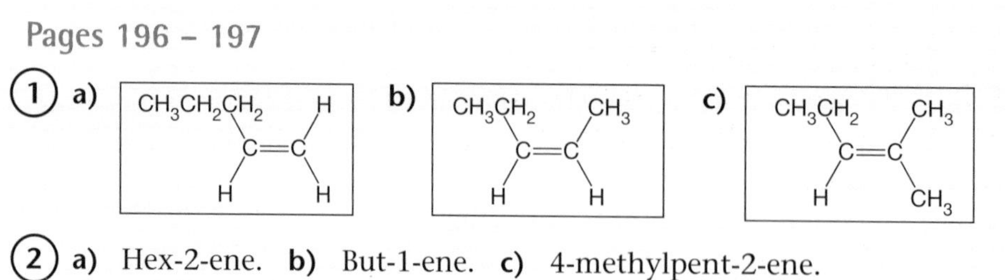

2 a) Hex-2-ene. b) But-1-ene. c) 4-methylpent-2-ene.
d) 2,3-dimethylbut-2-ene. a) and c) have geometrical isomers.

3 a) $C_4H_8 + 6O_2 \rightarrow 4CO_2 + 4H_2O$
b) $C_5H_{10} + 7\frac{1}{2}O_2 \rightarrow 5CO_2 + 5H_2O$
c) $C_4H_8 + 6O_2 \rightarrow 4CO_2 + 4H_2O$

Pages 198 – 199

1 Bromine is more electronegative than hydrogen, and attracts the pair of electrons in the H—Br bond.

2 The carbon—carbon double bond is a region of high electron density and is attacked by electron-deficient species.

3 The movement of a pair of electrons when a covalent bond is formed or broken.

Pages 200 – 201

1.
```
     H   H   H   H
     |   |   |   |
 H — C — C — C — C — H
     |   |   +   |
     H   H       H
```

2. An electron-deficient species. It can accept a pair of electrons.

3. **a)** $CH_3CH{=}CH_2 + Br_2 \rightarrow CH_3CHBrCH_2Br$
   b) $CH_3CH{=}CHCH_3 + HBr \rightarrow CH_3CH_2CHBrCH_3$
   c) $CH_3CH_2CH{=}CHCH_2CH_3 + H_2SO_4 \rightarrow CH_3CH_2CH_2CH(OSO_3H)CH_2CH_3$

Pages 202 – 203

1. **a)**
```
     H   H   H   H   H
     |   |   |   |   |
 H — C — C — C — C — C — H
     |   |   |   |   +
     H   H   H   H

     H   H   H   H   H
     |   |   |   |   |
 H — C — C — C — C — C — H
     |   |   |   +   |
     H   H   H       H
```
b)
```
     H   H   H   H   H
     |   |   |   |   |
 H — C — C — C — C — C — H
     |   |   |   +   |
     H   H   H       H

     H   H   H   H   H
     |   |   |   |   |
 H — C — C — C — C — C — H
     |   |   +   |   |
     H   H       H   H
```

2. **a)** Major: $CH_3CH_2CHBrCH_3$; minor: $CH_3CH_2CH_2CH_2Br$.
   b) Major: $(CH_3)_3CBr$; minor: $(CH_3)_2CHCH_2Br$.

3. **a)** Major: $CH_3CH_2CH_2CH(OH)CH_3$; minor: $CH_3CH_2CH_2CH_2CH_2OH$.
   b) Major: $CH_3CH_2C(OH)(CH_3)_2$; minor: $CH_3CH(OH)CH(CH_3)_2$.

4. **a)** But-1-ene, $CH_3CH_2CH{=}CH_2$, and but-2-ene, $CH_3CH{=}CHCH_3$.
   b) 2-methylpropene, $(CH_3)_2C{=}CH_2$.

Pages 204 – 205

1. **a)** $(CH_3)_2C{=}CH_2 + H_2 \rightarrow (CH_3)_2CHCH_3$
   b) $CH_3CH_2CH{=}CHCH_3 + H_2 \rightarrow CH_3CH_2CH_2CH_2CH_3$
   c) $(CH_3)_2C{=}CHCH_3 + H_2 \rightarrow (CH_3)_2CHCH_2CH_3$

2. But-1-ene and but-2-ene.

3. **a)**
```
  / CH3  CH3 \
 |  |     |   |
—|— C ——— C —|—
 |  |     |   |
  \ H     H  / n
```
b)
```
  / CH3  H \
 |  |    |  |
—|— C —— C —|—
 |  |    |  |
  \ CH3  H / n
```

Pages 206 – 207

1. The forward reaction involves reduction in the number of gaseous mole, so the forward reaction is favoured by an increase in pressure.

2.
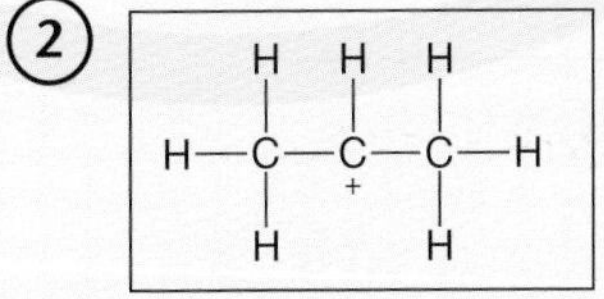

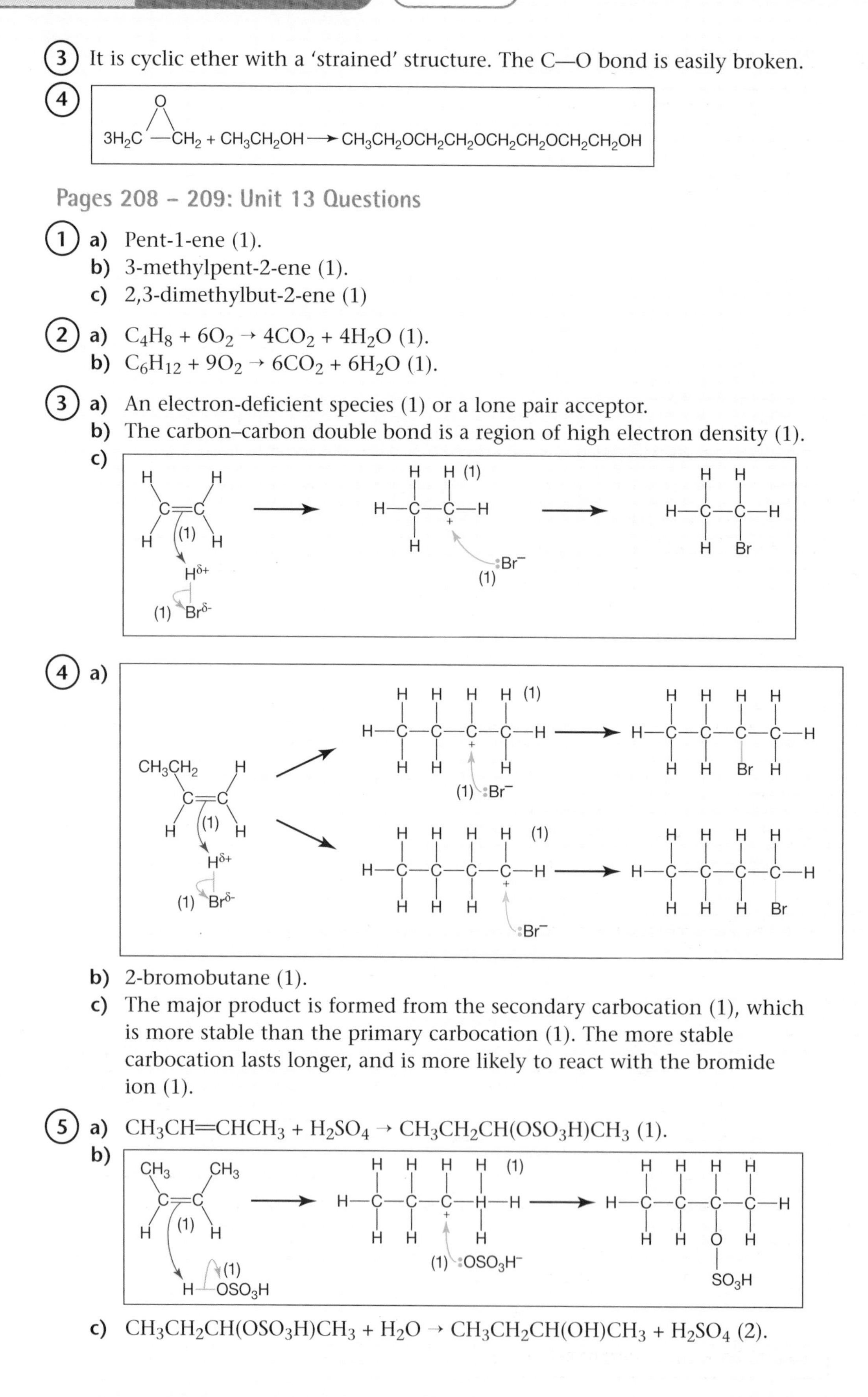

(3) It is cyclic ether with a 'strained' structure. The C—O bond is easily broken.

(4) $3H_2C(O)CH_2 + CH_3CH_2OH \rightarrow CH_3CH_2OCH_2CH_2OCH_2CH_2OCH_2CH_2OH$

Pages 208 – 209: Unit 13 Questions

(1) **a)** Pent-1-ene (1).
b) 3-methylpent-2-ene (1).
c) 2,3-dimethylbut-2-ene (1)

(2) **a)** $C_4H_8 + 6O_2 \rightarrow 4CO_2 + 4H_2O$ (1).
b) $C_6H_{12} + 9O_2 \rightarrow 6CO_2 + 6H_2O$ (1).

(3) **a)** An electron-deficient species (1) or a lone pair acceptor.
b) The carbon–carbon double bond is a region of high electron density (1).
c)

(4) **a)**

b) 2-bromobutane (1).
c) The major product is formed from the secondary carbocation (1), which is more stable than the primary carbocation (1). The more stable carbocation lasts longer, and is more likely to react with the bromide ion (1).

(5) **a)** $CH_3CH{=}CHCH_3 + H_2SO_4 \rightarrow CH_3CH_2CH(OSO_3H)CH_3$ (1).
b)

c) $CH_3CH_2CH(OSO_3H)CH_3 + H_2O \rightarrow CH_3CH_2CH(OH)CH_3 + H_2SO_4$ (2).

6 a)

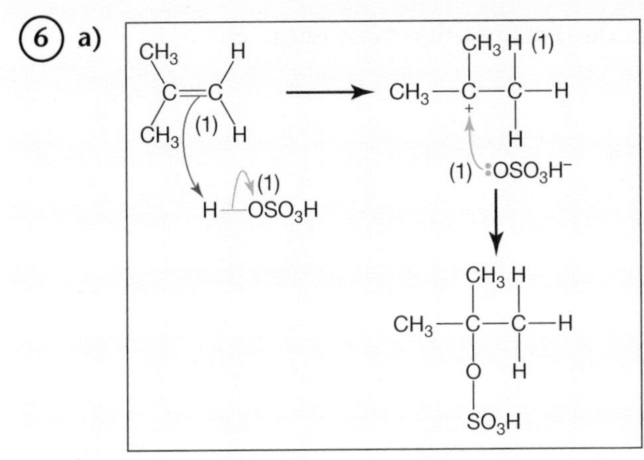

b) Two carbocations (1) are formed in the first step of the mechanism. The primary carbocation (1) is less stable than the tertiary carbocation, and it reacts to form the minor product, methylpropan-1-ol (1).

7 a) (i) Catalyst of concentrated H_3PO_4 (1), pressure of 7 MPa (1) and a temperature of 350 °C (1).

(ii) $CH_2{=}CH_2 + H_2O \rightarrow CH_3CH_2OH$ (1).

b)

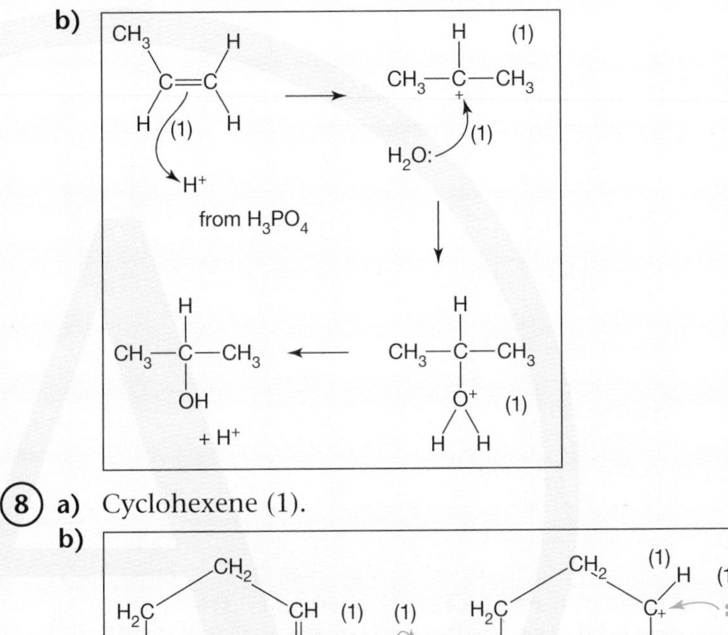

8 a) Cyclohexene (1).

b)

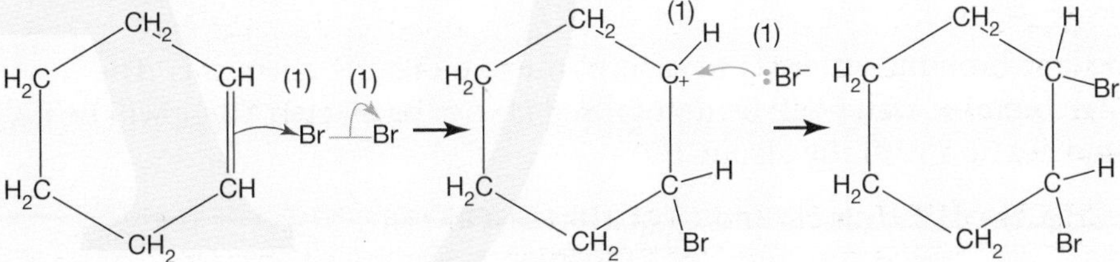

9 a) The carbon—carbon double bond is resistant (1) to rotation (1).

b)

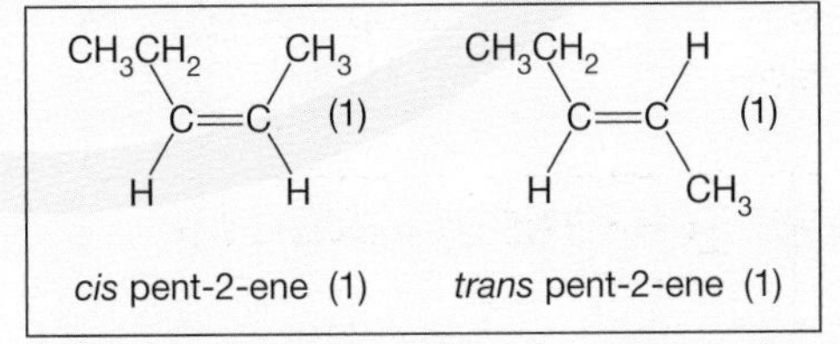

c) One of the carbon atoms (1) in the double bond has two hydrogen atoms (1) attached.

(10) a) $CH_2{=}CH_2 + \frac{1}{2}O_2 \rightarrow H_2C{-}CH_2$ (epoxide ring with O) (1)

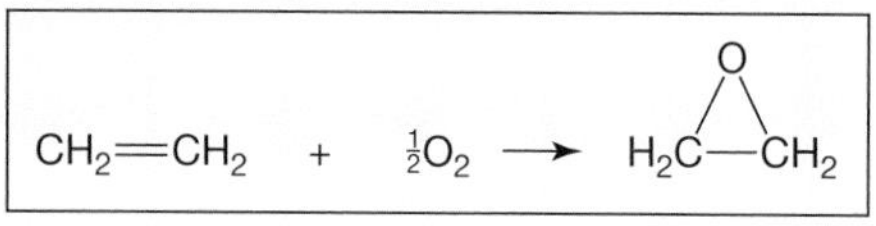

Silver catalyst (1); 180°C (1).

b) The cyclic structured is very strained (1), and the carbon-oxygen bond is easily broken (1).

c) $H_2C{-}CH_2$ (epoxide ring with O) $+ H_2O \rightarrow$ H–C(OH)(H)–C(OH)(H)–H (1)

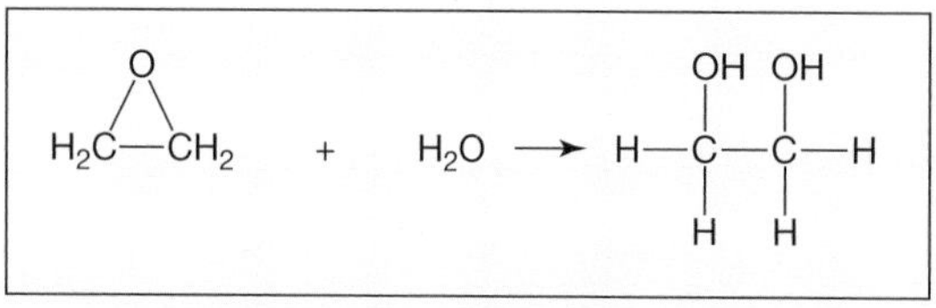

d) Antifreeze (1); making polyesters (1).

e) (i)

(ii)

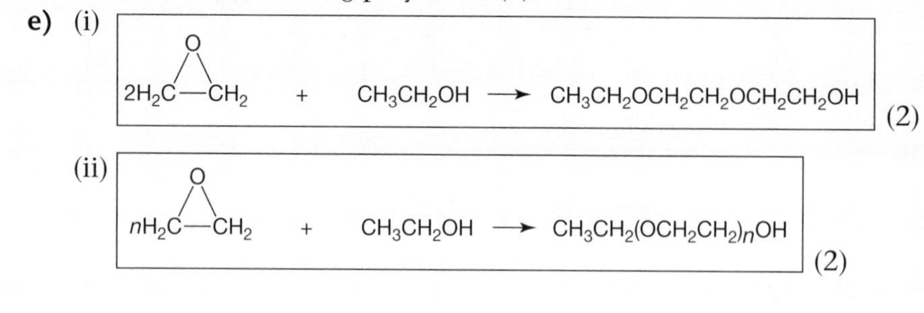

(11) a) **b)** **c)** **d)**

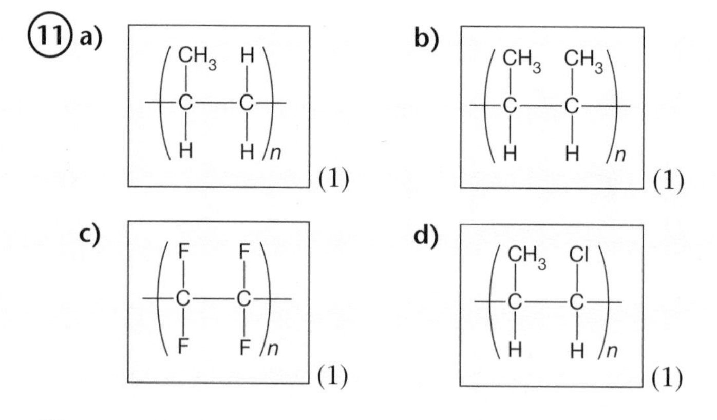

(12) Add some bromine water (1) to each of the two organic chemicals. The orange bromine water will be decolorised (1) by the alkene. There will be no visible reaction with the alkane (1).

(13) a) $CH_3CH(OH)CH_2Br$ (1) and $CH_3CHBrCH_2OH$ (1).

b) $CH_3(H)C{=}CH_2$ (1) attacks Br^+ → $CH_3{-}\overset{+}{C}H{-}CH_2Br$ (1); (1) $:OH^-$ attacks carbocation → $CH_3{-}CH(OH){-}CH_2Br$ (1)

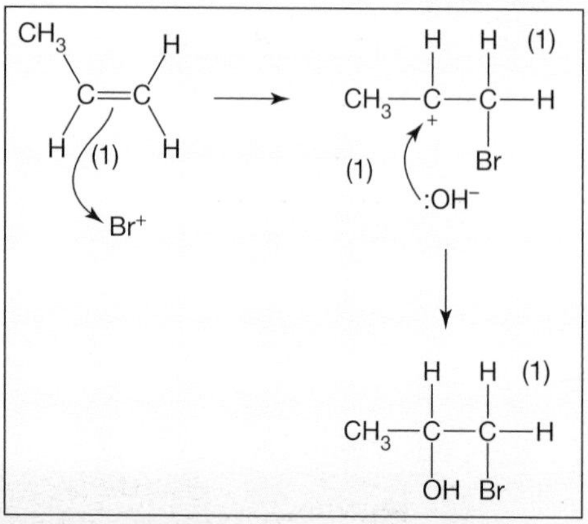

14 Haloalkanes

Pages 210 – 211

① **a)** Iodoethane. **b)** 2-chloropropane.
c) 2-bromopentane. **d)** 2,3-dibromobutane.

② **a)** $CH_3CH_2CHBrCH_2CH_3$ **b)** $CH_3CH_2CHCl_2$ **c)** $(CH_3)_3CI$

③ **a)** Primary. **b)** Secondary. **c)** Secondary. **d)** Secondary.
a) Secondary. **b)** Primary. **c)** Tertiary.

Pages 212 – 213

① **a)** $CH_3CH_2CH_2CH_2OH$ **b)** $CH_3CH_2CH(OH)CH_3$ **c)** CH_3CH_2OH

② **a)** $CH_3CH_2CH_2Br + KCN \rightarrow \underset{\text{butanenitrile}}{CH_3CH_2CH_2CN} + KBr$

b) $CH_3CH_2CH_2CH_2Cl + 2NH_3 \rightarrow \underset{\text{1-aminobutane}}{CH_3CH_2CH_2CH_2NH_2} + NH_4Cl$

c) $CH_3CH_2CH_2I + KOH \rightarrow \underset{\text{propan-1-ol}}{CH_3CH_2CH_2OH} + KI$

Pages 214 – 215

① **a)** $CH_3CH_2CH_2CH_2OH$ and $CH_3CH_2CH{=}CH_2$
b) $(CH_3)_3COH$ and $(CH_3)_2C{=}CH_2$

② **a)**

b)

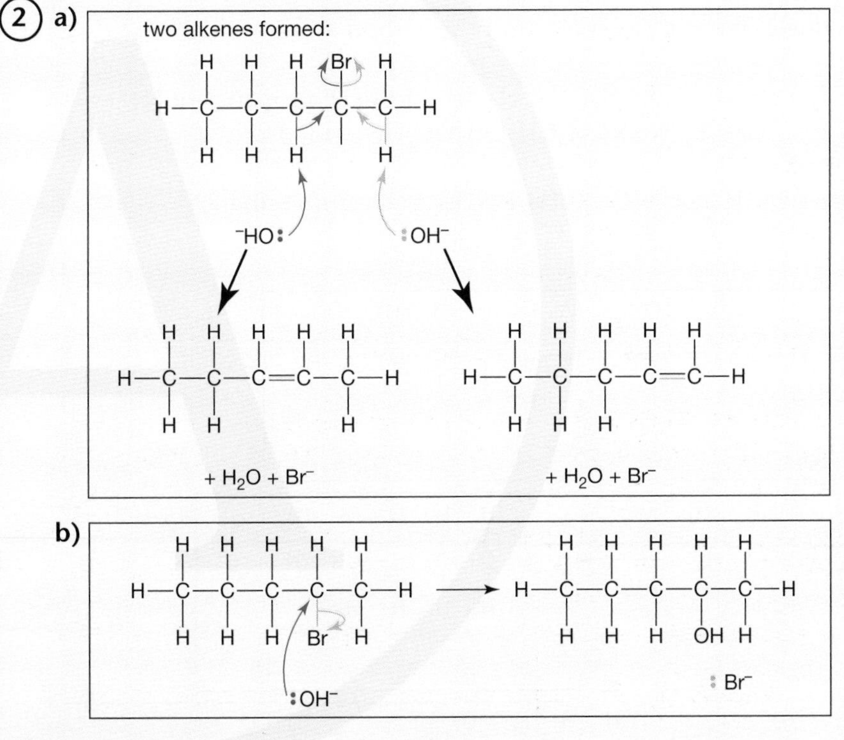

Pages 216 – 217: Unit 14 Questions

① **a)** 1-bromobutane (1).
b) 1-chloropropane (1).

c) 2,3-dibromobutane (1).
d) 1-iodo,2-methylbutane (1).

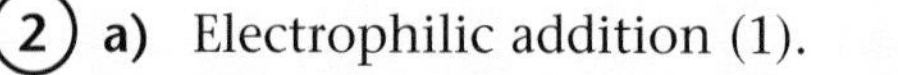

2 a) Electrophilic addition (1).
b) $CH_3CH{=}CHCH_3 + HBr \rightarrow CH_3CH_2CHBrCH_3$ (1); 2-bromobutane (1).
c)

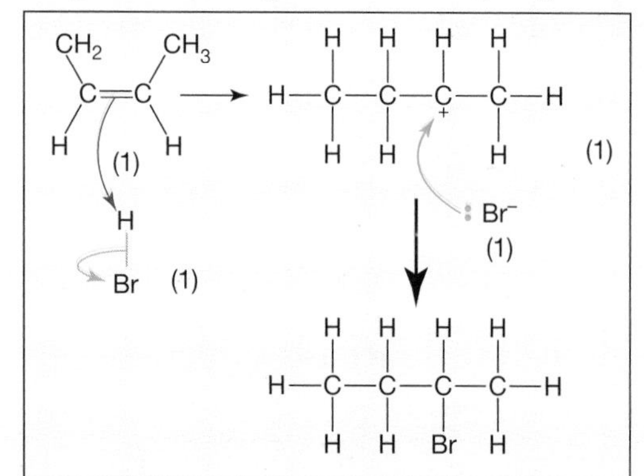

d) Free-radical substitution (1).
e) $CH_3CH_3 + Cl_2 \rightarrow CH_3CH_2Cl + HCl$ (1).
f) Initiation step: $Cl_2 \rightarrow Cl\cdot + Cl\cdot$ (1)
Propagation steps: $CH_3CH_3 + Cl\cdot \rightarrow CH_3CH_2\cdot + HCl$ (1)
$CH_3CH_2\cdot + Cl_2 \rightarrow CH_3CH_2Cl + Cl\cdot$ (1)
Termination step: $CH_3CH_2\cdot + Cl\cdot \rightarrow CH_3CH_2Cl$ (1)

3 a) A species with a lone pair of electrons that attacks an electron-deficient carbon atom (1).
b) The carbon–halogen bond is polar (1), and the carbon atom has a δ^+ charge (1).
c) (i) $CH_3CH_2CH_2Br + KOH \rightarrow CH_3CH_2CH_2OH + KBr$ (1); propan-1-ol (1).
(ii) $CH_3CH_2CH_2CH_2I + KCN \rightarrow CH_3CH_2CH_2CH_2CN + KI$ (1); pentanenitrile (1).
(iii) $CH_3CH_2Cl + 2NH_3 \rightarrow CH_3CH_2NH_2 + NH_4Cl$ (1); aminoethane (ethylamine) (1).

Mechanisms:

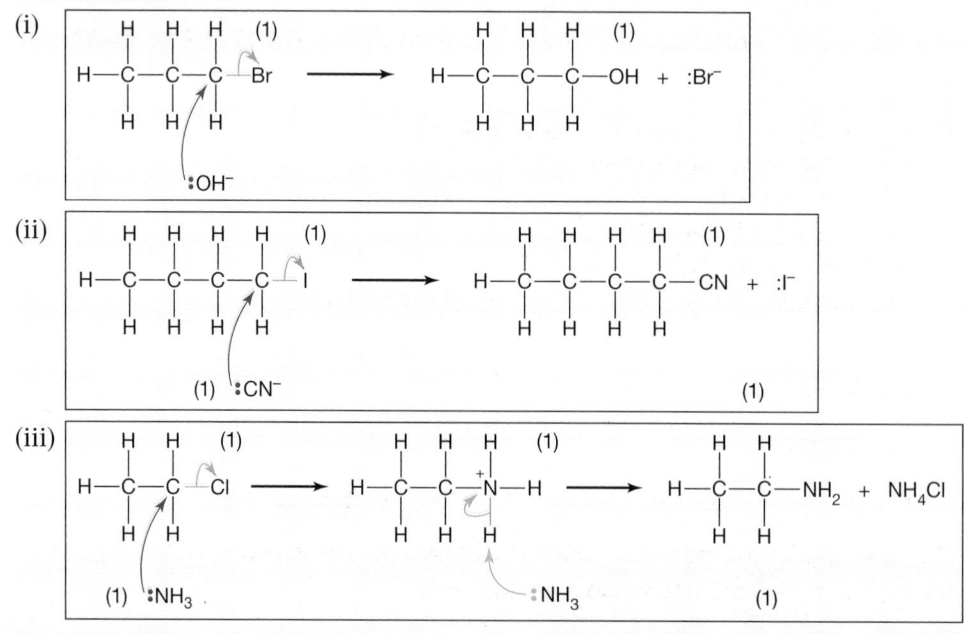

4 a) Nucleophilic substitution (1).

b)

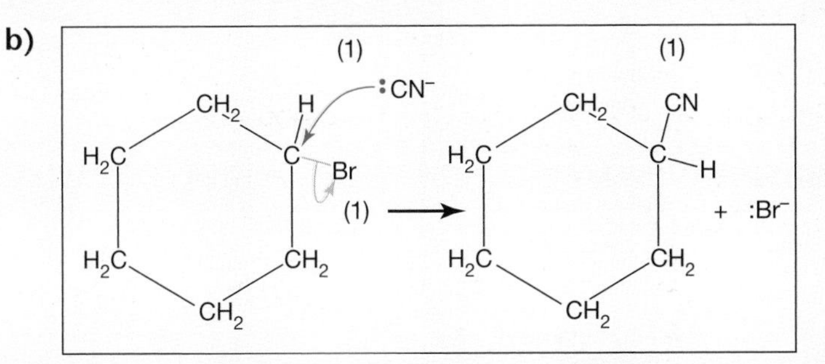

5 a) $CH_3CH_2CH_2CH_2Br + KOH \rightarrow CH_3CH_2CH_2CH_2OH + KBr$ (1); butan-1-ol (1).
$CH_3CH_2CH_2CH_2Br + KOH \rightarrow CH_3CH_2CH{=}CH_2 + H_2O + KBr$ (1);
but-1-ene (1).
Mechanisms:

(i)

(ii)

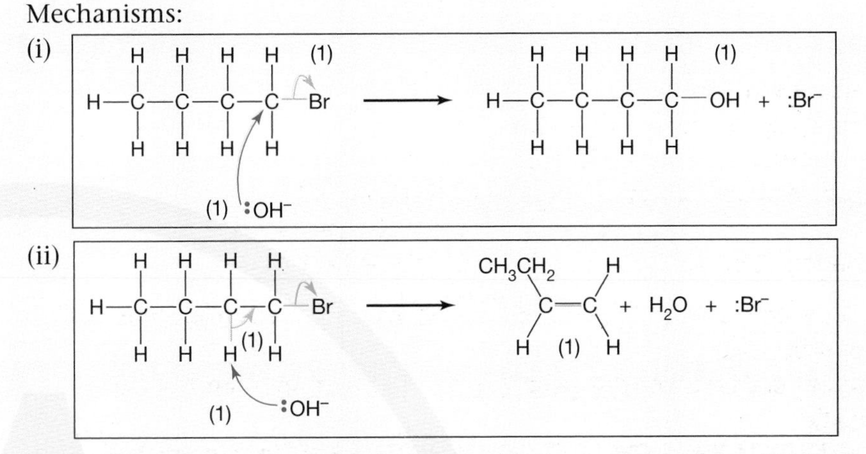

b) In the elimination reaction the hydroxide ion acts as a base (1); in the substitution reaction the hydroxide ions acts as a nucleophile (1).

6 a) $CH_3CH_2CH_2Cl + KCN \rightarrow CH_3CH_2CH_2CN + KCl$ (1).
b) Ethanol as the solvent (1); heat (1).
c) $CH_3CH_2CH_2CN + 2H_2O + HCl \rightarrow CH_3CH_2CH_2COOH + NH_4Cl$ (2).

7 The bond energy of the C—I bond is less than that of the C—Cl bond (1), so it is more easily broken (1).

8 a) (i) $Cl_2(g)$ (1); heat or ultraviolet light (1).
(ii) Free-radical substitution (1).
(iii) $C_6H_5CH_3 + Cl_2 \rightarrow C_6H_5CH_2Cl + HCl$ (1).
b) (i) KOH(aq) (1); heat (1).
(ii) Nucleophilic substitution (1).
(iii) $C_6H_5CH_2Cl + KOH \rightarrow C_6H_5CH_2OH + KCl$ (1).

15 Alcohols

Pages 218 – 219

1 a) Pentan-2-ol. b) Pentan-1-ol. c) 3-methylpentan-2-ol.

2 The alcohols have hydrogen bonding between the molecules. These are the strongest intermolecular forces.

3

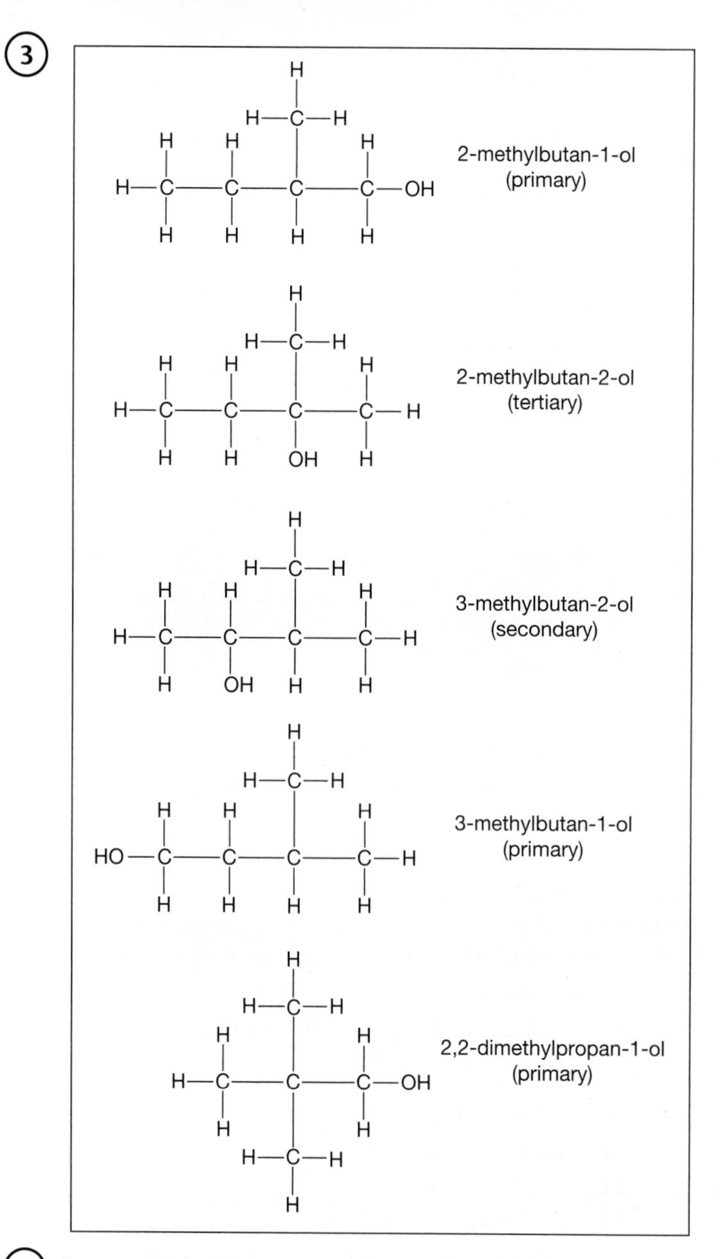

4 A renewable resource can be replaced (e.g. wood); a finite resource cannot be replaced once used (e.g. crude oil).

Pages 220 – 221

1 **a)** $CH_3CH_2CH_2CHO$ and $CH_3CH_2CH_2COOH$. **b)** $CH_3CH_2COCH_3$.
c) $(CH_3)_2CHCHO$ and $(CH_3)_2CHCOOH$. **d)** None – this is a tertiary alcohol.

2 *Propanal*: $CH_3CH_2CH_2OH + [O] \rightarrow CH_3CH_2CHO + H_2O$. Conditions: immediately distil off the aldehyde.
Propanoic acid: $CH_3CH_2CH_2OH + 2[O] \rightarrow CH_3CH_2COOH + H_2O$. Conditions: heat under reflux.

Pages 222 – 223

1 **a)** Pentanal. **b)** Pentan-3-one. **c)** Methylpropanal.

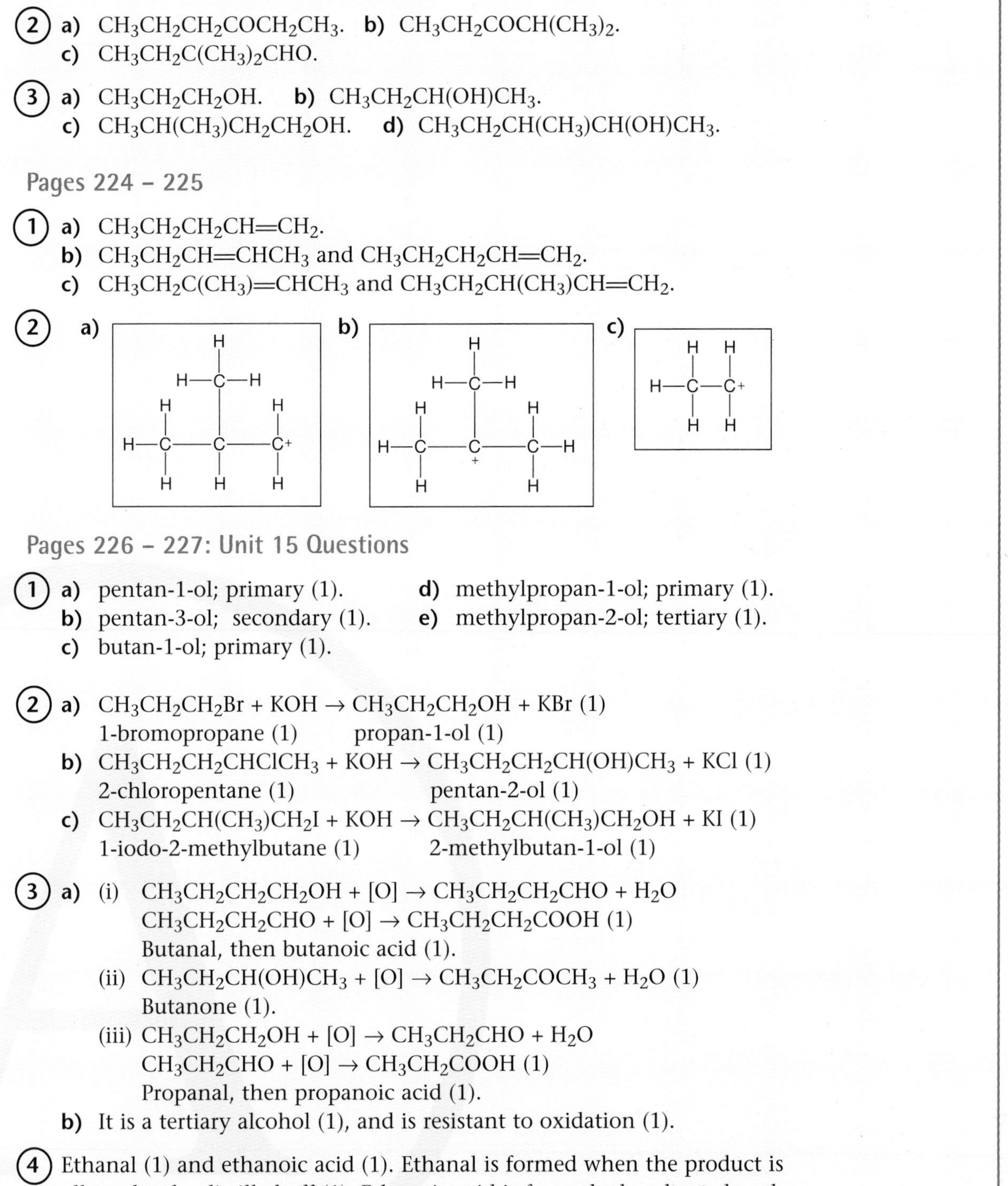

2 **a)** $CH_3CH_2CH_2COCH_2CH_3$. **b)** $CH_3CH_2COCH(CH_3)_2$.
c) $CH_3CH_2C(CH_3)_2CHO$.

3 **a)** $CH_3CH_2CH_2OH$. **b)** $CH_3CH_2CH(OH)CH_3$.
c) $CH_3CH(CH_3)CH_2CH_2OH$. **d)** $CH_3CH_2CH(CH_3)CH(OH)CH_3$.

Pages 224 – 225

1 **a)** $CH_3CH_2CH_2CH{=}CH_2$.
b) $CH_3CH_2CH{=}CHCH_3$ and $CH_3CH_2CH_2CH{=}CH_2$.
c) $CH_3CH_2C(CH_3){=}CHCH_3$ and $CH_3CH_2CH(CH_3)CH{=}CH_2$.

2 **a)** **b)** **c)**

Pages 226 – 227: Unit 15 Questions

1 **a)** pentan-1-ol; primary (1).
b) pentan-3-ol; secondary (1).
c) butan-1-ol; primary (1).
d) methylpropan-1-ol; primary (1).
e) methylpropan-2-ol; tertiary (1).

2 **a)** $CH_3CH_2CH_2Br + KOH \rightarrow CH_3CH_2CH_2OH + KBr$ (1)
1-bromopropane (1) propan-1-ol (1)
b) $CH_3CH_2CH_2CHClCH_3 + KOH \rightarrow CH_3CH_2CH_2CH(OH)CH_3 + KCl$ (1)
2-chloropentane (1) pentan-2-ol (1)
c) $CH_3CH_2CH(CH_3)CH_2I + KOH \rightarrow CH_3CH_2CH(CH_3)CH_2OH + KI$ (1)
1-iodo-2-methylbutane (1) 2-methylbutan-1-ol (1)

3 **a)** (i) $CH_3CH_2CH_2CH_2OH + [O] \rightarrow CH_3CH_2CH_2CHO + H_2O$
$CH_3CH_2CH_2CHO + [O] \rightarrow CH_3CH_2CH_2COOH$ (1)
Butanal, then butanoic acid (1).
(ii) $CH_3CH_2CH(OH)CH_3 + [O] \rightarrow CH_3CH_2COCH_3 + H_2O$ (1)
Butanone (1).
(iii) $CH_3CH_2CH_2OH + [O] \rightarrow CH_3CH_2CHO + H_2O$
$CH_3CH_2CHO + [O] \rightarrow CH_3CH_2COOH$ (1)
Propanal, then propanoic acid (1).
b) It is a tertiary alcohol (1), and is resistant to oxidation (1).

4 Ethanal (1) and ethanoic acid (1). Ethanal is formed when the product is allowed to be distilled off (1). Ethanoic acid is formed when heated under reflux (1).

5 **a)** Elimination *or* dehydration (1).

b)

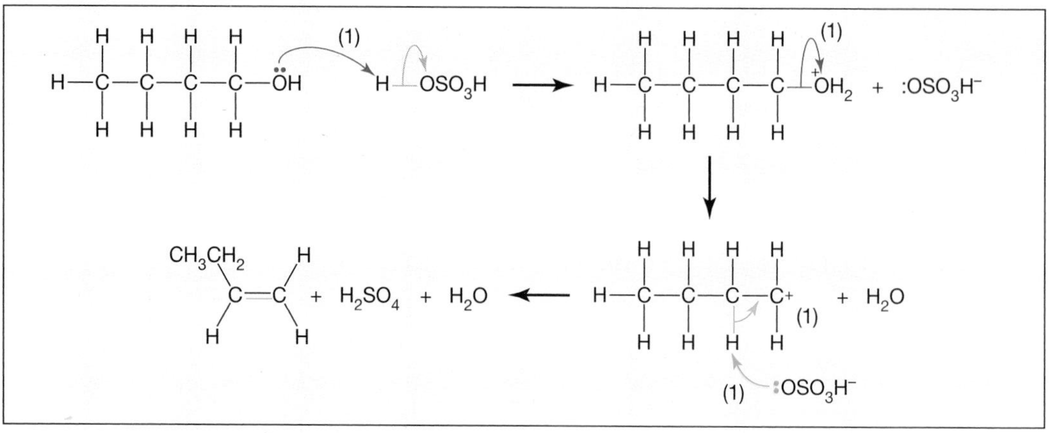

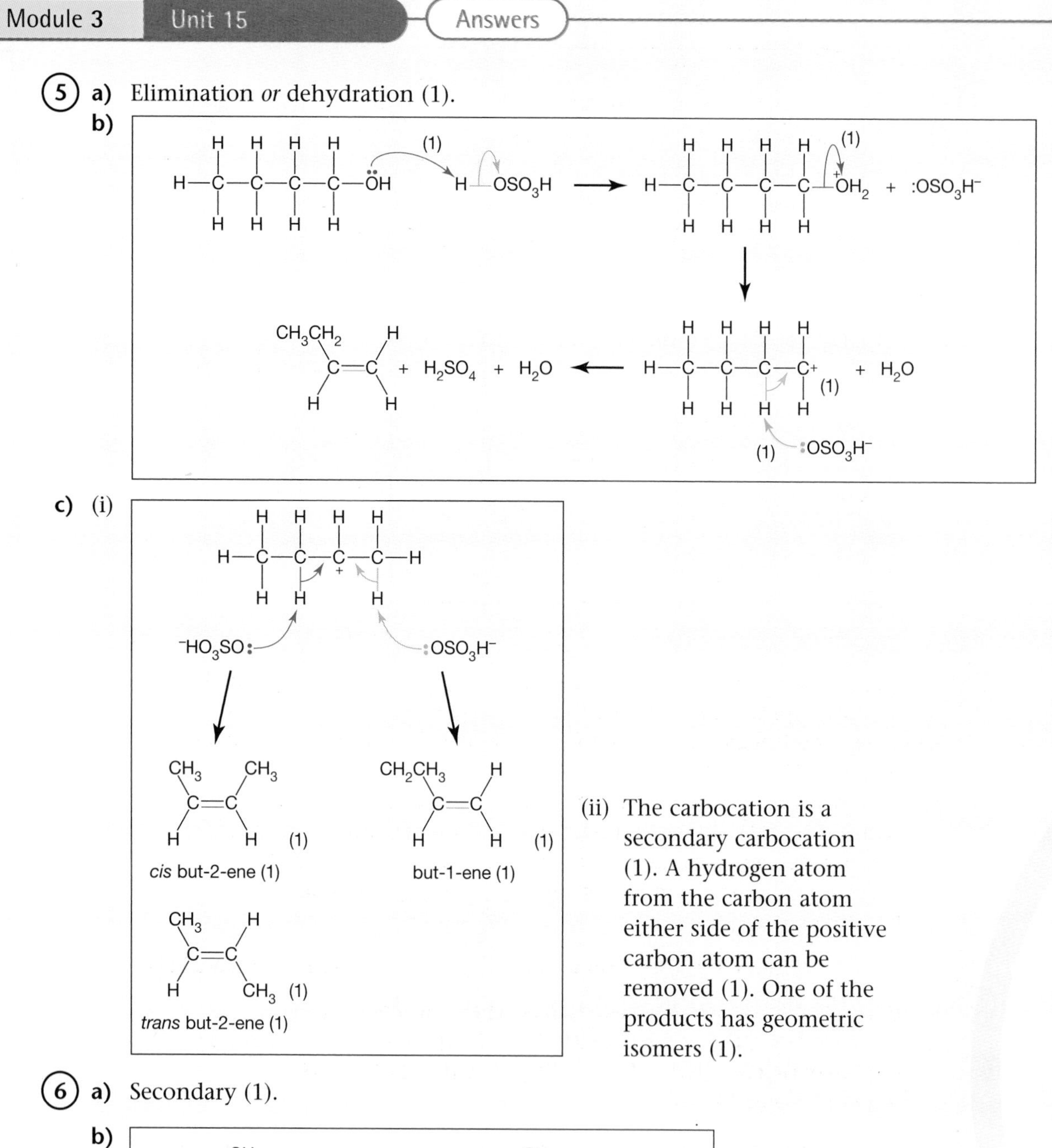

c) (i)

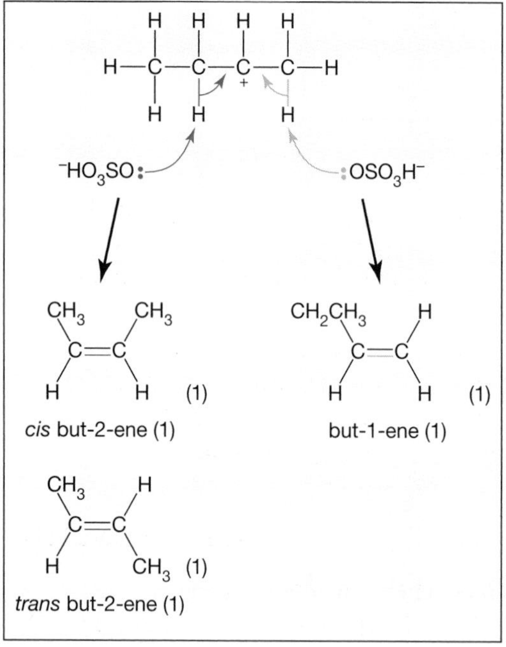

(ii) The carbocation is a secondary carbocation (1). A hydrogen atom from the carbon atom either side of the positive carbon atom can be removed (1). One of the products has geometric isomers (1).

6 **a)** Secondary (1).

b)

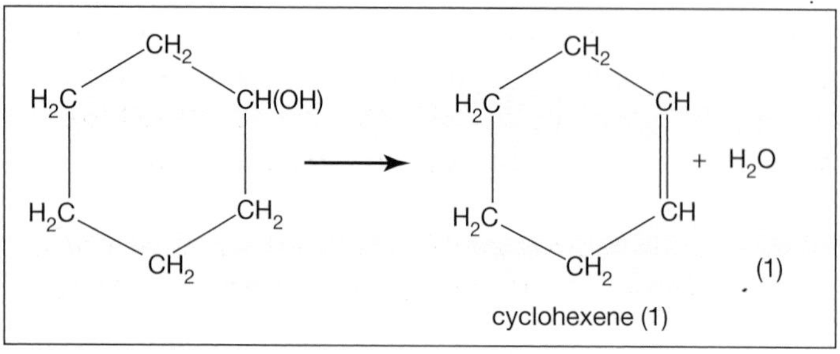

c)

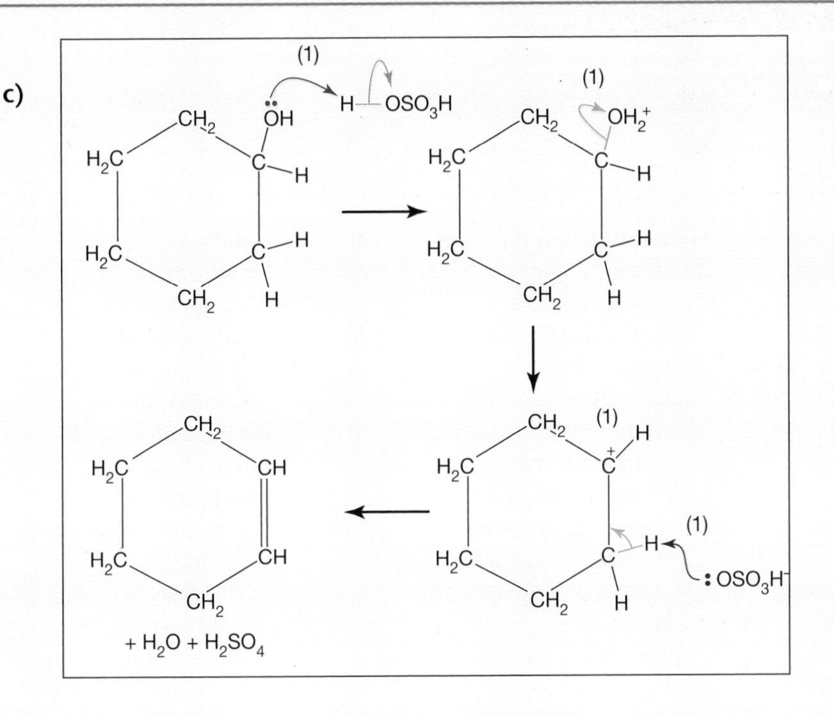

(7) **a)** Sugars dissolved in water (1), using the enzyme catalysts in yeast (1), at an optimum temperature of 37°C (1).

b) Ethene and steam passed over H_3PO_4 catalyst (1), at 70 atmospheres pressure (1), and 300°C (1).

c) *Fermentation*: advantages: uses renewable resources (1), low energy requirement (1); disadvantages: slow reaction (1), impure product (1). *Direct hydration*: advantages: fast reaction (1), pure product (1); disadvantages: uses non-renewable resources (1), high energy requirements (1).

(8) **a)** 3-methylbutanal (1). **b)** Hexan-3-one (1). **c)** 2,3-dimethylbutanal (1).

(9) **a)** $CH_3CH_2CHO + 2[H] \rightarrow CH_3CH_2CH_2OH$ (1); propan-1-ol (1).

b) $CH_3CH_2CH_2COCH_3 + 2[H] \rightarrow CH_3CH_2CH_2CH(OH)CH_3$ (1); pentan-2-ol (1).

c) $CH_3CH(CH_3)CH_2CHO + 2[H] \rightarrow CH_3CH(CH_3)CH_2CH_2OH$ (1); 3-methylbutan-1-ol (1).

(10) **a)** Warm with Tollen's Reagent (1): the aldehyde will form a silver mirror (1) on the inside of the test tube; there is no reaction with the ketone. Alternatively, heat with Fehling's solution: with the aldehyde the blue solution will turn a brick-red colour; there is no reaction with the ketone.

b) Warm with acidified potassium dichromate(VI) (1): the alcohol will turn the solution from orange to green (1); there will be no reaction with the ketone.

Index